W9-CZM-288

PROJECT PLANNING, SCHEDULING, AND CONTROL IN CONSTRUCTION

PROJECT PLANNING, SCHEDULING, AND CONTROL IN CONSTRUCTION

An Encyclopedia of Terms and Applications

CALIN M. POPESCU, Ph.D., PE

CHOTCHAI CHAROENNGAM, Ph.D.

A Wiley-Interscience Publication

JOHN WILEY & SONS, INC.

NEW YORK • CHICHESTER • BRISBANE • TORONTO • SINGAPORE

Library of Congress Cataloging-in-Publication Data
Popescu, Calin.
    Project planning, scheduling and control: encyclopedia of terms &
    applications / Calin Popescu, Chotchai Charoenngam. -- 1st ed.
        p.   cm.
    Includes bibliographical references and index.
    ISBN 0-471-02858-4 (acid-free paper)
    1. Engineering--Management--Terminology.   I. Charoenngam,
Chotchai.   II. Title.
TA190.P65 1995
624'.068'4--dc20   94-35464

PREFACE

This encyclopedia of terms in project planning, scheduling, and control is built on a solid foundation of hundreds of books, trade journals, research publications, and planning and scheduling software vendor literature. Most of them are listed in chronological order by type of publication at the end of the book, as an aid for future researchers.

The purpose of the encyclopedia is to bring together in one comprehensive source a useful collection of terms, definitions, and applications related to project time and cost control. The contents range over the narrow field of project time and cost management and certain terms and definitions that have evolved in the last four decades. The book should be an important research and quick reference tool, desktop information resource, and supplementary reading asset for those practicing, teaching, and learning the art and science of managing projects of various sites. It coordinates and unifies terms and definitions related to the critical path method (CPM) technique. With the present state of technological CPM software growth, this book should serve as an important tool not only for practitioners but also for educators, students, and salespeople here and abroad.

The information in this publication is reliable to the best of our knowledge. We would appreciate being informed of any errors or omissions so that these can be incorporated in subsequent editions. This book is dedicated to all present and future project managers concerned with completing projects on time and within budget, who will benefit from this pioneering reference, as a new standard established for all project management professionals.

This book could not have been published without the assistance of contributors who have given their time and guidance during the last two years. To all of them, our thanks.

Calin Popescu, Ph.D., PE
Chotchai Charoenngam, Ph.D.

Austin, Texas, 1994

CONTRIBUTORS LIST

John D. Borcherding, Ph.D., PE, The University of Texas at Austin, Civil Engineering Department, ECJ 5.424, Austin, Texas 78712.

Daniel M. Burns, Attorney, Assistant Attorney General of Texas, P.O. Box 12548, Austin, Texas 78711-2548.

Luh M. Chang, Ph.D., Purdue University, Civil Engineering School, West Lafayette, Indiana 47907.

Edward G. Gibson, Ph.D., PE, The University of Texas at Austin, Civil Engineering Department, ECJ 5.200, Austin, Texas 78712.

Daniel W. Halpin, Ph.D., PE, Purdue University, Civil Engineering School, West Lafayette, Indiana 47907.

Don Hancher, Ph.D., PE, University of Kentucky, College of Engineering, Lexington, Kentucky 40506-0043.

Jeff Jackmond, PMP, Vice President, Demand Construction Service Inc., 7430 East Caley Avenue, #350, Englewood, Colorado 80111.

Kyungrai Kim, 6-304 Hyundai Apt. Gunyang-Dong, Anyan-Si, Kyunggi-Do, South Korea.

Radhika V. Kulkarni, Ph.D., SAS Institute Inc., SAS Circle, Box 8000, Cary, North Carolina 27512-8000.

Dean Kyle, PE, President, Managing Associates, 325 Hector Road, Victoria, British Columbia, Canada V8X 3X1.

Garold D. Oberlender, Ph.D., PE, Oklahoma State University, School of Civil Engineering, Hester Street and Athletic Drive, Stillwater, Oklahoma 74078.

Suthi Pasiphol, Ph.D., 351 Soi Samtaharn, Sukumvit 50 Rd, Prakanong, Bangkok 10250, Thailand.

Anamaria Popescu, 8014 Landsman Drive, Austin, Texas 78736.

James E. Rowings, Jr., Ph.D., PE, Iowa State University, Construction Engineering, 2125 Country Club Boulevard, Ames, Iowa 50010.

Paul Stynchcomb, President, Brower & Company Inc., 6219 Executive Boulevard, Rockville, Maryland 20852.

Nicolae Tutus, Ph.D., Vice President, Stone & Webster, Advanced System Development Services Inc., 2045 Summer Street, Boston, Massachusetts 02210.

ENCYCLOPEDIA OF TERMS

A

ACCELERATION A compression of activity durations and/or logic changes modifying series work to concurrent work, such that a given quantity of work is performed in a time period shorter than the original planned performance period for that quantity of work.

ACCOUNT CODE STRUCTURE A system used to assign summary numbers to elements of the project breakdown structure and charge (account) numbers to individual work packages.

To create an account code structure for a construction project, the project breakdown structure and individual work packages should be established in advance. The project breakdown structure *(see Project breakdown structure)* is used to summarize cost information at any level, and the work packages *(see Work package)* are used to estimate and collect costs at the operating level *(see Level).* Based on these project breakdown structure and work packages, summary numbers are assigned to each element of the project breakdown structure using a hierarchical coding system. Charge numbers given to each item in the work packages are related to the account numbers *(see Account number).* The entire structure of these numbers is referred to as the *account code structure.* The account code structure is composed of charge numbers and summary numbers.

For example, the account code structure for a commercial building is developed as shown in Figure 1. In the account code structure, the cost data from "concrete" provide the information required to be summarized under the heading "concrete." Similarly, the "concrete" summary, plus other summaries on the same level, such as "form" and "re-bar," provide the information required for the "slab and beam" summary, and so on. In this example, 0331.001, 0332.002, and 0333.001 are the charge numbers for "concrete pouring," "concrete finishing," and "concrete curing," respectively, and 33222 is the summary number of the work packages.

The advantage of using the account code structure is that cost overruns can immediately be pointed to the particular activity responsible. It is used for the basic concepts of the PERT and CPM cost systems.

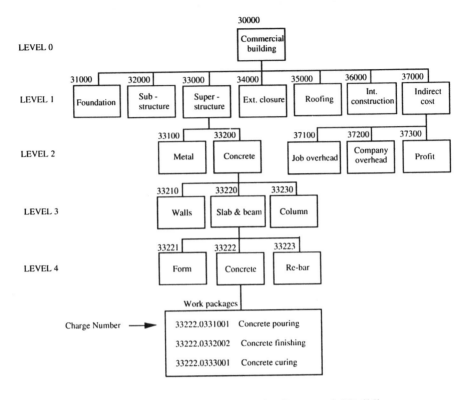

Figure 1 Account Code Structure for Commercial Building

2

ACCOUNTING VARIANCE (AV) The difference between the budgeted cost of work schedule (BCWS) and the actual cost of work performed (ACWP) at any point in time (time now) at a project level.

The accounting variance (AV) is one of the C/SCSC elements and represents a performance measure of whether the established plan for the work accomplished to date is over or under budget *(see Cost and schedule control system criteria)*. At a given time now, the AV is calculated using the formula

$$AV = BCWS - ACWP$$

(See Budgeted cost of work schedule and Actual cost of work performed.) Interpretation of the AV is as follows:

VARIANCE	-	0	+
AV	Over Budget	On Budget	Under Budget

The CV is a more accurate method of performance measurement than the AV because the AV is not necessary related to the earned value concept. In fact, any conclusion drawn as to the AV is of dubious value. For example, a large negative AV may reflect the fact that the project is way ahead of schedule. But it may also be a reflection of significant cost inefficiency and waste rather than speedy progress.

To develop the AV in a project, the following steps are generally followed:

Step 1: Use the original PBS of a project.
Step 2: Use the original cost estimation of the project based on the PBS.
Step 3: Use the latest updated CPM schedule by the data date *(see Data date)*.
Step 4: Calculate the BCWS at the point of the data date (time now) at the activity and project levels.
Step 5: Calculate the ACWP at the point of the data date (time now) at the activity and project levels.
Step 6: Calculate the AV by subtracting the ACWP from the BCWS. The results of steps 4 to 6 are shown in the monthly cost performance report.
Step 7: Draw the AV on the graph using the time-phased and cumulative values at the point of the data date (time now).

The preceding seven steps are explained using a residential house as an example.

1. *Step 1:* PBS

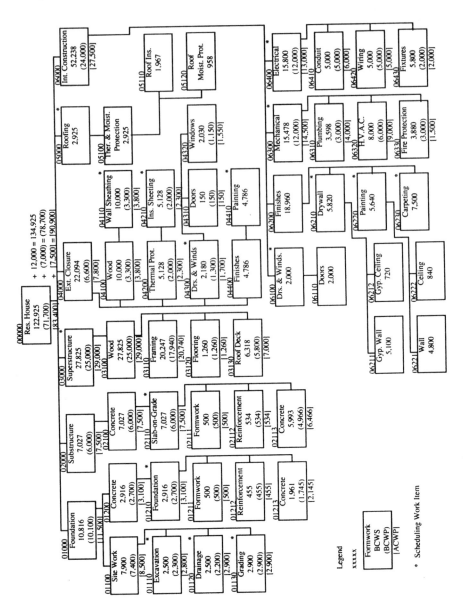

2. *Step 2:* Cost Estimation

DESCRIPTION	U/M	U/P	Q'TY	TOTAL
01 General Requirements				
Subtotal	L/S	12,000	1	12,000
02 Site Work				
01110 Excavation	C.Y.	1.89	1,323	2,500
01120 Drainage	L.F.	3.11	804	2,500
01130 Grading	C.Y.	1.06	2,736	2,900
Subtotal				7,900
03 Concrete				
01211 Foundation Form	SFCA	2.89	173	500
01212 Foundation Re-bar	Tons	910	0.5	455
01213 Foundation Concrete	C.Y.	103.2	19	1,961
02111 Slab-on-Grade Form	L.F.	1.38	362	500
02112 Slab-on-Grade Re-bar	Tons	890	0.6	534
02113 Slab-on-Grade Concrete	C.Y.	90.8	66	5,993
Subtotal				9,943
06 Wood				
03110 Superstructure Framing	M.B.F.	1,191	17	20,247
03120 Superstructure Flooring	S.F.	0.63	2,000	1,260
03130 Roof Decking	S.F.	5.01	1,261	6,318
04110 Wall Sheathing	S.F.	1.17	8,547	10,000
Subtotal				37,825
07 Thermal & Moisture Protection				
04210 Wall Insulating Sheeting	S.F.	0.6	8,547	5,128
05110 Roof Insulating	S.F.	1.56	1,261	1,967
05120 Roof Moisture Protection	S.F.	0.76	1,261	958
Subtotal				8,053
08 Doors & Windows				
04310 Ext. Doors	Ea.	150	1	150
04320 Ext. Windows	Ea.	203	10	2,030
06110 Int. Doors	Ea.	200	10	2,000
Subtotal				4,180
09 Finishes				
04410 Ext. Painting	S.F.	0.56	8,547	4,786
06211 Int. Gyp. Wallboard	S.F.	0.34	15,000	5,100
06212 Int. Gyp. Ceiling Board	S.F.	0.36	2,000	720
06311 Int. Wall Painting	S.F.	0.32	15,000	4,800
06312 Int. Ceiling Painting	S.F.	0.42	2,000	840
06230 Carpeting	S.Y.	5	1,500	7,500
Subtotal				23,746
15 Mechanical				
06310 Plumbing	L.F.	3.98	904	3,598
06320 H.V.A.C.	L/S	8,000	1	8,000
06330 Fire Protection	Ea.	776	5	3,880
Subtotal				15,478
16 Electrical				
06410 Electrical Conduit	L/S	5,000	1	5,000
06420 Electrical Wiring	L/S	5,000	1	5,000
06430 Electrical Fixtures	L/S	5,800	1	5,800
Subtotal				15,800
GRAND TOTAL				134,925

3. *Step 3:* Updated CPM Network Schedule

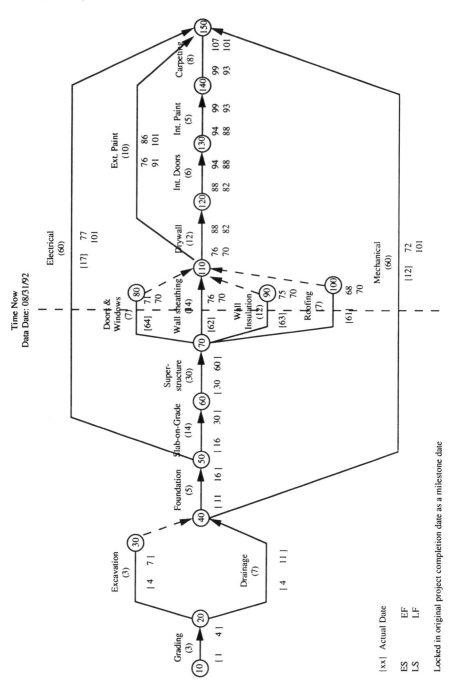

6

4. *Steps 4 to 6:* Monthly Cost Performance Report

MONTHLY COST PERFORMANCE REPORT: CPM ITEM

PROJECT : RESIDENTIAL HOUSE

REPORT DATE: 08/31/92

ITEM		CURRENT PERIOD									CUMULATIVE TO DATE									AT COMPLETION		
		BUDGET COST			VARIANCE			INDEX			BUDGET COST			VARIANCE			INDEX					
PBS	CPM	BCWS	BCWP	ACWP	SV	CV	AV	SPI	CPI	PC	BCWS	BCWP	ACWP	SV	CV	AV	SPI	CPI	PC	BAC	EAC	ACV
01130	10 - 20										2.900	2.900	2.900	0	0	0	1.00	1.00	1.00	2.900	2.900	
01110	20 - 30										2.500	2.300	2.800	-200	-500	-300	0.92	0.82	0.92	2.500	2.800	-300
01120	20 - 40										2.500	2.200	2.800	-300	-600	-300	0.88	0.79	0.88	2.500	2.800	-300
01210	40 - 50										2.916	2.700	3.100	-216	-400	-184	0.93	0.87	0.93	2.916	3.100	-184
06300	40 - 150	5.160	5.000	5.200	-160	-200	-40	0.97	0.96	0.32	13.416	12.000	14.500	-1.416	-2.500	-1.084	0.89	0.83	0.78	15.478	16.562	-1.084
02110	50 - 60										7.027	6.000	7.500	-1.027	-1.500	-473	0.85	0.80	0.85	7.027	7.500	-473
03000	60 - 70	14.903	13.900	15.000	-1.003	-1.100	-97	0.93	0.93	0.50	27.825	25.000	29.000	-2.825	-4.000	-1.175	0.90	0.86	0.90	27.825	29.000	-1.175
06400	50 - 150	5.260	5.100	5.300	-160	-200	-40	0.97	0.96	0.32	12.361	12.000	13.000	-361	-1.000	-639	0.97	0.92	0.76	15.800	16.439	-639
04300	70 - 80	1.555	1.300	1.700	-255	-400	-145	0.84	0.76	0.60	1.555	1.300	1.700	-255	-400	-145	0.84	0.76	0.60	2.180	2.325	-145
04210	70 - 90	2.135	2.000	2.300	-135	-300	-165	0.94	0.87	0.39	2.135	2.000	2.300	-135	-300	-165	0.94	0.87	0.39	5.128	5.293	-165
04110	70 - 110	3.570	3.300	3.800	-270	-500	-230	0.92	0.87	0.33	3.570	3.300	3.800	-270	-500	-230	0.92	0.87	0.33	10.000	10.230	-230
05000	70 - 100																			2.925	2.925	0
06210	110-120																			5.820	5.820	0
04410	110-150																			4.786	4.786	0
06100	120-130																			2.000	2.000	0
06220	130-140																			5.640	5.640	0
06230	140-150																			7.500	7.500	0
SUBTOTAL		32.583	30.600	33.300	-1.983	-2.700	-717	0.94	0.92	0.25	78.705	71.700	83.400	-7.005	-11.700	-4.695	0.91	0.86	0.58	122.925	127.620	-4.695
GEN. REQ.		2.400	2.400	2.500		-100	-100	1.00	0.96	0.20	7.200	7.000	7.500	-200	-500	-300	0.97	0.93	0.58	12.000	12.300	-300
TOTAL		34.983	33.000	35.800	-1.983	-2.800	-817	0.94	0.92	0.24	85.905	78.700	90.900	-7.205	-12.200	-4.995	0.92	0.87	0.58	134.925	139.920	-4.995

Note: The project at the data date has -817 in the current period and -4,995 cumulatively as the accounting variance (AV). It shows that the project is over budget because the AV is less than 0.
The same comment is valid for all work items except for item 01130, which is on schedule (AV=0).

5. *Step 7:* AV

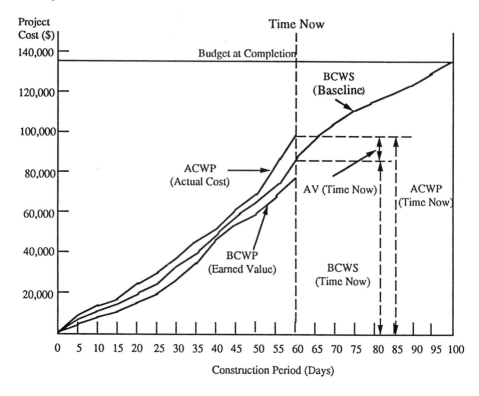

ACCOUNT NUMBER An alphanumeric identification of the work package (also known as the shop order number, charge number, or work order number).

The account number is one component of the account code structure *(see Account code structure)*. In the account code, a number is assigned to each work package item and called the account number. The actual cost of each work package item, such as material costs spent for acquiring materials, is collected and charged to the account number. This account number should be tied into the company's existing accounting system.

For example, Figure 1, the account code structure for a commercial building, shows the account number. When referring to the "concrete" work package in the account code structure, the account number 0331.001, 0332.002, and 0333.001 for "concrete pouring," "concrete finishing," and "concrete curing," respectively, are used for the collection of charges. For example, the actual labor cost related to "concrete finishing" should be charged to account number 0332.002. In addition to the labor cost, the material, equipment, and subcontract costs spent for "concrete finishing" should be charged to the same account number, 0332.002.

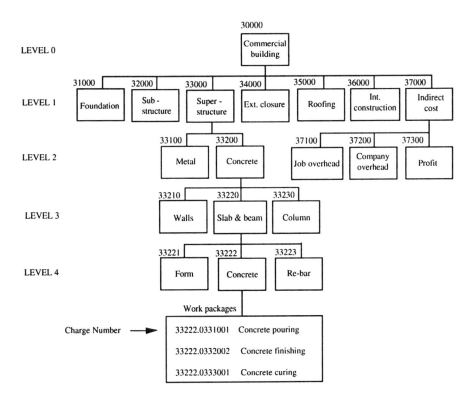

Figure 1 Account Code Structure for Commercial Building

Several coding systems can be used for the account number. One of them is the CSI master format coding system, which classifies work items into 16 divisions, such as Div.1: general requirements, Div.2: site work, Div.3: concrete, Div.4: masonry, Div.5: metals, Div.6: wood and plastics, Div.7: thermal and moisture protection, Div.8: doors and windows, Div.9: finishes, Div.10: specialties, Div.11: equipment, Div.12: furnishings, Div.13: special construction, Div.14: conveying systems, Div.15: mechanical, and Div.16: electrical.

In the account code structure described above, the CSI master format coding system is used for the account number and is explained in Figure 2. This coding system has the advantage of permitting easy comparison of actual charged costs with budget. In addition to the CSI coding system, the hierarchical coding system shown in Figure 3 can be used.

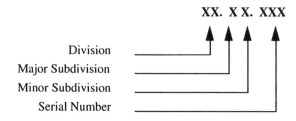

Figure 2 Account Number Based on CSI Master Format Coding System

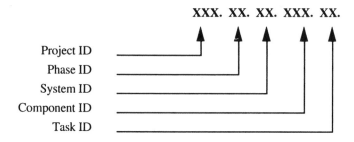

Figure 3 Account Number Based on Hierarchical Coding System

If the same account number were used, for both the initial estimate of costs and the collection of charges, a more efficient cost control system could be established. The account number is used for the basic concepts of the PERT and CPM cost control systems.

ACTIVITY The fundamental unit of work in a project plan and schedule. Each work activity has a defined geographical boundary, a concise description of the work to be performed, and an estimated duration during which the work can be performed. Activities can also have detailed resource and cost estimates.

An activity must have recognizable starting and ending points and may consume resources such as labor, materials, equipment, and paper work. Examples of activities are excavation, pouring concrete, and ordering materials. Activity representation in planning and scheduling can be an activity on arrow, on node, and on bar. For an activity on arrow as shown in Figure 1, each arrow represents one activity in the project. The tail of the arrow represents the starting point of the activity, and the head of the arrow represents completion. For an activity on node as shown in Figure 2, each box or node represents one activity in the project. The left side of the box (node) represents the starting point of the activity, and the right side of the box (node) represents completion. For an activity on bar as shown in Figure 3, each bar represents one activity. The starting point of the bar indicates the starting point of the activity, and the ending point of the bar indicates completion. The length of the bar represents the activity duration *(see Activity duration)*.

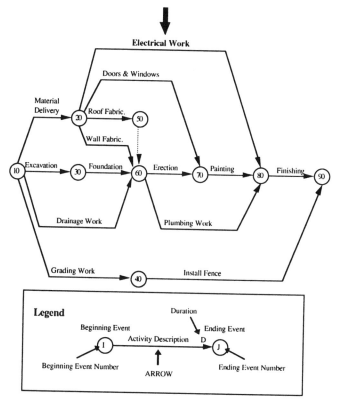

Figure 1 Arrow Diagram Network

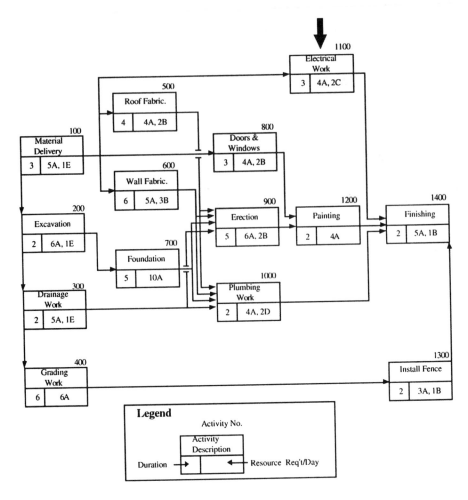

Figure 2 Precedence Diagram Network

Act. ID	Orig. Dur.	Rem. Dur.	Activity Description	March 17	18	19	20	21	22	23	24	25	26	27	28	29	30	31	April 1	2	3	4	5	6	7	8	9	10
100	3	0	Material Delivery	■	■	■																						
200	2	0	Excavation	■	■																							
300	2	0	Drainage	■	■																							
400	6	0	Grading Work	■	■				■	■																		
700	5	0	Foundation			■			■	■																		
600	6	0	Wall Fabrication						■	■	■																	
500	4	0	Roof Fabrication						■	■	■																	
800	3	0	Doors and Windows						■	■																		
1100	3	0	Electrical Work						■	■																		
1300	2	0	Install Fence								■	■																
900	5	0	Erection													■	■				■	■						
1000	2	0	Plumbing Work												■	■												
1200	2	0	Painting																						■	■		
1400	2	0	Finishing																							■	■	

■ Weekend

Figure 3 Bar Chart Schedule

ACTIVITY ACTUAL COSTS The actual expenditures incurred by a project activity at completion or accumulated up to the current reporting time.

Activity actual costs are comprised of activity direct cost *(see Activity direct costs)*, and activity indirect cost *(see Activity indirect costs)* spent at completion or up to a certain time. The activity actual costs are used for project cost control to evaluate as-planned and actual costs at a detailed level. The sum of activity actual costs for each activity at completion is the total actual cost of the project. Table 1 shows a sample of activity actual costs.

Table 1 Activity Actual Costs: Pouring Concrete for a Slab on Ground

Description	Unit	Actual
Quantity	C.Y.	27
Cost of concrete	$/C.Y.	55
Labor	Labor hours	12
Labor cost	$/L.H.	7
Direct cost		
Equipment cost	$	250
Material cost	$	27 x 55 = 1,485
Labor cost	$	42 x 7 = 294
Total direct cost	$	*2,029*
Indirect cost *		
Job overhead	$	100
General overhead	$	50
Total indirect cost	$	*150*
Total cost	$	**2,179**

* Difficult to estimate in practice.

ACTIVITY ALLOWABLE DURATION (ALLOWABLE DURATION) The estimated activity duration *(see Activity duration)* plus or minus the allocated total float *(see Allocated total float)* distributed evenly along the path containing the activity *(see also Activity performance window)*. For uneven float distribution, *see Independent float*.

Activity allowable duration is the maximum time duration allocated for a normal crew to carry out the work. The determination of the activity allowable duration is done after the logic diagram has been completed by adding the estimated activity

duration to the allocated total float distributed along the path with equal float. Figure 1 shows a sample of a nearly critical path with a total float of 8 units, and Figure 2 shows allowable duration and distributed total float along the same path.

Activity allowable duration = estimated activity duration + allocated total float

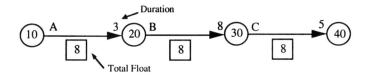

Figure 1 Nearly Critical Path with Total Float = 8 Units

Total duration = 3 + 8 + 5 = 16 units

Activity	Distributed Total Float	Allowable Duration
A	(3/16) x 8 = 1.5	3 + 1.5 = 4.5
B	(8/16) x 8 = 4.0	8 + 4.0 = 12.0
C	(5/16) x 8 = 2.5	5 + 2.5 = 7.5

• Distribution of total float is directly proportional to the estimated activity duration.

Figure 2 Allowable Duration and Distributed Total Float

Other choices of total float distribution are:

1. Distribution of total float along the path is indirectly proportional to the estimated activity duration.
2. Distribution of total float along the path is directly proportional to the estimated direct cost per time unit.
3. Distribution of total float along the path is indirectly proportional to the estimated direct cost per time unit.
4. Distribution of total float is based on a specific resource.

Note: None of the procedures described above have been implemented in existing commercial CPM-oriented software as of 1993.

15

ACTIVITY DESCRIPTION A concise explanation of the nature of the work to be performed which easily identifies an activity to any recipient of a schedule, report, or logic diagram.

Activity descriptions are included in a logic diagram to assist construction and management personnel in reading the schedule. The activity description length should be concise, due to the limited space on drawings and sometimes to limited software capacity. The descriptions should be unambiguous in order to communicate effectively among project personnel. Representation of activity descriptions in arrow and precedence diagram networks, respectively, is shown in Figures 1 and 2. Figure 3 is an activity description in a tabular schedule report.

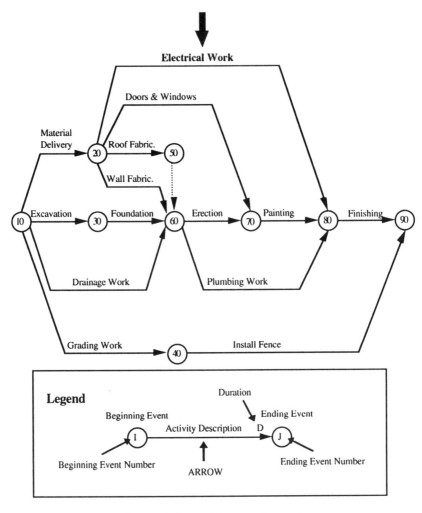

Figure 1 Arrow Diagram Network

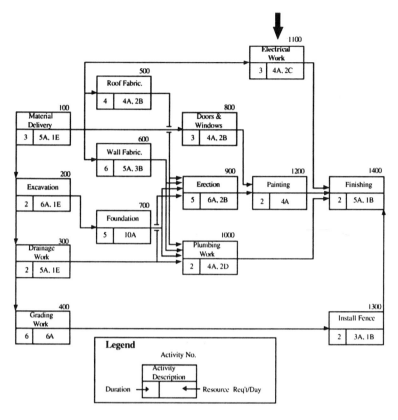

Figure 2 Precedence Diagram Network

CIVIL ENGINEERING DEPARTMENT   PRIMAVERA PROJECT PLANNER

REPORT DATE 23NOV93 RUN NO.   15   PROJECT SCHEDULE REPORT   START DATE 13MAR93 FIN DATE 9APR93
            20:31
CLASSIC SCHEDULE REPORT - SORT BY ES, TF   DATA DATE 17MAR93 PAGE NO.   1

ACTIVITY ID	ORIG DUR	REM DUR	%	CODE	ACTIVITY DESCRIPTION	EARLY START	EARLY FINISH	LATE START	LATE FINISH	TOTAL FLOAT
100	3	3	0		MATERIAL DELIVERY	17MAR93*	19MAR93	17MAR93*	19MAR93	0
200	2	2	0		EXCAVATION	17MAR93	18MAR93	19MAR93	22MAR93	2
300	2	2	0		DRAINAGE WORK	17MAR93	18MAR93	26MAR93	29MAR93	7
400	6	6	0		GRADING WORK	17MAR93	24MAR93	31MAR93	7APR93	10
700	5	5	0		FOUNDATION	19MAR93	25MAR93	23MAR93	29MAR93	2
600	6	6	0		WALL FABRICATION	22MAR93	29MAR93	22MAR93	29MAR93	0
500	4	4	0		ROOF FABRICATION	22MAR93	25MAR93	24MAR93	29MAR93	2
800	3	3	0		DOORS & WINDOWS	22MAR93	24MAR93	1APR93	5APR93	8
1100	3	3	0		ELECTRICAL WORK	22MAR93	24MAR93	5APR93	7APR93	10
1300	2	2	0		INSTALL FENCE	25MAR93	26MAR93	8APR93	9APR93	10
900	5	5	0		ERECTION	30MAR93	5APR93	30MAR93	5APR93	0
1000	2	2	0		PLUMBING WORK	30MAR93	31MAR93	6APR93	7APR93	5
1200	2	2	0		PAINTING	6APR93	7APR93	6APR93	7APR93	0
1400	2	2	0		FINISHING	8APR93	9APR93	8APR93	9APR93	0

Figure 3 Tabular Schedule Report in Precedence Diagramming Method

17

ACTIVITY DIRECT COSTS The cost of materials, equipment, and direct labor required to perform a specific activity.

By performing a quantity takeoff at an activity level, activity direct costs are directly assigned to each activity for installation or construction. The total resulting activity direct costs should equal the total project direct costs. The activity direct costs can be either as-planned or actual costs. A sample of activity direct costs is shown in Table 1.

Activity direct costs = material cost + labor cost + equipment cost

Table 1   Activity Direct Costs: Pouring Concrete for a Slab on Ground

Description	Unit	Planned	Actual
Quantity	C.Y.	25	27
Cost of concrete	$/C.Y.	50	55
Labor	Labor hours	20	12
Labor cost	$/L.H.	5	7
Equipment cost	$	300	250
Material cost	$	20 x 50 = 1,250	27 x 55 = 1,485
Labor cost	$	20 x 5 = 100	42 x 7 = 294
Total cost	$	**1,650**	**2,029**

ACTIVITY DURATION The length of time from the start to the finish of an activity, in any time units chosen for the project. The most common time units are minutes, hours, shifts, days, weeks, months, and years.

The duration of an activity depends on the quantity of work, the type of work, the type and quantity of available resources that may be used to complete an activity, working hours/day, and environmental factors that affect the work. An activity's duration can be determined by dividing the quantity of work obtained from the quantity takeoff at an activity level by the production rate estimated from an expected level of personnel, equipment, or other resources. Factors affecting the estimation of an activity's duration are:

1. Time needed to review required submittals
2. Contractor's personnel requirements and availability

3. Subcontractor's availability and resources
4. Type of material required
5. A/E and owners' time to review, inspect, and perform tasks
6. Equipment requirements and availability
7. Financial resources

An activity's duration can be estimated by the formula

Activity duration

$$= \frac{\text{estimated quantity of work required}}{\text{average production rate of crew or equipment assigned / time unit}}$$

$$= \text{time units}$$

Example: Activity = erection
Quantity = 10 tons of steel
Crew productivity rate = 2 tons/day
Activity duration = 10/2 = 5 days

Figure 1 shows activity duration in a tabular schedule report.

CIVIL ENGINEERING DEPARTMENT   PRIMAVERA PROJECT PLANNER

REPORT DATE 23NOV93 RUN NO.   15   PROJECT SCHEDULE REPORT
            20:31   START DATE 13MAR93 FIN DATE 9APR93
CLASSIC SCHEDULE REPORT - SORT BY ES, TF
                                                                       DATA DATE 17MAR93 PAGE NO.   1

ACTIVITY ID	ORIG DUR	REM DUR	%	CODE	ACTIVITY DESCRIPTION	EARLY START	EARLY FINISH	LATE START	LATE FINISH	TOTAL FLOAT
100	3	3	0		MATERIAL DELIVERY	17MAR93*	19MAR93	17MAR93*	19MAR93	0
200	2	2	0		EXCAVATION	17MAR93	18MAR93	19MAR93	22MAR93	2
300	2	2	0		DRAINAGE WORK	17MAR93	18MAR93	26MAR93	29MAR93	7
400	6	6	0		GRADING WORK	17MAR93	24MAR93	31MAR93	7APR93	10
700	5	5	0		FOUNDATION	19MAR93	25MAR93	23MAR93	29MAR93	2
600	6	6	0		WALL FABRICATION	22MAR93	29MAR93	22MAR93	29MAR93	0
500	4	4	0		ROOF FABRICATION	22MAR93	25MAR93	24MAR93	29MAR93	2
800	3	3	0		DOORS & WINDOWS	22MAR93	24MAR93	1APR93	5APR93	8
1100	3	3	0		ELECTRICAL WORK	22MAR93	24MAR93	5APR93	7APR93	10
1300	2	2	0		INSTALL FENCE	25MAR93	26MAR93	8APR93	9APR93	10
900	5	5	0		ERECTION	30MAR93	5APR93	30MAR93	5APR93	0
1000	2	2	0		PLUMBING WORK	30MAR93	31MAR93	6APR93	7APR93	5
1200	2	2	0		PAINTING	6APR93	7APR93	6APR93	7APR93	0
1400	2	2	0		FINISHING	8APR93	9APR93	8APR93	9APR93	0

Figure 1 Tabular Schedule Report

ACTIVITY IDENTIFICATION (ACTIVITY NUMBER OR ACTIVITY CODE)

The unique code used to identify various activities in the network for the purpose of organizing, storing, and retrieving information about activities.

There are two types of activity identification: (1) activity on nodes and (2) activity on arrows, as shown in Figures 1 to 4. For both types, the activity identification should be unique for each activity, or the same number should not be used more than once so that each activity cannot be mistaken for another activity. To accomplish this, the final activity identification should be assigned after the network logic has been completed. The activity-on-arrow identification is determined by the start and finish nodes (events). The start node should be smaller than the finish node so that it will be easy to locate the activity on a large logic drawing, although this is not a requirement.

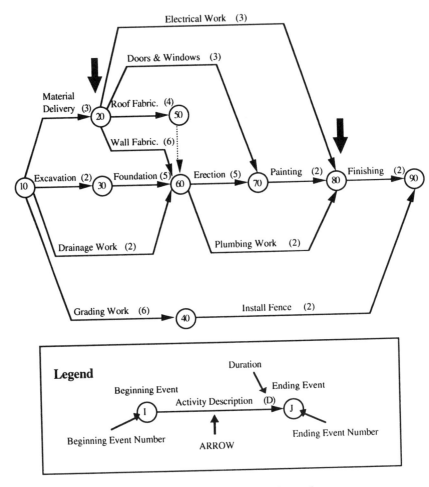

Figure 1 Arrow Diagram Network

PRED	SUCC	ORIG DUR	REM DUR	%	CODE	ACTIVITY DESCRIPTION	EARLY START	EARLY FINISH	LATE START	LATE FINISH	TOTAL FLOAT
10	20	3	3	0		MATERIAL DELIVERY	17MAR93*	19MAR93	17MAR93*	19MAR93	0
10	30	2	2	0		EXCAVATION	17MAR93	18MAR93	19MAR93	22MAR93	2
10	60	2	2	0		DRAINAGE WORK	17MAR93	18MAR93	26MAR93	29MAR93	7
10	40	6	6	0		GRADING WORK	17MAR93	24MAR93	31MAR93	7APR93	10
30	60	5	5	0		FOUNDATION	19MAR93	25MAR93	23MAR93	29MAR93	2
20	60	6	6	0		WALL FABRICATION	22MAR93	29MAR93	22MAR93	29MAR93	0
20	50	4	4	0		ROOF FABRICATION	22MAR93	25MAR93	24MAR93	29MAR93	2
20	70	3	3	0		DOORS AND WINDOWS	22MAR93	24MAR93	1APR93	5APR93	8
20	80	3	3	0		ELECTRICAL WORK	22MAR93	24MAR93	5APR93	7APR93	10
40	90	2	2	0		INSTALL FENCE	25MAR93	26MAR93	8APR93	9APR93	10
50	60	0	0	0		DUMMY 1	26MAR93	25MAR93	30MAR93	29MAR93	2
60	70	5	5	0		ERECTION	30MAR93	5APR93	30MAR93	5APR93	0
60	80	2	2	0		PLUMBING WORK	30MAR93	31MAR93	6APR93	7APR93	5
70	80	2	2	0		PAINTING	6APR93	7APR93	6APR93	7APR93	0
80	90	2	2	0		FINISHING	8APR93	9APR93	8APR93	9APR93	0

Figure 2   Tabular Schedule Report in Arrow Diagramming Method

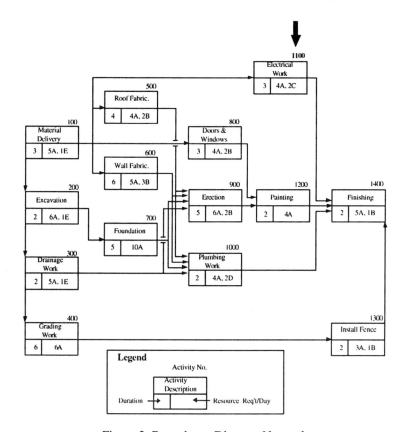

Figure 3   Precedence Diagram Network

CIVIL ENGINEERING DEPARTMENT   PRIMAVERA PROJECT PLANNER

REPORT DATE 23NOV93 RUN NO.   15   PROJECT SCHEDULE REPORT   START DATE 13MAR93 FIN DATE 9APR93
          20:31
CLASSIC SCHEDULE REPORT - SORT BY ES, TF   DATA DATE 17MAR93 PAGE NO.   1

ACTIVITY ID	ORIG DUR	REM DUR	%	CODE	ACTIVITY DESCRIPTION	EARLY START	EARLY FINISH	LATE START	LATE FINISH	TOTAL FLOAT
100	3	3	0		MATERIAL DELIVERY	17MAR93*	19MAR93	17MAR93*	19MAR93	0
200	2	2	0		EXCAVATION	17MAR93	18MAR93	19MAR93	22MAR93	2
300	2	2	0		DRAINAGE WORK	17MAR93	18MAR93	26MAR93	29MAR93	7
400	6	6	0		GRADING WORK	17MAR93	24MAR93	31MAR93	7APR93	10
700	5	5	0		FOUNDATION	19MAR93	25MAR93	23MAR93	29MAR93	2
600	6	6	0		WALL FABRICATION	22MAR93	29MAR93	22MAR93	29MAR93	0
500	4	4	0		ROOF FABRICATION	22MAR93	25MAR93	24MAR93	29MAR93	2
800	3	3	0		DOORS & WINDOWS	22MAR93	24MAR93	1APR93	5APR93	8
1100	3	3	0		ELECTRICAL WORK	22MAR93	24MAR93	5APR93	7APR93	10
1300	2	2	0		INSTALL FENCE	25MAR93	26MAR93	8APR93	9APR93	10
900	5	5	0		ERECTION	30MAR93	5APR93	30MAR93	5APR93	0
1000	2	2	0		PLUMBING WORK	30MAR93	31MAR93	6APR93	7APR93	5
1200	2	2	0		PAINTING	6APR93	7APR93	6APR93	7APR93	0
1400	2	2	0		FINISHING	8APR93	9APR93	8APR93	9APR93	0

Figure 4 Tabular Schedule Report in Precedence Diagramming Technique

ACTIVITY INDIRECT COSTS Costs other than direct costs which may be assigned to a specific activity. Activity indirect costs may consist of both job overhead and general overhead *(see Burden)*.

Difficulty and impracticality arise when relating the indirect cost to a specific activity unless that activity is a main activity and an accounting system is set up specifically for this purpose, such as site organization. The activity indirect costs can be classified into two categories: (1) general overhead, which is corporation general expenses allocated to a given project, and (2) job overhead, including supervision, site utilities and facilities, insurance, interest, penalty cost, and bonuses. A sample of activity indirect costs is shown in Table 1.

Activity indirect costs = activity general overhead + activity job overhead

Table 1 Activity Indirect Costs: Pouring Concrete for a Slab on Ground

Description	Planned	Actual
Activity general overhead	$50	$100
Activity job overhead	$100	$120
Activity indirect cost	50 + 100 = $150	100 + 120 = $220

Note:

$$\text{Activity general overhead} = \frac{\text{project total general overhead} \times \text{activity direct costs}}{\text{project total direct cost}}$$

$$\text{Activity job overhead} = \frac{\text{project total job overhead} \times \text{activity direct costs}}{\text{project total direct cost}}$$

22

ACTIVITY LIST (ACTIVITY REPORT) A list of activities involved in a project, or a portion of the project sorted by activity number, activity duration, activity dates, responsibilities, and so on.

An activity list is originated by a planner or scheduler to create a network. A planner or scheduler uses an activity list generated by a computer to review all activities or portions of activities to verify the validity of stored information and logic, making the interrelationships among activities before constructing a logic diagram and evaluating activities within the entire project. A computer-generated activity list can be obtained through several types of classification and sorting keys, such as activity dates, physical division, geographical division, area of responsibility, trades, and equipment or material usage. After network computation, an activity list can be sorted by the variety of sorting keys available within the CPM-oriented software.

ACTIVITY PERFORMANCE WINDOW The total time window during which an activity may be performed without delaying the project, measured as the time period between the early start date and the late finish date of the activity.

ACTIVITY SPLITTING Breaking down an activity with a long duration into smaller parts for the purpose of the allocated resources, for the technological purpose of accelerating the project, or to clarify the dependency logic.

Activity splitting can be triggered by two purposes: (1) meeting the resource restrictions (materials, equipment, and personnel) and (2) accelerating the project. To meet the resource restrictions the activity is subdivided into smaller parts that will be rescheduled to avoid the excessive use of the limited resources, resulting in a longer project duration. To accelerate the project, an activity is also broken down into smaller parts that will be rescheduled to occur concurrently by increasing the allocated resources in order to shorten the project duration. An example of activity splitting is shown in Figures 1 (arrow diagramming technique) and 2 (precedence diagramming technique). In this example, the activity with the longest activity duration, Trench Excavation, is split into three sections. Each section starts at the same time and requires 20 days to complete.

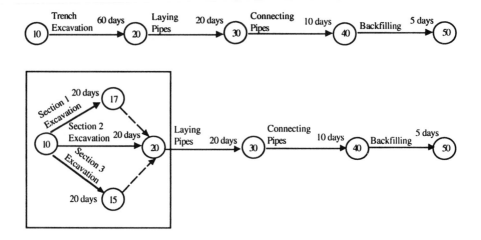

Figure 1 Activity Splitting in Arrow Diagramming Technique

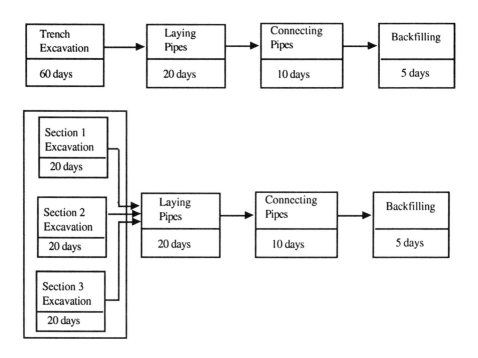

Figure 2 Activity Splitting in Precedence Diagramming Technique

ACTIVITY TIMES Time information generated through the CPM calculation that identifies the earliest and latest calculated start and finish times for each activity in the network.

The activity times are the points in time when the activity begins and ends. In precedence and arrow diagramming techniques *(see Precedence diagramming method and Arrow diagramming method)* the activity times are determined through the CPM time calculation methodology. In a bar chart not based on a CPM diagram, the activity start and finish times are defined by the estimate of the project planner based on mental computation. The activity times are early start, early finish, late start, and late finish. Figure 1 is a tabular schedule report with these four activity times.

```
----------------------------------------------------------------------------------------------------
CIVIL ENGINEERING DEPARTMENT                    PRIMAVERA PROJECT PLANNER

REPORT DATE 23NOV93  RUN NO.   16        PROJECT SCHEDULE REPORT              START DATE 13MAR93  FIN DATE  9APR93
            20:40                                                    /
CLASSIC SCHEDULE REPORT - SORT BY ES, TF                                    DATA DATE  17MAR93  PAGE NO.    1
```

ACTIVITY ID	ORIG DUR	REM DUR	%	CODE	ACTIVITY DESCRIPTION	EARLY START	EARLY FINISH	LATE START	LATE FINISH	TOTAL FLOAT
100	3	3	0		MATERIAL DELIVERY	17MAR93*	19MAR93	17MAR93*	19MAR93	0
200	2	2	0		EXCAVATION	17MAR93	18MAR93	19MAR93	22MAR93	2
300	2	2	0		DRAINAGE WORK	17MAR93	18MAR93	26MAR93	29MAR93	7
400	6	6	0		GRADING WORK	17MAR93	24MAR93	31MAR93	7APR93	10
700	5	5	0		FOUNDATION	19MAR93	25MAR93	23MAR93	29MAR93	2
600	6	6	0		WALL FABRICATION	22MAR93	29MAR93	22MAR93	29MAR93	0
500	4	4	0		ROOF FABRICATION	22MAR93	25MAR93	24MAR93	29MAR93	2
800	3	3	0		DOORS & WINDOWS	22MAR93	24MAR93	1APR93	5APR93	8
1100	3	3	0		ELECTRICAL WORK	22MAR93	24MAR93	5APR93	7APR93	10
1300	2	2	0		INSTALL FENCE	25MAR93	26MAR93	8APR93	9APR93	10
900	5	5	0		ERECTION	30MAR93	5APR93	30MAR93	5APR93	0
1000	2	2	0		PLUMBING WORK	30MAR93	31MAR93	6APR93	7APR93	5
1200	2	2	0		PAINTING	6APR93	7APR93	6APR93	7APR93	0
1400	2	2	0		FINISHING	8APR93	9APR93	8APR93	9APR93	0

Figure 1 Tabular Schedule Report

ACTIVITY TOTAL FLOAT (ACTIVITY TOTAL SLACK) The latest allowable finish minus the earliest allowable finish and/or the latest allowable start minus the earliest allowable start. It is the possible maximum number of time units (hours, days, etc.) that an activity can be delayed from its earliest start or finish without extending the project completion time.

Late finish minus early finish = finish total float (slack)
Late start minus early start = start total float (slack)

The activity total float denotes the amount of time (number of working time units) that an activity path is away from late finish of a project. The amount of time that an activity's actual start and finish times can exceed its earliest expected start

and finish times without causing the project overall duration to exceed its scheduled completion time is called the *activity total float.* Figures 1 and 2 show activity total floats in arrow and precedence diagram networks. A tabular schedule report for a precedence diagramming technique is shown in Figure 3.

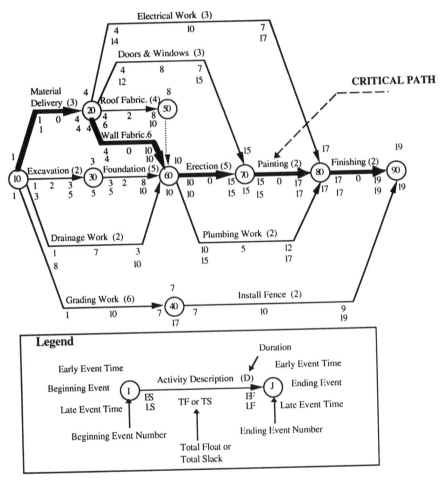

NOTE: The schedule computation in time units is based on the following assumption:

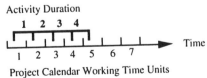

Project Calendar Working Time Units

Activity with 4 time units starts at working time unit 1 and is finished at working time unit 5.

Figure 1 Arrow Diagram Network with Schedule Computation

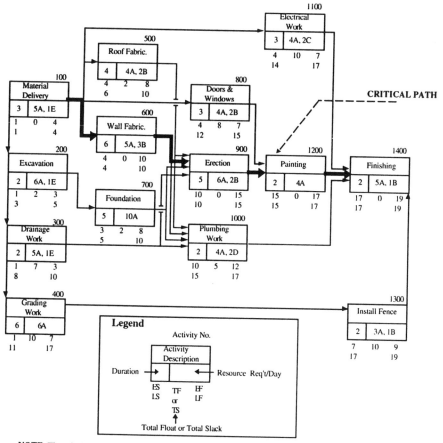

Legend

Activity No.

Activity Description

Duration → ← Resource Req't/Day

ES TF EF
LS or LF
TS
↑
Total Float or Total Slack

NOTE: The schedule computation in time units is based on the following assumption:

Activity Duration

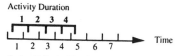

Time

Project Calendar Working Time Units

Activity with 4 time units starts at working time unit 1 and is finished at working time unit 5.

Figure 2 Precedence Diagram Network with Schedule Computation

CIVIL ENGINEERING DEPARTMENT   PRIMAVERA PROJECT PLANNER

REPORT DATE 23NOV93 RUN NO.   16   PROJECT SCHEDULE REPORT   START DATE 13MAR93 FIN DATE 9APR93
       20:40
CLASSIC SCHEDULE REPORT - SORT BY ES, TF   DATA DATE 17MAR93 PAGE NO.   1

ACTIVITY ID	ORIG DUR	REM DUR	%	CODE	ACTIVITY DESCRIPTION	EARLY START	EARLY FINISH	LATE START	LATE FINISH	TOTAL FLOAT
100	3	3	0		MATERIAL DELIVERY	17MAR93*	19MAR93	17MAR93*	19MAR93	0
200	2	2	0		EXCAVATION	17MAR93	18MAR93	19MAR93	22MAR93	2
300	2	2	0		DRAINAGE WORK	17MAR93	18MAR93	26MAR93	29MAR93	7
400	6	6	0		GRADING WORK	17MAR93	24MAR93	31MAR93	7APR93	10
700	5	5	0		FOUNDATION	19MAR93	25MAR93	23MAR93	29MAR93	2
600	6	6	0		WALL FABRICATION	22MAR93	29MAR93	22MAR93	29MAR93	0
500	4	4	0		ROOF FABRICATION	22MAR93	25MAR93	24MAR93	29MAR93	2
800	3	3	0		DOORS & WINDOWS	22MAR93	24MAR93	1APR93	5APR93	8
1100	3	3	0		ELECTRICAL WORK	22MAR93	24MAR93	5APR93	7APR93	10
1300	2	2	0		INSTALL FENCE	25MAR93	26MAR93	8APR93	9APR93	10
900	5	5	0		ERECTION	30MAR93	5APR93	30MAR93	5APR93	0
1000	2	2	0		PLUMBING WORK	30MAR93	31MAR93	6APR93	7APR93	5
1200	2	2	0		PAINTING	6APR93	7APR93	6APR93	7APR93	0
1400	2	2	0		FINISHING	8APR93	9APR93	8APR93	9APR93	0

Figure 3 Tabular Schedule Report in Precedence Diagramming Technique

ACTUAL COMPLETION DATE (ACTUAL FINISH DATE) The calendar date on which an activity was completed. (This date should be no later than the data date or time now.)

An actual completion date is recorded when an in-progress activity since last update, or activity started since last update, is finished. A planner may need this information to update the schedule and compare the progress of the project to the as-planned project schedule *(see As-planned project schedule)*. The actual completion date cannot be specified for any activity not yet started. Note that the actual completion date cannot be earlier than its actual start date *(see Actual start date)* or later than the data date (time now). The actual completion dates can be recorded according to various diagrams and reports as follows:

1. *Arrow diagram technique.* Figure 1 shows the original arrow diagram network schedule, and Figure 2 shows the network with the actual finish date written down on the right of the activity description. The two diagonal lines across an arrow indicate that the activity has been completed.

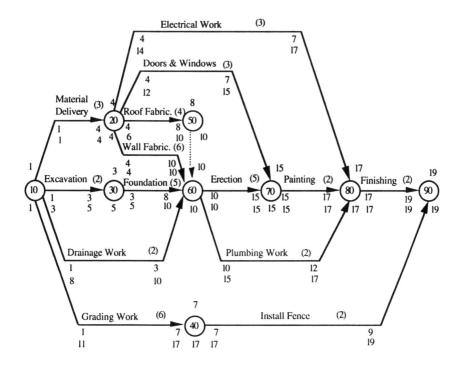

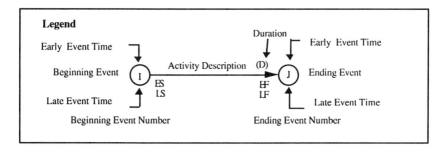

NOTE: The schedule computation in time units is based on the following assumption:

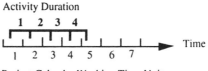

Activity with 4 time units starts at working time unit 1 and is finished at working time unit 5.

Figure 1 Planned Schedule on Arrow Diagram Network

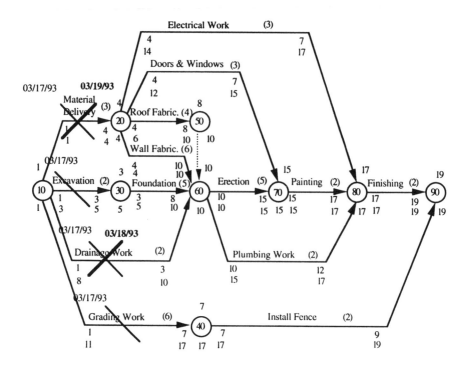

Electrical Work (3)

Doors & Windows (3)

03/17/93   03/19/93
Material
Delivery (3)   Roof Fabric. (4)

Wall Fabric. (6)

03/17/93
Excavation (2)   Foundation (5)   Erection (5)   Painting (2)   Finishing (2)

03/17/93   03/18/93
Drainage Work (2)   Plumbing Work (2)

03/17/93
Grading Work (6)   Install Fence (2)

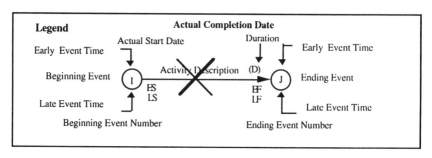

Legend

Actual Completion Date

Early Event Time

Beginning Event

Late Event Time

Beginning Event Number

Actual Start Date

Activity Description (D)

ES
LS

Duration

Early Event Time

Ending Event

Late Event Time

EF
LF

Ending Event Number

NOTE: The schedule computation in time units is based on the following assumption:

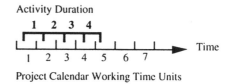

Activity Duration

Time

Project Calendar Working Time Units

Activity with 4 time units starts at working time unit 1 and is finished at working time unit 5.

Figure 2 Updated (Progressed) Arrow Diagram
with Record of Actual Completion Dates

2. *Precedence technique.* *Figure* 3 illustrates the original precedence diagram schedule, whereas Figure 4 illustrates the precedence diagram with the actual finish date recorded on the top right of the activity's box. The two diagonal lines inside the box indicates the completion of the activity.

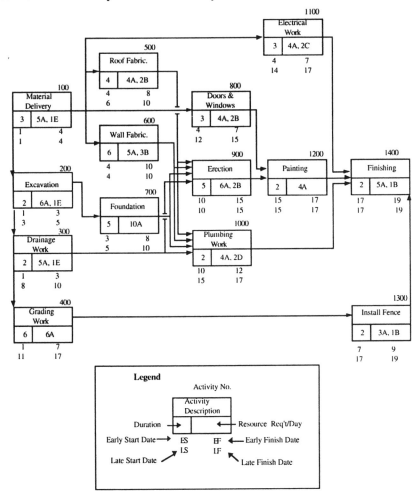

NOTE: The schedule computation in time units is based on the following assumption:

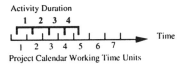

Activity with 4 time units starts at working time unit 1 and is finished at working time unit 5.

Figure 3 Planned Schedule on Precedence Diagram Network

31

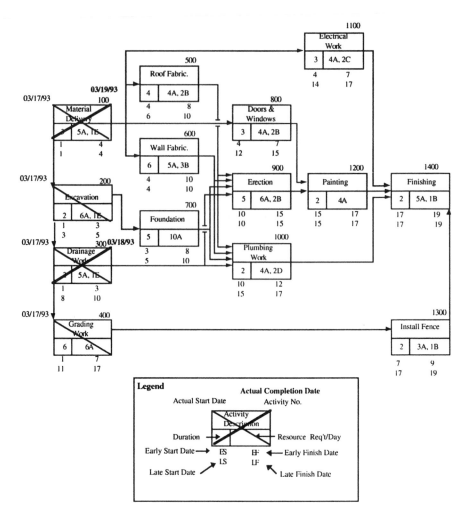

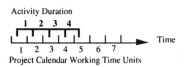

NOTE: The schedule computation in time units is based on the following assumption:

Activity Duration

Activity with 4 time units starts at working time unit 1 and is finished at working time unit 5.

Figure 4 Updated (Progressed) Precedence Diagram
with Record of Actual Completion Dates

32

3. *Bar chart technique (activity codes as in arrow diagram technique).* After specifying the actual finish date to the computer, it will generate a bar chart diagram with all c's (or other symbol) as a bar of the activity, as shown in Figure 5.

```
-------------------------------------------------------------------------------------------------------
CIVIL ENGINEERING DEPARTMENT                  PRIMAVERA PROJECT PLANNER

REPORT DATE 23NOV93  RUN NO.   13             PROJECT SCHEDULE REPORT              START DATE 17MAR93  FIN DATE 12APR93
          19:33
ACTUAL VS. PLANNED BAR CHART                                                      DATA DATE 19MAR93    PAGE NO.    1

                                                                                          DAILY-TIME PER.   1
-------------------------------------------------------------------------------------------------------
     ............ACTIVITY DESCRIPTION.............       22    29    05    12    19    26    03    10    17    24
   PRED  SUCC  OD  RD  PCT  CODES  FLOAT  SCHEDULE       MAR   MAR   APR   APR   APR   APR   MAY   MAY   MAY   MAY
   ----  ----  --  --  ---  -----  -----  --------       93    93    93    93    93    93    93    93    93    93
                                                    ---------------------------------------------------------------
DRAINAGE WORK                            CURRENT    AA*  .      .     .     .     .     .     .     .     .     .
      10   60   2   0 100                PLANNED    BB*  .      .     .     .     .     .     .     .     .     .

MATERIAL DELIVERY                        CURRENT    AA*  .      .     .     .     .     .     .     .     .     .
      10   20   3   0 100                PLANNED    BBB  .      .     .     .     .     .     .     .     .     .

EXCAVATION                               CURRENT    AAB..BB    .     .     .     .     .     .     .     .     .
      10   30   2   3   0            0    PLANNED    BB*  .      .     .     .     .     .     .     .     .     .

GRADING WORK                             CURRENT    AAB..BBB   .     .     .     .     .     .     .     .     .
      10   40   6   4  33           11    PLANNED    BBB..BBB   .     .     .     .     .     .     .     .     .

DOORS AND WINDOWS                        CURRENT    B..BB      .     .     .     .     .     .     .     .     .
      20   70   3   3   0           10    PLANNED    *   BBB    .     .     .     .     .     .     .     .     .

ELECTRICAL WORK                          CURRENT    B..BB      .     .     .     .     .     .     .     .     .
      20   80   3   3   0           12    PLANNED    *   BBB    .     .     .     .     .     .     .     .     .

ROOF FABRICATION                         CURRENT    B..BBB     .     .     .     .     .     .     .     .     .
      20   50   4   4   0            4    PLANNED    *   BBBB   .     .     .     .     .     .     .     .     .

WALL FABRICATION                         CURRENT    B..BBBBB   .     .     .     .     .     .     .     .     .
      20   60   6   6   0            2    PLANNED    *   BBBBB..B     .     .     .     .     .     .     .     .

FOUNDATION                               CURRENT    *   .  BBB..BB    .     .     .     .     .     .     .     .
      30   60   5   5   0            0    PLANNED    B..BBBB    .     .     .     .     .     .     .     .     .

DUMMY 1                                  CURRENT    *   .  B    .     .     .     .     .     .     .     .     .
      50   60   0   0   0            4    PLANNED    *   .  B   .     .     .     .     .     .     .     .     .

INSTALL FENCE                            CURRENT    *   .  BB   .     .     .     .     .     .     .     .     .
      40   90   2   2   0           11    PLANNED    *   .  BB  .     .     .     .     .     .     .     .     .

PLUMBING WORK                            CURRENT    *   .     .  BB   .     .     .     .     .     .     .     .
      60   80   2   2   0            5    PLANNED    *   .     .BB    .     .     .     .     .     .     .     .

ERECTION                                 CURRENT    *   .     .  BBB..BB    .     .     .     .     .     .     .
      60   70   5   5   0            0    PLANNED    *   .     .BBBB..B     .     .     .     .     .     .     .

PAINTING                                 CURRENT    *   .     .     .  BB   .     .     .     .     .     .     .
      70   80   2   2   0            0    PLANNED    *   .     .     .BB    .     .     .     .     .     .     .

FINISHING                                CURRENT    *   .     .     .  B..B      .     .     .     .     .     .
      80   90   2   2   0            0    PLANNED    *   .     .     .  BB   .     .     .     .     .     .     .
```

Figure 5 Updated Bar Chart Based on Arrow Diagram Technique

4. *Bar chart technique (activity codes as in precedence diagram technique).* After specifying the actual finish date to the computer, it will generate a bar chart diagram with all c's (or other symbol) as a bar of the activity, as shown in Figure 6.

```
------------------------------------------------------------------------------------------------------------------
CIVIL ENGINEERING DEPARTMENT              PRIMAVERA PROJECT PLANNER

REPORT DATE 23NOV93  RUN NO.   12         PROJECT SCHEDULE REPORT                  START DATE 13MAR93  FIN DATE 12APR93
              19:39
ACTUAL VS. PLANNED BAR CHART                                                       DATA DATE 19MAR93    PAGE NO.    1

                                                                                                DAILY-TIME PER.    1
------------------------------------------------------------------------------------------------------------------
.............ACTIVITY DESCRIPTION.............           22    29    05    12    19    26    03    10    17    24
ACTIVITY ID  OD   RD  PCT    CODES   FLOAT   SCHEDULE    MAR   MAR   APR   APR   APR   APR   MAY   MAY   MAY   MAY
-----------  ---- --- --- ------------ ----- --------    93    93    93    93    93    93    93    93    93    93
------------------------------------------------------------------------------------------------------------------
DRAINAGE WORK                                CURRENT   AA* .     .     .     .     .     .     .     .     .     .
        300   2    0 100                      PLANNED   EE* .     .     .     .     .     .     .     .     .     .

MATERIAL DELIVERY                            CURRENT   AA* .     .     .     .     .     .     .     .     .     .
        100   3    0 100                      PLANNED   EEE .     .     .     .     .     .     .     .     .     .

EXCAVATION                                   CURRENT   AAE..EE   .     .     .     .     .     .     .     .     .
        200   2    3   0              0       PLANNED   EE* .     .     .     .     .     .     .     .     .     .

GRADING WORK                                 CURRENT   AAE..EEE  .     .     .     .     .     .     .     .     .
        400   6    4  33             11       PLANNED   EEE..EEE  .     .     .     .     .     .     .     .     .

DOORS & WINDOWS                              CURRENT   E..EE     .     .     .     .     .     .     .     .     .
        800   3    3   0             10       PLANNED   *   EEE   .     .     .     .     .     .     .     .     .

ELECTRICAL WORK                              CURRENT   E..EE     .     .     .     .     .     .     .     .     .
       1100   3    3   0             12       PLANNED   *   EEE   .     .     .     .     .     .     .     .     .

ROOF FABRICATION                            CURRENT   E..EEE    .     .     .     .     .     .     .     .     .
        500   4    4   0              4       PLANNED   *   EEEE  .     .     .     .     .     .     .     .     .

WALL FABRICATION                            CURRENT   E..EEEEE  .     .     .     .     .     .     .     .     .
        600   6    6   0              2       PLANNED   *   EEEEE..E .     .     .     .     .     .     .     .

FOUNDATION                                   CURRENT   *   . EEE..EE  .     .     .     .     .     .     .     .
        700   5    5   0              0       PLANNED   E..EEEE   .     .     .     .     .     .     .     .     .

INSTALL FENCE                                CURRENT   *   . EE  .     .     .     .     .     .     .     .     .
       1300   2    2   0             11       PLANNED   *   . EE  .     .     .     .     .     .     .     .     .

PLUMBING WORK                               CURRENT   *   .   . EE  .     .     .     .     .     .     .     .
       1000   2    2   0              5       PLANNED   *   .   .EE  .     .     .     .     .     .     .     .

ERECTION                                     CURRENT   *   .   . EEE..EE  .     .     .     .     .     .     .
        900   5    5   0              0       PLANNED   *   .   .EEEE..E   .     .     .     .     .     .     .

PAINTING                                     CURRENT   *   .   .   . EE  .     .     .     .     .     .     .
       1200   2    2   0              0       PLANNED   *   .   .   .EE  .     .     .     .     .     .     .

FINISHING                                    CURRENT   *   .   .   . E..E    .     .     .     .     .     .
       1400   2    2   0              0       PLANNED   *   .   .   . EE  .     .     .     .     .     .     .
```

Figure 6 Updated Bar Chart Based on Precedence Diagram Technique

5. *Computerized reports (activity codes as in arrow diagram technique).* The underlined dates in the column of the earliest finish date indicate the actual finish date of activities (Figure 7).

CIVIL ENGINEERING DEPARTMENT   PRIMAVERA PROJECT PLANNER

REPORT DATE 23NOV93 RUN NO.   10   PROJECT SCHEDULE REPORT   START DATE 17MAR93 FIN DATE 12APR93
           19:22
CLASSIC SCHEDULE REPORT - SORT BY ES, TF   DATA DATE 19MAR93 PAGE NO.   1

PRED	SUCC	ORIG DUR	REM DUR	%	CODE	ACTIVITY DESCRIPTION	EARLY START	EARLY FINISH	LATE START	LATE FINISH	TOTAL FLOAT
10	20	3	0	100		MATERIAL DELIVERY	17MAR93A	19MAR93A			
10	60	2	0	100		DRAINAGE WORK	17MAR93A	18MAR93A			
10	30	2	3	0		EXCAVATION	17MAR93A	23MAR93		23MAR93	0
10	40	6	4	33		GRADING WORK	17MAR93A	24MAR93		8APR93	11
20	60	6	6	0		WALL FABRICATION	19MAR93	26MAR93	23MAR93	30MAR93	2
20	50	4	4	0		ROOF FABRICATION	19MAR93	24MAR93	25MAR93	30MAR93	4
20	70	3	3	0		DOORS AND WINDOWS	19MAR93	23MAR93	2APR93	6APR93	10
20	80	3	3	0		ELECTRICAL WORK	19MAR93	23MAR93	6APR93	8APR93	12
30	60	5	5	0		FOUNDATION	24MAR93	30MAR93	24MAR93	30MAR93	0
50	60	0	0	0		DUMMY 1	25MAR93	24MAR93	31MAR93	30MAR93	4
40	90	2	2	0		INSTALL FENCE	25MAR93	26MAR93	9APR93	12APR93	11
60	70	5	5	0		ERECTION	31MAR93	6APR93	31MAR93	6APR93	0
60	80	2	2	0		PLUMBING WORK	31MAR93	1APR93	7APR93	8APR93	5
70	80	2	2	0		PAINTING	7APR93	8APR93	7APR93	8APR93	0
80	90	2	2	0		FINISHING	9APR93	12APR93	9APR93	12APR93	0

Figure 7 Computerized Report with Underlined Actual Dates
Based on Arrow Diagram Technique

6. *Computerized reports (activity codes as in precedence diagram technique).*
The underlined dates in the column of the earliest finish date indicate the actual
finish date of activities (Figure 8).

CIVIL ENGINEERING DEPARTMENT   PRIMAVERA PROJECT PLANNER

REPORT DATE 23NOV93 RUN NO.   13   PROJECT SCHEDULE REPORT   START DATE 13MAR93 FIN DATE 12APR93
           19:43
CLASSIC SCHEDULE REPORT - SORT BY ES, TF   DATA DATE 19MAR93 PAGE NO.   1

ACTIVITY ID	ORIG DUR	REM DUR	%	CODE	ACTIVITY DESCRIPTION	EARLY START	EARLY FINISH	LATE START	LATE FINISH	TOTAL FLOAT
100	3	0	100		MATERIAL DELIVERY	17MAR93A	19MAR93A			
300	2	0	100		DRAINAGE WORK	17MAR93A	18MAR93A			
200	2	3	0		EXCAVATION	17MAR93A	23MAR93		23MAR93	0
400	6	4	33		GRADING WORK	17MAR93A	24MAR93		8APR93	11
600	6	6	0		WALL FABRICATION	19MAR93	26MAR93	23MAR93	30MAR93	2
500	4	4	0		ROOF FABRICATION	19MAR93	24MAR93	25MAR93	30MAR93	4
800	3	3	0		DOORS & WINDOWS	19MAR93	23MAR93	2APR93	6APR93	10
1100	3	3	0		ELECTRICAL WORK	19MAR93	23MAR93	6APR93	8APR93	12
700	5	5	0		FOUNDATION	24MAR93	30MAR93	24MAR93	30MAR93	0
1300	2	2	0		INSTALL FENCE	25MAR93	26MAR93	9APR93	12APR93	11
900	5	5	0		ERECTION	31MAR93	6APR93	31MAR93	6APR93	0
1000	2	2	0		PLUMBING WORK	31MAR93	1APR93	7APR93	8APR93	5
1200	2	2	0		PAINTING	7APR93	8APR93	7APR93	8APR93	0
1400	2	2	0		FINISHING	9APR93	12APR93	9APR93	12APR93	0

Figure 8 Computerized Report with Underlined Actual Dates
Based on Precedence Diagram Technique

ACTUAL COST OF WORK PERFORMED (ACWP) The actual cost for completed work items, including the completed part of work items in progress at a given data date.

The actual cost of work performed (ACWP) is one element of the C/SCSC, and represents the actual cost of the performance measurement. The ACWP is acquired by multiplying the actual quantities in place times the actual unit price of each work item *(see Cost and schedule control system criteria)*. To develop the ACWP in a project, five steps are generally followed:

Step 1: Use the original PBS of a project.
Step 2: Use the original cost estimation of the project based on the PBS.
Step 3: Use the latest updated CPM by the data date *(see Data date)*.
Step 4: Measure the works completed, and calculate the ACWP periodically by multiplying the actual quantities in place by the actual unit price of each work item. This measurement and calculation are shown in the monthly cost performance report.
Step 5: Develop the ACWP curve on the graph using the cumulative and time-phased actual cost by the data date.

The preceding five steps are explained using a residential house as an example.

1. *Step 1:* PBS

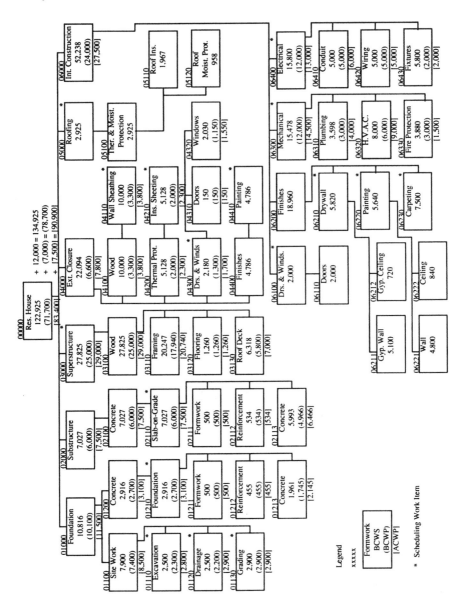

2. *Step 2:* Cost Estimation

DESCRIPTION	U/M	U/P	QTY	TOTAL
01 General Requirements				
Subtotal	L/S	12,000	1	12,000
02 Site Work				
01110 Excavation	C.Y.	1.89	1,323	2,500
01120 Drainage	L.F.	3.11	804	2,500
01130 Grading	C.Y.	1.06	2,736	2,900
Subtotal				7,900
03 Concrete				
01211 Foundation Form	SFCA	2.89	173	500
01212 Foundation Re-bar	Tons	910	0.5	455
01213 Foundation Concrete	C.Y.	103.2	19	1,961
02111 Slab-on-Grade Form	L.F.	1.38	362	500
02112 Slab-on-Grade Re-bar	Tons	890	0.6	534
02113 Slab-on-Grade Concrete	C.Y.	90.8	66	5,993
Subtotal				9,943
06 Wood				
03110 Superstructure Framing	M.B.F.	1,191	17	20,247
03120 Superstructure Flooring	S.F.	0.63	2,000	1,260
03130 Roof Decking	S.F.	5.01	1,261	6,318
04110 Wall Sheathing	S.F.	1.17	8,547	10,000
Subtotal				37,825
07 Thermal & Moisture Protection				
04210 Wall Insulating Sheeting	S.F.	0.6	8,547	5,128
05110 Roof Insulating	S.F.	1.56	1,261	1,967
05120 Roof Moisture Protection	S.F.	0.76	1,261	958
Subtotal				8,053
08 Doors & Windows				
04310 Ext. Doors	Ea.	150	1	150
04320 Ext. Windows	Ea.	203	10	2,030
06110 Int. Doors	Ea.	200	10	2,000
Subtotal				4,180
09 Finishes				
04410 Ext. Painting	S.F.	0.56	8,547	4,786
06211 Int. Gyp. Wallboard	S.F.	0.34	15,000	5,100
06212 Int. Gyp. Ceiling Board	S.F.	0.36	2,000	720
06311 Int. Wall Painting	S.F.	0.32	15,000	4,800
06312 Int. Ceiling Painting	S.F.	0.42	2,000	840
06230 Carpeting	S.Y.	5	1,500	7,500
Subtotal				23,746
15 Mechanical				
06310 Plumbing	L.F.	3.98	904	3,598
06320 H.V.A.C.	L/S	8,000	1	8,000
06330 Fire Protection	Ea.	776	5	3,880
Subtotal				15,478
16 Electrical				
06410 Electrical Conduit	L/S	5,000	1	5,000
06420 Electrical Wiring	L/S	5,000	1	5,000
06430 Electrical Fixtures	L/S	5,800	1	5,800
Subtotal				15,800
GRAND TOTAL				134,925

3. *Step 3:* Updated CPM Network Schedule

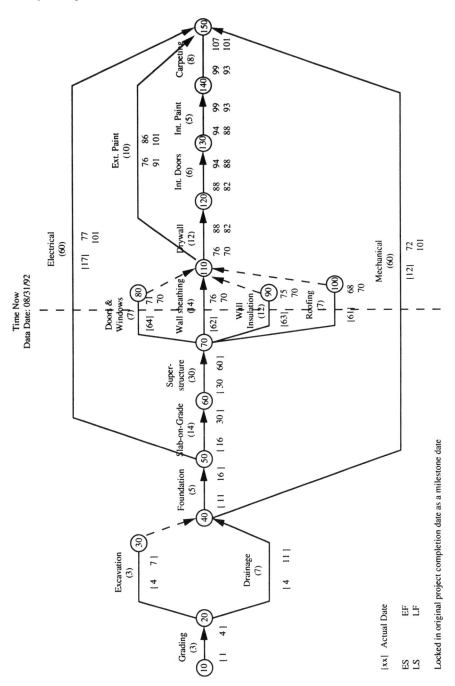

4. *Step 4:* Monthly Cost Performance Report

MONTHLY COST PERFORMANCE REPORT: CPM ITEM

PROJECT : RESIDENTIAL HOUSE   REPORT DATE: 08/31/92

ITEM		CURRENT PERIOD									CUMULATIVE TO DATE									AT COMPLETION		
		BUDGET COST		ACWP	VARIANCE			INDEX			BUDGET COST		ACWP	VARIANCE			INDEX					
PBS	CPM	BCWS	BCWP		SV	CV	AV	SPI	CPI	PC	BCWS	BCWP		SV	CV	AV	SPI	CPI	PC	BAC	EAC	ACV
01130	10 - 20										2.900	2.900	2.900	0	0	0	1.00	1.00	1.00	2.900	2.900	0
01110	20 - 30										2.500	2.300	2.800	-200	-500	-300	0.92	0.82	0.92	2.500	2.800	-300
01120	20 - 40										2.500	2.200	2.800	-300	-600	-300	0.88	0.79	0.88	2.500	2.800	-300
01210	40 - 50										2.916	2.700	3.100	-216	-400	-184	0.93	0.87	0.93	2.916	3.100	-184
06300	40 - 150	5.160	5.000	5.200	-160	-200	-40	0.97	0.96	0.32	13.416	12.000	14.500	-1.416	-2.500	-1.084	0.89	0.83	0.78	15.478	16.562	-1.084
02110	50 - 60										7.027	6.000	7.500	-1.027	-1.500	-473	0.85	0.80	0.85	7.027	7.500	-473
03000	60 - 70	14.903	13.900	15.000	-1.003	-1.100	-97	0.93	0.93	0.50	27.825	25.000	29.000	-2.825	-4.000	-1.175	0.90	0.86	0.90	27.825	29.000	-1.175
06400	50 - 150	5.260	5.100	5.300	-160	-200	-40	0.97	0.96	0.32	12.361	12.000	13.000	-361	-1.000	-639	0.97	0.92	0.76	15.800	16.439	-639
04300	70 - 80	1.555	1.300	1.700	-255	-400	-145	0.84	0.76	0.60	1.555	1.300	1.700	-255	-400	-145	0.84	0.76	0.60	2.180	2.325	-145
04210	70 - 90	2.135	2.000	2.300	-135	-300	-165	0.94	0.87	0.39	2.135	2.000	2.300	-135	-300	-165	0.94	0.87	0.39	5.128	5.293	-165
04110	70 - 110	3.570	3.300	3.800	-270	-500	-230	0.92	0.87	0.33	3.570	3.300	3.800	-270	-500	-230	0.92	0.87	0.33	10.000	10.230	-230
05000	70 - 100																			2.925	2.925	0
06210	110-120																			5.820	5.820	0
06410	110-150																			4.786	4.786	0
06100	120-130																			2.000	2.000	0
06220	130-140																			5.640	5.640	0
06230	140-150																			7.500	7.500	0
SUB-TOTAL		32.583	30.600	33.300	-1.983	-2.700	-717	0.94	0.92	0.25	78.705	71.700	83.400	-7.005	-11.700	-4.695	0.91	0.86	0.58	122.925	127.620	-4.695
GEN. REQ.		2.400	2.400	2.500		-100	-100	1.00	0.96	0.20	7.200	7.000	7.500	-200	-500	-300	0.97	0.93	0.58	12.000	12.300	-300
TOTAL		34.983	33.000	35.800	-1.983	-2.800	-817	0.94	0.92	0.24	85.905	78.700	90.900	-7.205	-12.200	-4.995	0.92	0.87	0.58	134.925	139.920	-4.995

5. *Step 5:* ACWP (Actual Cost, S Curve)

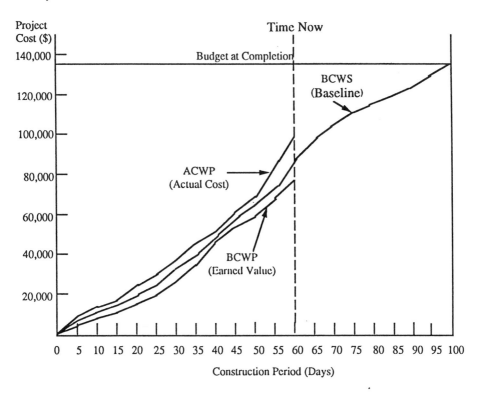

ACTUAL DATE *(see Actual start date* and *Actual completion date)*

ACTUAL FINISH DATE *(see Actual completion date)*

ACTUAL PROJECT COST The actual expenditures incurred by a program or project. The sum of actual costs of all activities and other project costs incurred not represented as activities.

The actual project cost is collected from the work package level using account numbers and is reported in the monthly cost reports. In the reports, the actual costs of work items are calculated by multiplying the actual quantities times the actual unit price *(see Actual cost of work performed).* The actual project cost is the sum of the actual cost of work items.

The actual project cost is shown in the monthly cost reports (Figure 1), which work items are derived from the lower levels of the project breakdown structure *(see*

41

Project breakdown structure). Based on the cost reports, the actual project cost is shown using PBS (Figure 2) or a project cost/schedule progress report (Figure 3).

Data Date: Dec. 31, 1993

Item Code	Description	Planned Cost			Actual Cost		
		U. Price	Q'ty	Cost	U. Price	Q'ty	Cost
3.1000	Foundation Subtotal	-	-	20,000	-	-	30,000
3.2000	Substruc. Subtotal	-	-	40,000	-	-	50,000
3.3221	Form	20	250	5,000	25	240	6,000
3.3222	Concrete	136.67	30	4,100	140	30	4,200
3.3223	Re-bar	800	25	2,000	900	3	2,700
3.7100	Job Overhead			2,000			2,000
Total				73,100			94,900

Figure 1 Monthly Cost Report

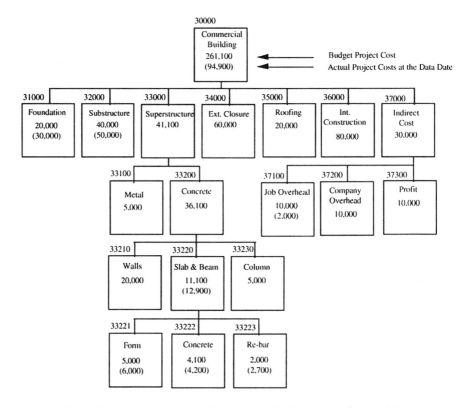

Figure 2 Project Breakdown Structure with Planned and Actual Cost
at the Data Date (Dec. 31, 1993)

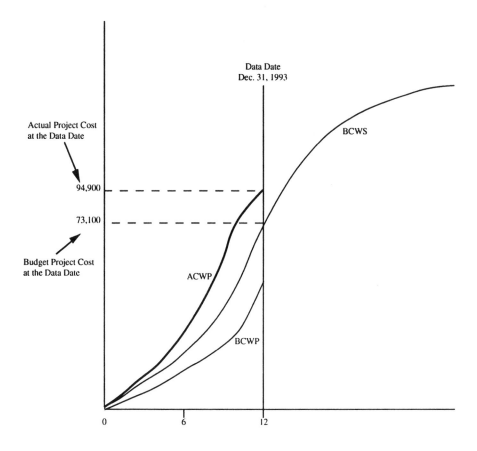

Figure 3 Project Cost Curves: ACWP, BCWS, and BCWP

ACTUAL START DATE The calendar date on which the activity actually begins. It must be prior to or equal to the data date (time now).

An actual start date should be recorded once an activity begins for comparison with the planned project schedule *(see As-planned project schedule).* The actual start date is used for monitoring progress. Without that, the computer assumes that the activity has not yet started and that the result calculated will not reflect the real project status.

The actual start dates can be recorded in various ways:

1. *Arrow diagram technique.* Figure 1 shows the original arrow diagram network schedule, and Figure 2 shows the network with the actual start date written down on the left of the activity description. A diagonal line across an arrow indicates that the activity has already started.

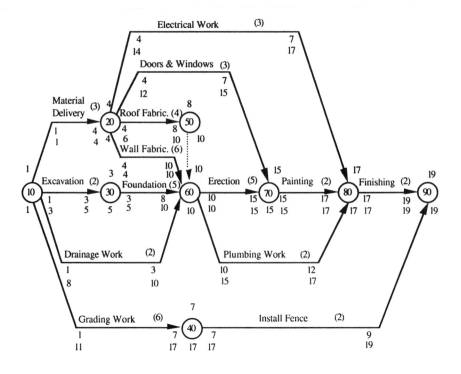

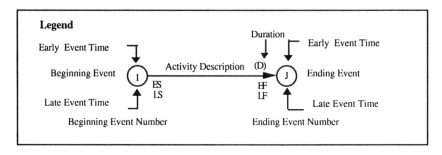

NOTE: The schedule computation in time units is based on the following assumption:

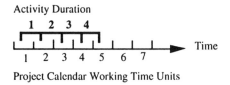

Activity with 4 time units starts at working time unit 1 and is finished at working time unit 5.

Figure 1 Planned Schedule on Arrow Diagram Network

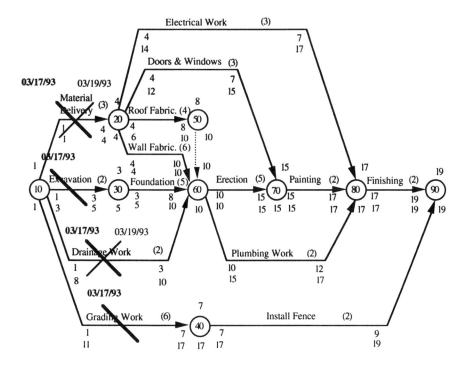

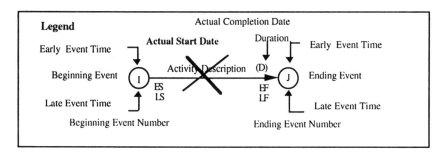

NOTE: The schedule computation in time units is based on the following assumption:

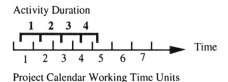

Activity Duration

Project Calendar Working Time Units

Activity with 4 time units starts at working time unit 1 and is finished at working time unit 5.

Figure 2 Updated (Progressed) Arrow Diagram
with Record of Actual Start Dates

2. *Precedence diagram technique.* Figure 3 illustrates the original precedence diagram schedule, and Figure 4 illustrates the precedence diagram with the actual start date recorded on the top left of the activity's box. The diagonal line inside the box indicates that the activity has already started.

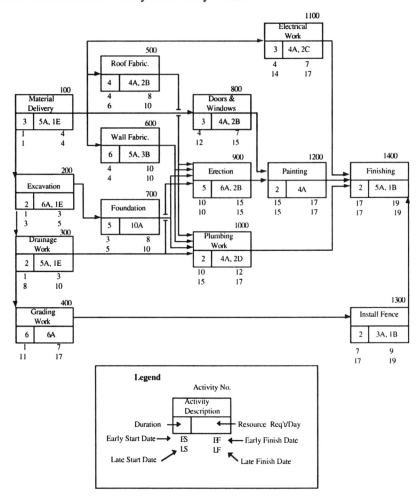

NOTE: The schedule computation in time units is based on the following assumption:

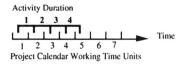

Activity with 4 time units starts at working time unit 1 and is finished at working time unit 5.

Figure 3 Planned Schedule on Precedence Diagram Technique

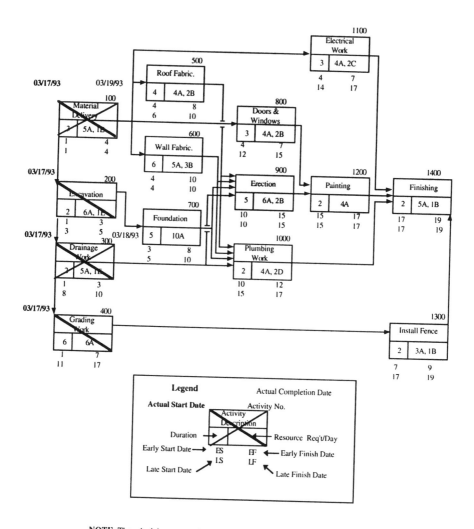

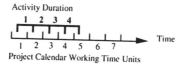

NOTE: The schedule computation in time units is based on the following assumption:

Activity Duration

Time

Project Calendar Working Time Units

Activity with 4 time units starts at working time unit 1 and is finished at working time unit 5.

Figure 4 Updated (Progressed) Precedence Diagram
with Record of Actual Start Dates

47

3. *Bar chart technique (activity codes as in arrow diagram technique).* Once the actual start date is specified to the computer, it will generate a bar chart diagram with an "A" (or other symbol) on the date, as shown in Figure 5.

```
---------------------------------------------------------------------------------------
CIVIL ENGINEERING DEPARTMENT              PRIMAVERA PROJECT PLANNER

REPORT DATE 23NOV93  RUN NO.  13          PROJECT SCHEDULE REPORT           START DATE 17MAR93  FIN DATE 12APR93
            19:33                                                           DATA DATE 19MAR93   PAGE NO.    1
ACTUAL VS. PLANNED BAR CHART
                                                                               DAILY-TIME PER.  1
---------------------------------------------------------------------------------------
                                                      22   29   05   12   19   26   03   10   17   24
..........ACTIVITY DESCRIPTION..........              MAR  MAR  APR  APR  APR  APR  MAY  MAY  MAY  MAY
PRED  SUCC  OD  RD  PCT   CODES   FLOAT   SCHEDULE     93   93   93   93   93   93   93   93   93   93
----------  ----  ---  -----------  -----  --------
```

			Bar
DRAINAGE WORK	CURRENT	AA* .	. . . .
10 60 2 0 100	PLANNED	EE* .	. . . .
MATERIAL DELIVERY	CURRENT	AA* .	. . . .
10 20 3 0 100	PLANNED	EEE .	. . . .
EXCAVATION	CURRENT	AAE..EE	. . . .
10 30 2 3 0 0	PLANNED	EE* .	. . . .
GRADING WORK	CURRENT	AAE..EEE	. . . .
10 40 6 4 33 11	PLANNED	EEE..EEE	. . . .
DOORS AND WINDOWS	CURRENT	E..EE	. . . .
20 70 3 3 0 10	PLANNED	* EEE	. . . .
ELECTRICAL WORK	CURRENT	E..EE	. . . .
20 80 3 3 0 12	PLANNED	* EEE	. . . .
ROOF FABRICATION	CURRENT	E..EEE	. . . .
20 50 4 4 0 4	PLANNED	* EEEE	. . . .
WALL FABRICATION	CURRENT	E..EEEEE .	. . . .
20 60 6 6 0 2	PLANNED	* EEEEE..E	. . . .
FOUNDATION	CURRENT	* . EEE..EE	. . . .
30 60 5 5 0 0	PLANNED	E..EEEE .	. . . .
DUMMY 1	CURRENT	* . E	. . . .
50 60 0 0 0 4	PLANNED	* . E .	. . . .
INSTALL FENCE	CURRENT	* . EE	. . . .
40 90 2 2 0 11	PLANNED	* . EE .	. . . .
PLUMBING WORK	CURRENT	* . . EE	. . . .
60 80 2 2 0 5	PLANNED	* . .EE	. . . .
ERECTION	CURRENT	* . . EEE..EE	. . . . . .
60 70 5 5 0 0	PLANNED	* . .EEEE..E	. . . .
PAINTING	CURRENT	* . . . EE	. . . . . .
70 80 2 2 0 0	PLANNED	* . . .EE .	. . . .
FINISHING	CURRENT	* . . . E..E	. . . . .
80 90 2 2 0 0	PLANNED	* . . . EE .	. . . . . .

Figure 5 Updated Bar Chart Based on Arrow Diagram Technique

48

4. Bar chart technique (activity codes as in precedence diagram technique).
Once the actual start date is specified to the computer, it will generate a bar chart
diagram with an "A" (or other symbol) on the date, as shown in Figure 6.

```
------------------------------------------------------------------------------------------------------------------
CIVIL ENGINEERING DEPARTMENT              PRIMAVERA PROJECT PLANNER

REPORT DATE 23NOV93  RUN NO.   12         PROJECT SCHEDULE REPORT                    START DATE 13MAR93  FIN DATE 12APR93
           19:39
ACTUAL VS. PLANNED BAR CHART                                                         DATA DATE 19MAR93   PAGE NO.    1

                                                                                           DAILY-TIME PER.    1

............ACTIVITY DESCRIPTION.............          22    29    05    12    19    26    03    10    17    24
ACTIVITY ID  OD   RD  PCT   CODES   FLOAT   SCHEDULE   MAR   MAR   APR   APR   APR   APR   MAY   MAY   MAY   MAY
------------ ----  ---- --- ----------- -----  --------   93    93    93    93    93    93    93    93    93    93
                                                       ------------------------------------------------------------
DRAINAGE WORK                              CURRENT  AA* .     .     .     .     .     .     .     .     .     .
          300    2   0 100                 PLANNED  EE* .     .     .     .     .     .     .     .     .     .

MATERIAL DELIVERY                          CURRENT  AA* .     .     .     .     .     .     .     .     .     .
          100    3   0 100                 PLANNED  EEE .     .     .     .     .     .     .     .     .     .

EXCAVATION                                 CURRENT  AAE..EE   .     .     .     .     .     .     .     .     .
          200    2   3   0          0      PLANNED  EE* .     .     .     .     .     .     .     .     .     .

GRADING WORK                               CURRENT  AAE..EEE  .     .     .     .     .     .     .     .     .
          400    6   4  33         11      PLANNED  EEE..EEE  .     .     .     .     .     .     .     .     .

DOORS & WINDOWS                            CURRENT  E..EE     .     .     .     .     .     .     .     .     .
          800    3   3   0         10      PLANNED  *  EEE    .     .     .     .     .     .     .     .     .

ELECTRICAL WORK                            CURRENT  E..EE     .     .     .     .     .     .     .     .     .
         1100    3   3   0         12      PLANNED  *  EEE    .     .     .     .     .     .     .     .     .

ROOF FABRICATION                           CURRENT  E..EEE    .     .     .     .     .     .     .     .     .
          500    4   4   0          4      PLANNED  *  EEEE   .     .     .     .     .     .     .     .     .

WALL FABRICATION                           CURRENT  E..EEEEE  .     .     .     .     .     .     .     .     .
          600    6   6   0          2      PLANNED  *  EEEEE..E .    .     .     .     .     .     .     .     .

FOUNDATION                                 CURRENT  *  . EEE..EE  .    .     .     .     .     .     .     .     .
          700    5   5   0          0      PLANNED  E..EEEE   .     .     .     .     .     .     .     .     .

INSTALL FENCE                              CURRENT  *  . EE   .     .     .     .     .     .     .     .     .
         1300    2   2   0         11      PLANNED  *  . EE   .     .     .     .     .     .     .     .     .

PLUMBING WORK                              CURRENT  *  .   . EE   .     .     .     .     .     .     .     .
         1000    2   2   0          5      PLANNED  *  .   .EE    .     .     .     .     .     .     .     .

ERECTION                                   CURRENT  *  .   . EEE..EE  .    .     .     .     .     .     .     .
          900    5   5   0          0      PLANNED  *  .   .EEEE..E   .    .     .     .     .     .     .     .

PAINTING                                   CURRENT  *  .     .   . EE   .     .     .     .     .     .     .
         1200    2   2   0          0      PLANNED  *  .     .   .EE    .     .     .     .     .     .     .

FINISHING                                  CURRENT  *  .     .   . E..E    .     .     .     .     .     .
         1400    2   2   0          0      PLANNED  *  .     .   . EE   .     .     .     .     .     .     .
```

Figure 6 Updated Bar Chart Based on Precedence Diagram Technique

5. Computerized reports (activity codes as in arrow diagram technique). The
underlined dates in the column of the earliest start date indicate the actual start date
of activities (Figure 7).

REPORT DATE 23NOV93 RUN NO. 10   PROJECT SCHEDULE REPORT   START DATE 17MAR93 FIN DATE 12APR93
19:22
CLASSIC SCHEDULE REPORT - SORT BY ES, TF   DATA DATE 19MAR93 PAGE NO. 1

PRED	SUCC	ORIG DUR	REM DUR	%	CODE	ACTIVITY DESCRIPTION	EARLY START	EARLY FINISH	LATE START	LATE FINISH	TOTAL FLOAT
10	20	3	0	100		MATERIAL DELIVERY	17MAR93A	19MAR93A			
10	60	2	0	100		DRAINAGE WORK	17MAR93A	18MAR93A			
10	30	2	3	0		EXCAVATION	17MAR93A	23MAR93		23MAR93	0
10	40	6	4	33		GRADING WORK	17MAR93A	24MAR93		8APR93	11
20	60	6	6	0		WALL FABRICATION	19MAR93	26MAR93	23MAR93	30MAR93	2
20	50	4	4	0		ROOF FABRICATION	19MAR93	24MAR93	25MAR93	30MAR93	4
20	70	3	3	0		DOORS AND WINDOWS	19MAR93	23MAR93	2APR93	6APR93	10
20	80	3	3	0		ELECTRICAL WORK	19MAR93	23MAR93	6APR93	8APR93	12
30	60	5	5	0		FOUNDATION	24MAR93	30MAR93	24MAR93	30MAR93	0
50	60	0	0	0		DUMMY 1	25MAR93	24MAR93	31MAR93	30MAR93	4
40	90	2	2	0		INSTALL FENCE	25MAR93	26MAR93	9APR93	12APR93	11
60	70	5	5	0		ERECTION	31MAR93	6APR93	31MAR93	6APR93	0
60	80	2	2	0		PLUMBING WORK	31MAR93	1APR93	7APR93	8APR93	5
70	80	2	2	0		PAINTING	7APR93	8APR93	7APR93	8APR93	0
80	90	2	2	0		FINISHING	9APR93	12APR93	9APR93	12APR93	0

Figure 7 Computerized Report with Underlined Actual Dates
Based on Arrow Diagram Technique

6. *Computerized reports (activity codes as in precedence diagram technique).*
The underlined dates in the column of the earliest start date indicate the actual start
date of activities (Figure 8).

REPORT DATE 23NOV93 RUN NO. 13   PROJECT SCHEDULE REPORT   START DATE 13MAR93 FIN DATE 12APR93
19:43
CLASSIC SCHEDULE REPORT - SORT BY ES, TF   DATA DATE 19MAR93 PAGE NO. 1

ACTIVITY ID	ORIG DUR	REM DUR	%	CODE	ACTIVITY DESCRIPTION	EARLY START	EARLY FINISH	LATE START	LATE FINISH	TOTAL FLOAT
100	3	0	100		MATERIAL DELIVERY	17MAR93A	19MAR93A			
300	2	0	100		DRAINAGE WORK	17MAR93A	18MAR93A			
200	2	3	0		EXCAVATION	17MAR93A	23MAR93		23MAR93	0
400	6	4	33		GRADING WORK	17MAR93A	24MAR93		8APR93	11
600	6	6	0		WALL FABRICATION	19MAR93	26MAR93	23MAR93	30MAR93	2
500	4	4	0		ROOF FABRICATION	19MAR93	24MAR93	25MAR93	30MAR93	4
800	3	3	0		DOORS & WINDOWS	19MAR93	23MAR93	2APR93	6APR93	10
1100	3	3	0		ELECTRICAL WORK	19MAR93	23MAR93	6APR93	8APR93	12
700	5	5	0		FOUNDATION	24MAR93	30MAR93	24MAR93	30MAR93	0
1300	2	2	0		INSTALL FENCE	25MAR93	26MAR93	9APR93	12APR93	11
900	5	5	0		ERECTION	31MAR93	6APR93	31MAR93	6APR93	0
1000	2	2	0		PLUMBING WORK	31MAR93	1APR93	7APR93	8APR93	5
1200	2	2	0		PAINTING	7APR93	8APR93	7APR93	8APR93	0
1400	2	2	0		FINISHING	9APR93	12APR93	9APR93	12APR93	0

Figure 8 Computerized Report with Underlined Actual Dates
Based on Precedence Diagram Technique

ACWP *(see Actual cost of work performed)*

ADM *(see Arrow diagramming method)*

AGGREGATION (RESOURCE) The method of calculating resource requirements at the project level to obtain the needed resource quantities to implement the schedule selected for each project time unit, without imposing any limitation on resource availability.

Resource aggregation is a method of calculating needed resources at the project or multiproject level for each time unit to enable implementation of the plan. During this process, no limitation on available resources is considered. An illustrative aggregation example for a precedence network and the resources estimated for each activity are shown in Figure 1.

Resources needed for the progress of any activity from the network may be constant (the same quantity is needed every day of that activity for the entire duration) or it may be a variable requirement. The intensity fluctuates over the activity duration. Example: For the first 2 days, the requirements are three pipefitters/day, and for the last 2 days, only one pipefitter/day with a break of 2 days (Figure 2).

To simplify the process, it is possible to assume that for the same activity, the demand is constant over the activity duration. In this situation, Figure 3 shows the resource pipe-fitters' distribution over the activity duration.

The third approach in estimating needed resources at the activity level is to figure the total amount needed to complete the activity under normal working conditions. In this case, we specify that six pipefitters are needed each day for activity X. To figure the requirements during every activity day, the total amount estimated is divided by the activity's estimated duration and a constant level will be considered for further analysis. Figure 4 is a standard schedule report indicating resources/activity.

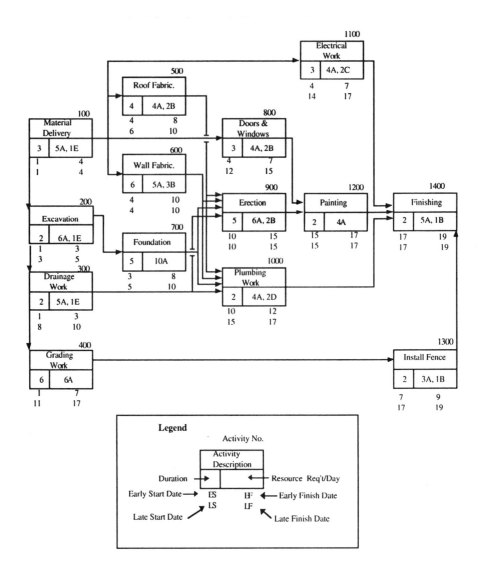

NOTE: The schedule computation in time units is based on the following assumption:

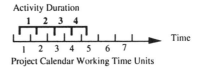

Activity Duration

Project Calendar Working Time Units

Activity with 4 time units starts at working time unit 1 and is finished at working time unit 5.

Figure 1 Required Resources Shown in Precedence Diagram Network

52

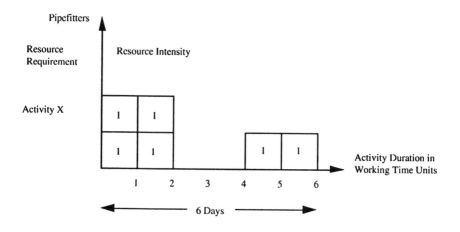

Figure 2 Requirement of Resource Pipefitters

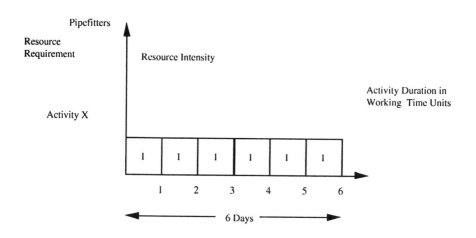

Figure 3 Distribution of Resource Pipefitters

```
-----------------------------------------------------------------------------------------------------------------------------
CIVIL ENGINEERING DEPARTMENT                    PRIMAVERA PROJECT PLANNER

REPORT DATE 24NOV93  RUN NO.    29                   PROJECT SCHEDULE                     START DATE 13MAR93  FIN DATE 09APR93
             05:04
CLASSIC SCHEDULE REPORT (RESOURCE)                                                       DATA DATE  17MAR93  PAGE NO.   1
-----------------------------------------------------------------------------------------------------------------------------
ACTIVITY   ORIG REM  CAL              ACTIVITY                  EARLY     EARLY     LATE      LATE     TOTAL
   ID      DUR  DUR  ID  %   CODE     DESCRIPTION               START     FINISH    START     FINISH   FLOAT RESOURC QUNTY
---------- ---- ---- --- --- -------- ------------------------- --------- --------- --------- -------- ----- ------- -----

       100    3    3   1   0           MATERIAL DELIVERY         17MAR93   19MAR93   17MAR93   19MAR93     0   AA      15
                                                                                                              BB       3

       200    2    2   1   0           EXCAVATION                17MAR93   18MAR93   19MAR93   22MAR93     2   AA      12
                                                                                                              BB       2

       300    2    2   1   0           DRAINAGE WORK             17MAR93   18MAR93   26MAR93   29MAR93     7   AA      10
                                                                                                              BB       2

       400    6    6   1   0           GRADING WORK              17MAR93   24MAR93   31MAR93   07APR93    10   AA      36

       700    5    5   1   0           FOUNDATION                19MAR93   25MAR93   23MAR93   29MAR93     2   AA      50

       600    6    6   1   0           WALL FABRICATION          22MAR93   29MAR93   22MAR93   29MAR93     0   AA      30
                                                                                                              BB      18

       500    4    4   1   0           ROOF FABRICATION          22MAR93   25MAR93   24MAR93   29MAR93     2   AA      16
                                                                                                              BB       8

       800    3    3   1   0           DOORS & WINDOWS           22MAR93   24MAR93   01APR93   05APR93     8   AA      12
                                                                                                              BB       6

      1100    3    3   1   0           ELECTRICAL WORK           22MAR93   24MAR93   05APR93   07APR93    10   AA      12
                                                                                                              CC       6

      1300    2    2   1   0           INSTALL FENCE             25MAR93   26MAR93   08APR93   09APR93    10   AA       6
                                                                                                              BB       2

       900    5    5   1   0           ERECTION                  30MAR93   05APR93   30MAR93   05APR93     0   AA      30
                                                                                                              BB      10

      1000    2    2   1   0           PLUMBING WORK             30MAR93   31MAR93   06APR93   07APR93     5   AA       8
                                                                                                              DD       4

      1200    2    2   1   0           PAINTING                  06APR93   07APR93   06APR93   07APR93     0   AA       8

      1400    2    2   1   0           FINISHING                 08APR93   09APR93   08APR93   09APR93     0   AA      10
                                                                                                              BB       2
```

Figure 4 Standard Report Including Resources/Activity in Early Start Schedule

54

A printed computer-generated histogram for each resource type is shown in Figure 5 for activities considered to start at the early start schedule. Similar reports can be generated for activities starting at the late start schedule (Figure 6, late start schedule sorting histogram; Figure 7, late schedule resource histogram).

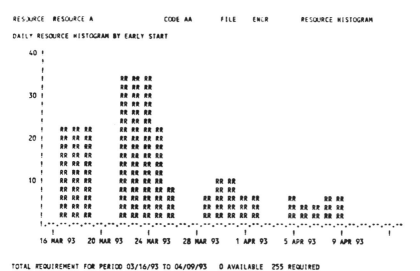

Figure 5 Resource Histogram of Activities in Early Start Schedule

```
------------------------------------------------------------------------------------------
CIVIL ENGINEERING DEPARTMENT                    PRIMAVERA PROJECT PLANNER

REPORT DATE 24NOV93  RUN NO.   52                   PROJECT SCHEDULE              START DATE 13MAR93  FIN DATE 09APR93
           07:05
CLASSIC SCHEDULE REPORT (RESOURCE)                                                DATA DATE  17MAR93  PAGE NO.    1
------------------------------------------------------------------------------------------
ACTIVITY  ORIG REM CAL        ACTIVITY               EARLY    EARLY    LATE     LATE     TOTAL
   ID     DUR  DUR ID  %  CODE DESCRIPTION           START    FINISH   START    FINISH   FLOAT RESOURC QUNTY
--------- ---- ---- --- --- ---------- ----------------------  -------- -------- -------- -------- ----- ------- -----

     100    3    3   1   0       MATERIAL DELIVERY    17MAR93  19MAR93  17MAR93  19MAR93    0    AA      15
                                                                                                BB       3

     200    2    2   1   0       EXCAVATION           17MAR93  18MAR93  19MAR93  22MAR93    2    AA      12
                                                                                                BB       2

     600    6    6   1   0       WALL FABRICATION     22MAR93  29MAR93  22MAR93  29MAR93    0    AA      30
                                                                                                BB      18

     700    5    5   1   0       FOUNDATION           19MAR93  25MAR93  23MAR93  29MAR93    2    AA      50

     500    4    4   1   0       ROOF FABRICATION     22MAR93  25MAR93  24MAR93  29MAR93    2    AA      16
                                                                                                BB       8

     300    2    2   1   0       DRAINAGE WORK        17MAR93  18MAR93  26MAR93  29MAR93    7    AA      10
                                                                                                BB       2

     900    5    5   1   0       ERECTION             30MAR93  05APR93  30MAR93  05APR93    0    AA      30
                                                                                                BB      10

     400    6    6   1   0       GRADING WORK         17MAR93  24MAR93  31MAR93  07APR93   10    AA      36

     800    3    3   1   0       DOORS & WINDOWS      22MAR93  24MAR93  01APR93  05APR93    8    AA      12
                                                                                                BB       6

    1100    3    3   1   0       ELECTRICAL WORK      22MAR93  24MAR93  05APR93  07APR93   10    AA      12
                                                                                                CC       6

    1200    2    2   1   0       PAINTING             06APR93  07APR93  06APR93  07APR93    0    AA       8

    1000    2    2   1   0       PLUMBING WORK        30MAR93  31MAR93  06APR93  07APR93    5    AA       8
                                                                                                DD       4

    1400    2    2   1   0       FINISHING            08APR93  09APR93  08APR93  09APR93    0    AA      10
                                                                                                BB       2

    1300    2    2   1   0       INSTALL FENCE        25MAR93  26MAR93  08APR93  09APR93   10    AA       6
                                                                                                BB       2
```

Figure 6 Standard Report Including Resources / Activity in Late Start Schedule

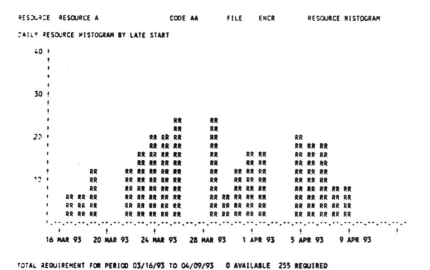

Figure 7 Resource Histogram of Activities in Late Start Schedule

Another useful managerial report is a presentation of needed resources at the project level in tabular format. Figure 8 shows the early start schedule tabular format, and Figure 9 shows the late start schedule in the same format.

57

CIVIL ENGINEERING DEPARTMENT   PRIMAVERA PROJECT PLANNER

REPORT DATE 24NOV93 RUN NO.   35   TABULAR RESOURCE REPORT-DAILY   START DATE 13MAR93 FIN DATE   9APR93
             5:42
TABULAR RESOURCE USAGE   DATA DATE 17MAR93   PAGE NO.   1

| PERIOD | ----AVAILABLE---- | | -----EARLY SCHEDULE---- | | -----LATE SCHEDULE----- | | ---TARGET 1 SCHEDULE--- | |
BEGINNING	NORMAL	MAXIMUM	USAGE	CUMULATIVE	USAGE	CUMULATIVE	USAGE	CUMULATIVE
	AA	-			UNIT OF MEASURE =			
13MAR93	15	15	.00	.00	.00	.00	.00	.00
14MAR93	15	15	.00	.00	.00	.00	.00	.00
15MAR93	15	15	.00	.00	.00	.00	.00	.00
16MAR93	15	15	.00	.00	.00	.00	.00	.00
DATA DATE								
17MAR93	15	15	22.00	22.00	.00	.00	.00	.00
18MAR93	15	15	22.00	44.00	.00	.00	.00	.00
19MAR93	15	15	21.00	65.00	.00	.00	.00	.00
20MAR93	15	15	.00	65.00	.00	.00	.00	.00
21MAR93	15	15	.00	65.00	.00	.00	.00	.00
22MAR93	15	15	33.00	98.00	.00	.00	.00	.00
23MAR93	15	15	33.00	131.00	.00	.00	.00	.00
24MAR93	15	15	33.00	164.00	.00	.00	.00	.00
25MAR93	15	15	22.00	186.00	.00	.00	.00	.00
26MAR93	15	15	8.00	194.00	.00	.00	.00	.00
27MAR93	15	15	.00	194.00	.00	.00	.00	.00
28MAR93	15	15	.00	194.00	.00	.00	.00	.00
29MAR93	15	15	5.00	199.00	.00	.00	.00	.00
30MAR93	15	15	10.00	209.00	.00	.00	.00	.00
31MAR93	15	15	10.00	219.00	.00	.00	.00	.00
1APR93	15	15	6.00	225.00	.00	.00	.00	.00
2APR93	15	15	6.00	231.00	.00	.00	.00	.00
3APR93	15	15	.00	231.00	.00	.00	.00	.00
4APR93	15	15	.00	231.00	.00	.00	.00	.00
5APR93	15	15	6.00	237.00	.00	.00	.00	.00
6APR93	15	15	4.00	241.00	.00	.00	.00	.00
7APR93	15	15	4.00	245.00	.00	.00	.00	.00
8APR93	15	15	5.00	250.00	.00	.00	.00	.00
9APR93	15	15	5.00	255.00	.00	.00	.00	.00

Figure 8 Required Resources in Tabular Format of Early Start Schedule

58

```
--------------------------------------------------------------------------------------------------------------------
CIVIL ENGINEERING DEPARTMENT                          PRIMAVERA PROJECT PLANNER

REPORT DATE  24NOV93  RUN NO.   36                TABULAR RESOURCE REPORT-DAILY              START DATE 13MAR93  FIN DATE  9APR93
             5:44
TABULAR RESOURCE USAGE                                                                     DATA DATE  17MAR93   PAGE NO.    1

--------------------------------------------------------------------------------------------------------------------
    PERIOD     ----AVAILABLE----     -----EARLY SCHEDULE----     -----LATE SCHEDULE-----      ---TARGET 1 SCHEDULE---
   BEGINNING   NORMAL   MAXIMUM      USAGE     CUMULATIVE        USAGE      CUMULATIVE        USAGE      CUMULATIVE
  ------------- -------- --------   ---------- -----------      ---------- -----------      ---------- -----------
               AA        -                                   UNIT OF MEASURE =

    13MAR93       15       15          .00        .00            .00          .00              .00          .00
    14MAR93       15       15          .00        .00            .00          .00              .00          .00
    15MAR93       15       15          .00        .00            .00          .00              .00          .00
    16MAR93       15       15          .00        .00            .00          .00              .00          .00
  ***DATA DATE***
    17MAR93       15       15          .00        .00           5.00         5.00              .00          .00
    18MAR93       15       15          .00        .00           5.00        10.00              .00          .00
    19MAR93       15       15          .00        .00          11.00        21.00              .00          .00
    20MAR93       15       15          .00        .00            .00        21.00              .00          .00
    21MAR93       15       15          .00        .00            .00        21.00              .00          .00
    22MAR93       15       15          .00        .00          11.00        32.00              .00          .00
    23MAR93       15       15          .00        .00          15.00        47.00              .00          .00
    24MAR93       15       15          .00        .00          19.00        66.00              .00          .00
    25MAR93       15       15          .00        .00          19.00        85.00              .00          .00
    26MAR93       15       15          .00        .00          24.00       109.00              .00          .00
    27MAR93       15       15          .00        .00            .00       109.00              .00          .00
    28MAR93       15       15          .00        .00            .00       109.00              .00          .00
    29MAR93       15       15          .00        .00          24.00       133.00              .00          .00
    30MAR93       15       15          .00        .00           6.00       139.00              .00          .00
    31MAR93       15       15          .00        .00          12.00       151.00              .00          .00
    1APR93        15       15          .00        .00          16.00       167.00              .00          .00
    2APR93        15       15          .00        .00          16.00       183.00              .00          .00
    3APR93        15       15          .00        .00            .00       183.00              .00          .00
    4APR93        15       15          .00        .00            .00       183.00              .00          .00
    5APR93        15       15          .00        .00          20.00       203.00              .00          .00
    6APR93        15       15          .00        .00          18.00       221.00              .00          .00
    7APR93        15       15          .00        .00          18.00       239.00              .00          .00
    8APR93        15       15          .00        .00           8.00       247.00              .00          .00
    9APR93        15       15          .00        .00           8.00       255.00              .00          .00
```

Figure 9 Required Resources in Tabular Format of Late Start Schedule

Most CPM-oriented software on the market is capable of performing aggregation or finding daily (time unit) project requirements for all resources estimated as needed to implement the plan.

ALLOCATED TOTAL FLOAT (AF) [DISTRIBUTED TOTAL FLOAT (DTF)]

The amount of total float distributed pro rata to activities along the same path based on the ratio of activity duration to path duration:

$$\text{AF of an activity} = \frac{\text{total float of the path x activity duration}}{\text{total duration of the path}}$$

Allocated total float is the portion of total float *(see Total float)* of any path that is shared among all activities of the path based on a criterion selected by the project management team. The allocated total float concept is a very new concept developed to assist reducing numbers of arguments of total float ownership. Recently, there have been a large number of construction claims regarding total float ownership. All participants demand the use of total float to allow them to extend their time (without penalty) when the project is delayed.

To distribute the total float to an activity appropriately, some agreement needs to be made prior to project scheduling. The project management team has to select and agree with the criterion used for total float distribution. For example, the criterion selected is activity duration, which means that the more duration an activity requires to perform the work, the more total float will be distributed to that activity. The criterion chosen can also be activity direct cost, activity resource demand, and so on.

Figure 1 is an example of allocated total float calculation based on the activity duration criterion mentioned earlier. The allocated total float can be computed from the following formula:

$$\text{Allocated total float of an activity} = \frac{\text{total float of the path x activity duration}}{\text{total duration of the path}}$$

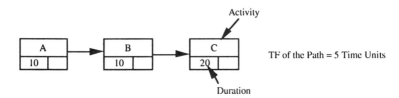

Figure 1 Example Path of Three Activities with 5 Time Units
of Total Float to Be Distributed

$$\text{AF of activity A} = \frac{5 \times 10}{10+10+20} = 1.25 = 1$$

$$\text{AF of activity B} = \frac{5 \times 10}{10+10+20} = 1.25 = 1$$

$$\text{AF of activity C} = \frac{5 \times 20}{10+10+20} = 2.50 = 3$$

The distribution procedures will start on the path with the least total float. After the first distribution, the process is repeated until all remaining total floats *(see Remaining total float)* are distributed. Figures 2 and 3 show the allocated total float for each activity in arrow and precedence diagrams, respectively.

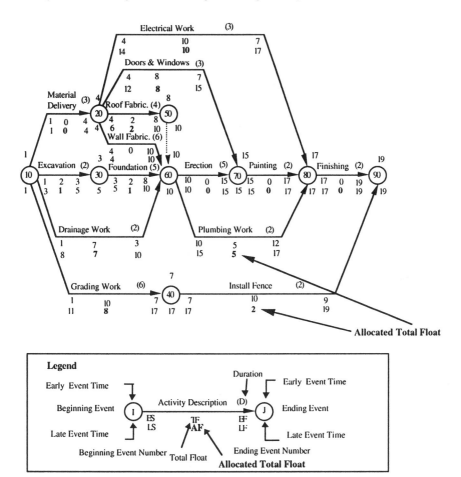

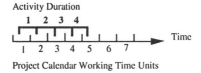

NOTE: The schedule computation in time units is based on the following assumption:

Activity Duration

Project Calendar Working Time Units

Activity with 4 time units starts at working time unit 1 and is finished at working time unit 5.

Figure 2 Allocated Total Float Shown in Arrow Diagram Network

61

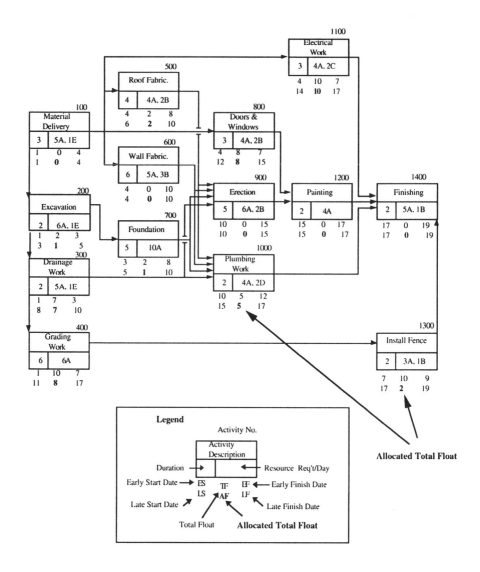

Legend

Activity No.

Activity Description	
Duration →	← Resource Req't/Day

Early Start Date → ES   TF   EF ← Early Finish Date
LS   AF   LF
Late Start Date ↗   ↘ Late Finish Date

Total Float   **Allocated Total Float**

Allocated Total Float

NOTE: The schedule computation in time units is based on the following assumption:

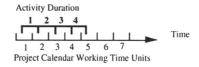

Activity Duration

Time

Project Calendar Working Time Units

Activity with 4 time units starts at working time unit 1 and is finished at working time unit 5.

Figure 3 Allocated Total Float Shown in Precedence Diagram

ALLOCATION (RESOURCE) The assignment of project resources to each project activity. Project activity scheduling with limited resources and with knowledge of certain constraints imposed on resource availability.

Project schedule considering limited available resources and with knowledge of certain constraints imposed on resources is the most realistic approach to project management. The allocation of available resources to project activities provides a way to produce a schedule from which resources required by the project during each working day (working time unit) will not exceed the available resources.

The availability level for each resource considered can be set constant over the activity duration, or variable, such as intensity over shorter time intervals, as shown in Figure 1. The availability of resources controls activity, project duration, and project final cost.

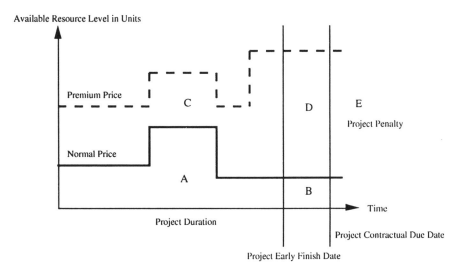

Figure 1 Variable Intensity of Available Resources over Project Duration

When analyzing resource availability, which affects project cost and project completion date, the following steps should be considered:

1. Obtain the early start schedule and project early finish date based on a CPM logic diagram. (All activities are scheduled to start at the early start date.)
2. Obtain the resources needed to implement the early start schedule based on the aggregation *(see Aggregation)*.
3. Determine the levels of resource availability to be allocated to various project activities:

a. Available resource levels at the normal procurement price (e.g., regular hours for the labor force).

b. Available resource levels at premium prices (e.g., extended shift: 10 hours/ day, 12 hours/day, and extended week, 6 or 7 days; the price of overtime hours will constitute premium price resources).

4. If, based on project resource requirements, availability at the normal price level does not satisfy, the delayed project completion date and cost of delay are determined.

5. Consider the premium price resources for a limited period, when project requirements exceed the normal resource price level for implementing the early start schedule. Compare the additional cost of premium-price resources with the cost of project delay.

Resource requirement/availability and economics of project completion date

1. The project's required resources can be satisfied at normal prices. The recommendation is to adopt the early start schedule and target the project early completion date (Figure 2).

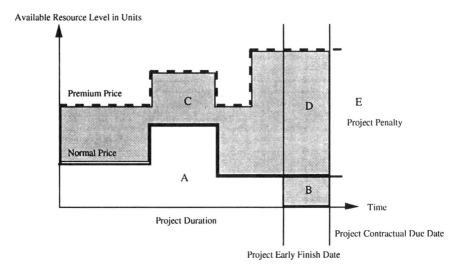

Figure 2 Required Resource Satisfied at Normal Prices

2. If the project requirements cannot be satisfied at the normal price level, maintain the project completion date, reschedule the activities, and target the project contractual due date (project late finish date) (Figure 3).

64

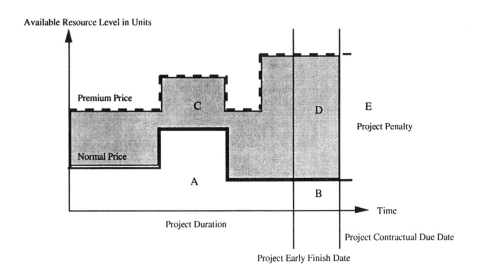

Figure 3 Required Resource Satisfied at Normal Prices Within Late Finish Date

3. If the early start schedule resource requirements can be satisfied using the resource premium-price level (if economical), the planner may choose this solution, targeting early completion (Figure 4).

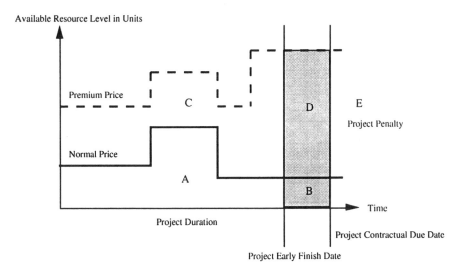

Figure 4 Required Resource at Premium Price Within Early Finish Date

4. If project resource requirements cannot be satisfied at the premium-price level by implementing an activity early start schedule but it will be possible to complete the project by the contractual due date, there is no alternative (Figure 5).

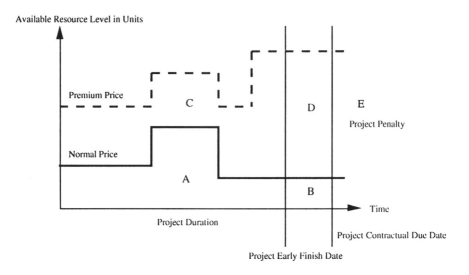

Figure 5 Required Resource at Premium Price Within Late Finish Date

Allocation impact on project schedule (computerized example)

1. *Precedence diagram.* With resources estimated for each activity, makes it possible to consider contract requirements during each activity's duration (Figure 6).

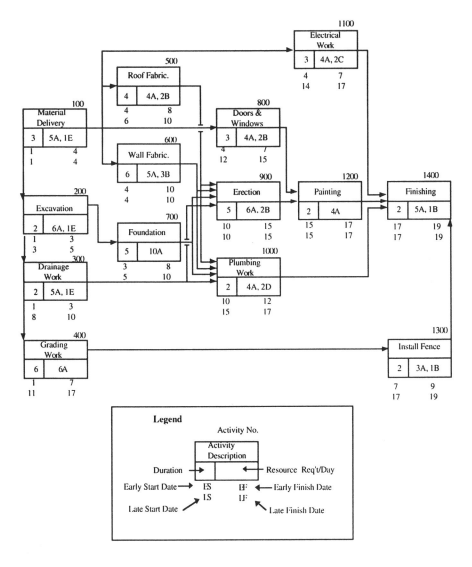

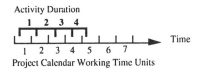

NOTE: The schedule computation in time units is based on the following assumption:

Activity with 4 time units starts at working time unit 1 and is finished at working time unit 5.

Figure 6 Precedence Diagram With Resources Estimated for Activities

2. *Computerized early start schedule* (Figure 7) and *resource histogram* (daily requirement for resource type AA) (Figure 8).

```
-----------------------------------------------------------------------------------------------------
CIVIL ENGINEERING DEPARTMENT                     PRIMAVERA PROJECT PLANNER

REPORT DATE 24NOV93  RUN NO.   39                   PROJECT SCHEDULE                 START DATE 13MAR93  FIN DATE 09APR93
            06:13
CLASSIC SCHEDULE REPORT (RESOURCE)                                               DATA DATE  17MAR93  PAGE NO.   1
-----------------------------------------------------------------------------------------------------
ACTIVITY   ORIG REM CAL              ACTIVITY              EARLY    EARLY    LATE     LATE    TOTAL
  ID       DUR  DUR  ID  %   CODE    DESCRIPTION           START    FINISH   START    FINISH  FLOAT RESOURC QUNTY
---------- ---- ---- --- --- -------- ------------------- -------- -------- -------- -------- ----- ------- -----

     100     3    3   1   0          MATERIAL DELIVERY    17MAR93  19MAR93  17MAR93  19MAR93    0     AA     15
                                                                                                     BB      3

     200     2    2   1   0          EXCAVATION           17MAR93  18MAR93  19MAR93  22MAR93    2     AA     12
                                                                                                     BB      2

     300     2    2   1   0          DRAINAGE WORK        17MAR93  18MAR93  26MAR93  29MAR93    7     AA     10
                                                                                                     BB      2

     400     6    6   1   0          GRADING WORK         17MAR93  24MAR93  31MAR93  07APR93   10     AA     36

     700     5    5   1   0          FOUNDATION           19MAR93  25MAR93  23MAR93  29MAR93    2     AA     50

     600     6    6   1   0          WALL FABRICATION     22MAR93  29MAR93  22MAR93  29MAR93    0     AA     30
                                                                                                     BB     18

     500     4    4   1   0          ROOF FABRICATION     22MAR93  25MAR93  24MAR93  29MAR93    2     AA     16
                                                                                                     BB      8

     800     3    3   1   0          DOORS & WINDOWS      22MAR93  24MAR93  01APR93  05APR93    8     AA     12
                                                                                                     BB      6

    1100     3    3   1   0          ELECTRICAL WORK      22MAR93  24MAR93  05APR93  07APR93   10     AA     12
                                                                                                     CC      6

    1300     2    2   1   0          INSTALL FENCE        25MAR93  26MAR93  08APR93  09APR93   10     AA      6
                                                                                                     BB      2

     900     5    5   1   0          ERECTION             30MAR93  05APR93  30MAR93  05APR93    0     AA     30
                                                                                                     BB     10

    1000     2    2   1   0          PLUMBING WORK        30MAR93  31MAR93  06APR93  07APR93    5     AA      8
                                                                                                     DD      4

    1200     2    2   1   0          PAINTING             06APR93  07APR93  06APR93  07APR93    0     AA      8

    1400     2    2   1   0          FINISHING            08APR93  09APR93  08APR93  09APR93    0     AA     10
                                                                                                     BB      2
```

Figure 7 Report with Required Resources for Activities

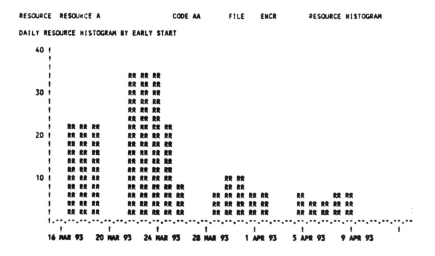

Figure 8 Resource Histogram for Resource AA

3. *Limited resource schedule.* The resource AA allocation is considered to be available at a constant rate of 15 units/project time unit. Figure 9 shows the resulting computerized schedule, and Figure 10 shows the new resource histogram for resource AA.

```
-----------------------------------------------------------------------------------------------------
CIVIL ENGINEERING DEPARTMENT                    PRIMAVERA PROJECT PLANNER

REPORT DATE 24NOV93  RUN NO.  43                  PROJECT SCHEDULE               START DATE 13MAR93  FIN DATE 15APR93
           06:36
CLASSIC SCHEDULE REPORT (RESOURCE)                                              DATA DATE  17MAR93  PAGE NO.   1
-----------------------------------------------------------------------------------------------------
```

ACTIVITY ID	ORIG DUR	REM DUR	CAL ID	%	CODE	ACTIVITY DESCRIPTION	EARLY START	EARLY FINISH	LATE START	LATE FINISH	TOTAL FLOAT	RESOURC	QUNTY
100	3	3	1	0		MATERIAL DELIVERY	17MAR93	19MAR93	17MAR93	19MAR93	0	AA	15
												BB	3
200	2	2	1	0		EXCAVATION	17MAR93	18MAR93	25MAR93	26MAR93	6	AA	12
												BB	2
300	2	2	1	0		DRAINAGE WORK	19MAR93	22MAR93	01APR93	02APR93	9	AA	10
												BB	2
600	6	6	1	0		WALL FABRICATION	22MAR93	29MAR93	26MAR93	02APR93	4	AA	30
												BB	18
500	4	4	1	0		ROOF FABRICATION	22MAR93	25MAR93	30MAR93	02APR93	6	AA	16
												BB	8
800	3	3	1	0		DOORS & WINDOWS	23MAR93	25MAR93	07APR93	09APR93	11	AA	12
												BB	6
700	5	5	1	0		FOUNDATION	26MAR93	01APR93	29MAR93	02APR93	1	AA	50
1100	3	3	1	0		ELECTRICAL WORK	30MAR93	01APR93	09APR93	13APR93	8	AA	12
												CC	6
900	5	5	1	0		ERECTION	02APR93	08APR93	05APR93	09APR93	1	AA	30
												BB	10
1000	2	2	1	0		PLUMBING WORK	02APR93	05APR93	12APR93	13APR93	6	AA	8
												DD	4
400	6	6	1	0		GRADING WORK	06APR93	13APR93	06APR93	13APR93	0	AA	36
1200	2	2	1	0		PAINTING	09APR93	12APR93	12APR93	13APR93	1	AA	8
1300	2	2	1	0		INSTALL FENCE	14APR93	15APR93	14APR93	15APR93	0	AA	6
												BB	2
1400	2	2	1	0		FINISHING	14APR93	15APR93	14APR93	15APR93	0	AA	10
												BB	2

Figure 9 Schedule When Resource AA Is Limited to 15 Units/Project Time Unit

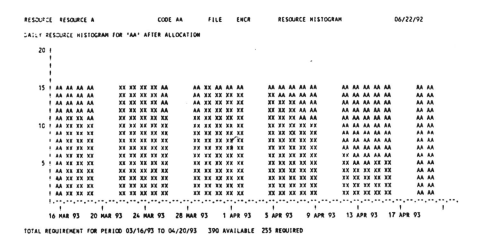

DAILY RESOURCE HISTOGRAM FOR 'AA' AFTER ALLOCATION

TOTAL REQUIREMENT FOR PERIOD 03/16/93 TO 04/20/93   390 AVAILABLE 255 REQUIRED

Legend: XX: Project time unit when required resources have been satisfied
              from availability pool
         AA: Available resources in excess of required resources

Figure 10 Resource Histogram When Resource AA Is Limited
to 15 Units/Project Time Unit

4. Limited resource schedule/allocation considering the availability of resource
    AA during project duration is shown in Table 1.

Table 1 Availability of Resource AA

Period	Duration	Resource	Intensity
1	8	AA	15 Units/working time unit
2	7	AA	6 Units/working time unit
3	6	AA	20 Units/working time unit

Figure 11 shows the project schedule of all activities sorted in ascending order
of early start, and Figure 12 shows the histogram for resource AA.

71

CIVIL ENGINEERING DEPARTMENT   PRIMAVERA PROJECT PLANNER

REPORT DATE 24NOV93 RUN NO.   45   PROJECT SCHEDULE   START DATE 13MAR93 FIN DATE 19APR93
            06:47
CLASSIC SCHEDULE REPORT (RESOURCE)   DATA DATE 17MAR93 PAGE NO.   1

ACTIVITY ID	ORIG DUR	REM DUR	CAL ID	%	CODE	ACTIVITY DESCRIPTION	EARLY START	EARLY FINISH	LATE START	LATE FINISH	TOTAL FLOAT	RESOURC	QUNTY
100	3	3	1	0		MATERIAL DELIVERY	17MAR93	19MAR93	17MAR93	19MAR93	0	AA	15
												BB	3
200	2	2	1	0		EXCAVATION	17MAR93	18MAR93	29MAR93	30MAR93	8	AA	12
												BB	2
300	2	2	1	0		DRAINAGE WORK	19MAR93	22MAR93	05APR93	06APR93	11	AA	10
												BB	2
600	6	6	1	0		WALL FABRICATION	22MAR93	29MAR93	30MAR93	06APR93	6	AA	30
												BB	18
800	3	3	1	0		DOORS & WINDOWS	23MAR93	25MAR93	09APR93	13APR93	13	AA	12
												BB	6
500	4	4	1	0		ROOF FABRICATION	30MAR93	02APR93	01APR93	06APR93	2	AA	16
												BB	8
1100	3	3	1	0		ELECTRICAL WORK	30MAR93	01APR93	13APR93	15APR93	10	AA	12
												CC	6
700	5	5	1	0		FOUNDATION	31MAR93	06APR93	31MAR93	06APR93	0	AA	50
400	6	6	1	0		GRADING WORKS	06APR93	13APR93	08APR93	15APR93	2	AA	36
900	5	5	1	0		ERECTION	07APR93	13APR93	07APR93	13APR93	0	AA	30
												BB	10
1000	2	2	1	0		PLUMBING WORK	07APR93	08APR93	14APR93	15APR93	5	AA	8
												DD	4
1200	2	2	1	0		PAINTING	14APR93	15APR93	14APR93	15APR93	0	AA	8
1300	2	2	1	0		INSTALL FENCE	14APR93	15APR93	16APR93	19APR93	2	AA	6
												BB	2
1400	2	2	1	0		FINISHING	16APR93	19APR93	16APR93	19APR93	0	AA	10
												BB	2

Figure 11 Project Schedule After Allocating Resource AA Based on Table 1

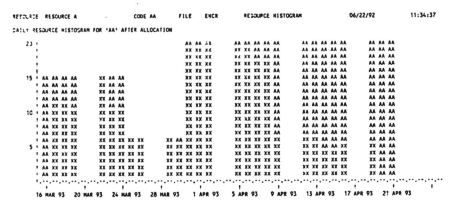

RESOURCE RESOURCE A   CODE AA   FILE   ENCR   RESOURCE HISTOGRAM   06/22/92   11:34:37

Legend: XX: Project time unit when required resources have been satisfied
from availability pool
AA: Available resources in excess of required resources

Figure 12 Histogram of Resource AA After Allocation
Based on Availability from Table 1

APPORTIONED CONCURRENCY An apportionment of time and/or cost impacts arising from two or more delays that occur during the same general time window within a project schedule—wherein one delay may overlap with another delay event, each caused by different parties. For example, if several subcontractors delay a contractor's CPM schedule, there may be apportioned concurrency whereby the contractor assesses a portion of the total delay period to each of the contributing parties if there is no single party to which the overriding critical path delay can be assigned.

ARROW The graphical representation of an activity in the CPM network arrow diagramming method or PERT. One arrow represents one activity. Unlike the time-scaled diagram, the arrow is not a vector quantity and is not drawn to scale. It is defined uniquely by two event nodes *(see Event)*.

Each arrow represents an activity in the network with a defined start and end. It is drawn in a solid line, normally a straight line except for a dummy activity *(see Dummy activity)*, in which the activity representation is a dashed line. The beginning and end of an arrow show the start and finish of one activity in the CPM network. The direction of the arrow shows the flow of time during which a project must

be processed. The arrow size and spacing are quite important. If the arrows in the CPM network are too long and widely spread apart, the diagram will become too large. Conversely, if the arrows are too tight and small, the diagram will be crowded and difficult to read, review, and amend. An example of an arrow network is shown in Figure 1.

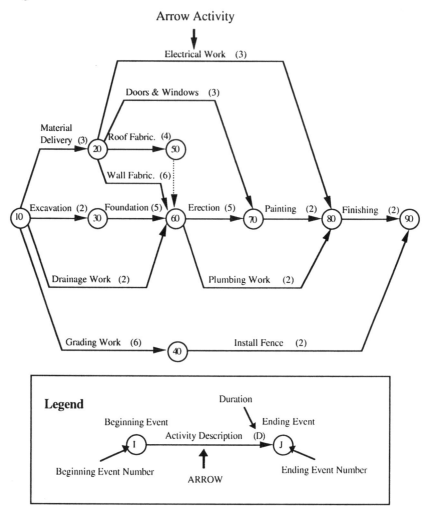

Figure 1 Arrow Diagram Network

ARROW DIAGRAM *(see Arrow diagramming method)*

ARROW DIAGRAMMING METHOD (ADM) A network on which activities are represented by an arrow between event nodes *(see Event)*.

The arrow diagramming method is a network scheduling technique that uses arrows to show the construction sequences and relationships of activities to be performed in a project. The diagram is drawn after the activities and their interrelationships have been defined. In an arrow diagram each line or arrow represents one activity, and the sequential start and finish between activities represent their relationship. A circle represents an event that is the start of the activity when all activities preceding it have been completed. Normally, an arrow diagram is developed before the activity data are input in a computer and a network computation is performed. Construction of the arrow diagram depends on the technical and logical constraints of how work will be performed. A sample arrow diagram is shown in Figure 1.

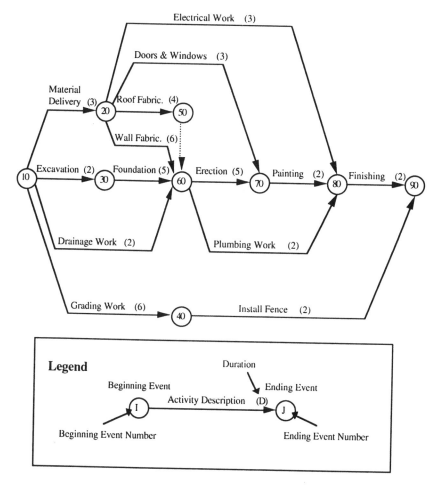

Figure 1 Arrow Diagram Network

When an arrow diagram cannot be drawn on one page, a connection from one page to another has to be made. Figure 2 shows how to make that connection.

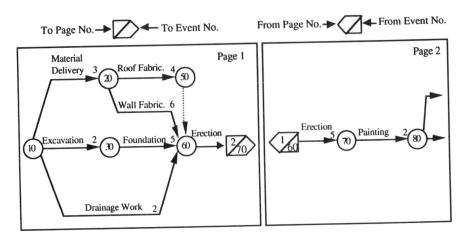

Figure 2 Connection Between Two Pages in Arrow Diagram

ARROW NETWORK *(see Arrow diagramming method)*

AS-BUILT PROJECT SCHEDULE (AS-BUILT SCHEDULE) An interim or final project schedule that depicts for each completed activity the actual start and completion dates, actual duration, cost, resources consumed, and actual logic relations with other activities.

An as-built project schedule demonstrates what occurred historically in a project, from start to completion. The as-built schedule represents project progress by proper updating of the project schedule and related information such as cost and resources. During the updating process, the completed as-planned activity, *(see As-planned project schedule)* is documented, becoming the as-built information, as the project proceeds. Moreover, any occurrence that affects the as-planned project schedule should be recorded to reflect the causes of project delays. The last updating of the project will be used as an as-built project schedule.

If contractors neglect to update the project schedule adequately and delay or acceleration disputes arise, the as-built schedule will be prepared from daily logs or other contract documents. The creation of an accurate as-built schedule after project completion is difficult, since sequencing or relationships of work activities may have changed from the as-planned schedule. When delay or acceleration disputes arise, the as-built and as-planned schedule will be compared to determine the causes and impacts. Figures 1, 2, and 3 show an as-built precedence diagram network, tabular schedule report, and bar chart schedule report, respectively.

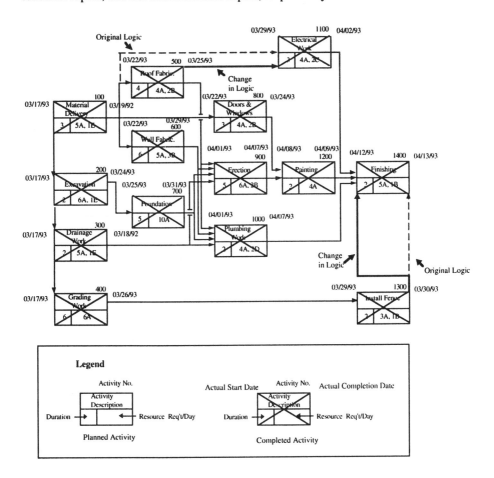

Figure 1 As-built Precedence Diagram Network

CIVIL ENGINEERING DEPARTMENT   PRIMAVERA PROJECT PLANNER

REPORT DATE 23NOV93 RUN NO.   7   PROJECT SCHEDULE REPORT   START DATE 13MAR93 FIN DATE 16APR93
            21:11
CLASSIC SCHEDULE REPORT - SORT BY ES, TF   DATA DATE 17APR93 PAGE NO.   1

ACTIVITY ID	ORIG DUR	REM DUR	%	CODE	ACTIVITY DESCRIPTION	EARLY START	EARLY FINISH	LATE START	LATE FINISH	TOTAL FLOAT
100	3	0	100		MATERIAL DELIVERY	17MAR93A	19MAR93A			
200	2	0	100		EXCAVATION	17MAR93A	24MAR93A			
300	2	0	100		DRAINAGE WORK	17MAR93A	18MAR93A			
400	6	0	100		GRADING WORK	17MAR93A	26MAR93A			
500	4	0	100		ROOF FABRICATION	22MAR93A	25MAR93A			
600	6	0	100		WALL FABRICATION	22MAR93A	29MAR93A			
800	3	0	100		DOORS & WINDOWS	22MAR93A	24MAR94A			
700	5	0	100		FOUNDATION	25MAR93A	31MAR93A			
1100	3	0	100		ELECTRICAL WORK	29MAR93A	2APR93A			
1300	2	0	100		INSTALL FENCE	29MAR93A	30MAR93A			
900	5	0	100		ERECTION	1APR93A	7APR93A			
1000	2	0	100		PLUMBING WORK	1APR93A	7APR93A			
1200	2	0	100		PAINTING	8APR93A	9APR93A			
1400	2	0	100		FINISHING	12APR93A	13APR93A			

Figure 2 As-built Tabular Schedule Report

```
-------------------------------------------------------------------------------------------------------
CIVIL ENGINEERING DEPARTMENT              PRIMAVERA PROJECT PLANNER

REPORT DATE 23NOV93  RUN NO.   6          PROJECT SCHEDULE REPORT                    START DATE 13MAR93  FIN DATE 16APR93
         21:08
BAR CHART BY ES, EF, TF                                                             DATA DATE 17APR93   PAGE NO.   1

                                                                                    DAILY-TIME PER.   1

-------------------------------------------------------------------------------------------------------
..........ACTIVITY DESCRIPTION............    15    22    29    05    12    19    26    03    10    17
ACTIVITY ID  OD   RD  PCT   CODES   FLOAT   SCHEDULE   MAR   MAR   MAR   APR   APR   APR   APR   MAY   MAY   MAY
-----------  ----  ---- ---  ------------  -----   --------   93    93    93    93    93    93    93    93    93    93
-------------------------------------------------------------------------------------------------------
DRAINAGE WORK                              EARLY    . AA   .     .     .     .    * .    .     .     .     .
      300    2    0 100                              .     .     .     .     .    * .    .     .     .     .

MATERIAL DELIVERY                          EARLY    . AAA  .     .     .     .    * .    .     .     .     .
      100    3    0 100                              .     .     .     .     .    * .    .     .     .     .

EXCAVATION                                 EARLY    . AAA..AAA   .     .     .    * .    .     .     .     .
      200    2    0 100                              .     .     .     .     .    * .    .     .     .     .

GRADING WORK                               EARLY    . AAA..AAAAA  .    .     .    * .    .     .     .     .
      400    6    0 100                              .     .     .     .     .    * .    .     .     .     .

ROOF FABRICATION                           EARLY    .    AAAA   .     .     .    * .    .     .     .     .
      500    4    0 100                              .     .     .     .     .    * .    .     .     .     .

WALL FABRICATION                           EARLY    .    AAAAA..A  .    .     .    * .    .     .     .     .
      600    6    0 100                              .     .     .     .     .    * .    .     .     .     .

DOORS & WINDOWS                            EARLY    .    AAAAA..AAAAA..AAAAA..AAAAA* .    .     .     .     .
      800    3    0 100                              .     .     .     .     .    * .    .     .     .     .

FOUNDATION                                 EARLY    .     . AA..AAA   .     .    * .    .     .     .     .
      700    5    0 100                              .     .     .     .     .    * .    .     .     .     .

INSTALL FENCE                              EARLY    .     .    AA    .     .    * .    .     .     .     .
     1300    2    0 100                              .     .     .     .     .    * .    .     .     .     .

ELECTRICAL WORK                            EARLY    .     .    AAAAA  .     .    * .    .     .     .     .
     1100    3    0 100                              .     .     .     .     .    * .    .     .     .     .

ERECTION                                   EARLY    .     .    . AA..AAA   .    * .    .     .     .     .
      900    5    0 100                              .     .     .     .     .    * .    .     .     .     .

PLUMBING WORK                              EARLY    .     .    . AA..AAA   .    * .    .     .     .     .
     1000    2    0 100                              .     .     .     .     .    * .    .     .     .     .

PAINTING                                   EARLY    .     .     .     . AA   .    * .    .     .     .     .
     1200    2    0 100                              .     .     .     .     .    * .    .     .     .     .

FINISHING                                  EARLY    .     .     .     .    AA   * .    .     .     .     .
     1400    2    0 100                              .     .     .     .     .    * .    .     .     .     .
```

Figure 3 As-built Bar Chart Schedule Report

79

AS-PLANNED PROJECT SCHEDULE (AS-PLANNED SCHEDULE) (1) The original or baseline schedule, generally developed prior to or soon after project construction is begun, demonstrating the anticipated sequence, durations, and interdependencies for the activities constituting the contract work. (2) An interim project schedule demonstrating the anticipated sequence, durations, and interdependencies for all incomplete activities as of the data date.

An as-planned schedule is the schedule that a contractor submits and that the owner approves at the beginning of the project. The most important part of this schedule is that it represents the contractor's plan and initiation of the work in order to meet the requirements of the contract before the project is executed. This schedule will form the basis for project schedule control. If it is not created thoughtfully and carefully, poor project execution results. Conversely, a well-prepared project schedule will provide a sound working plan for daily activities.

In preparing an as-planned schedule, a contractor should establish the project breakdown structure (PBS) *(see Project breakdown structure)* to serve as a framework for establishing a project schedule. The detailed project as-planned schedule should be established at the work package level of the PBS *(see Work package),* as shown in Figure 1.

An as-planned project schedule can be changed or modified during the course of construction due to change orders or unexpected occurrences. A good as-planned schedule will accurately assess the project program and readily identify problem areas during construction. To minimize project disputes and have better project control, the as-planned schedule should be updated regularly so that potential problem areas can be located rapidly before they take place. A poor as-planned schedule, on the other hand, will not allow adequate project evaluation and will produce unidentifiable problems. In analyzing construction delays, the as-planned schedule will be used together with an as-built schedule *(see As-built project schedule)* as an analysis tool to identify who and what is causing the delays.

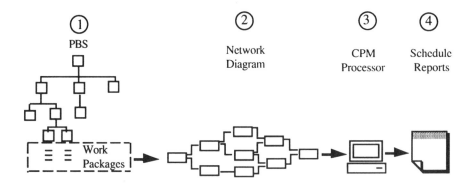

Figure 1 As-planned Project Schedule Development

Figures 2 and 3 show arrow and precedence diagram networks for an as-planned project schedule. Figures 4 and 5 show as-planned tabular and bar chart schedule reports for a precedence diagram network.

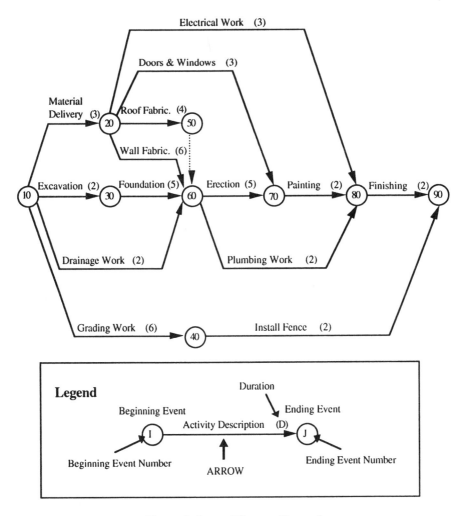

Figure 2 Arrow Diagram Network

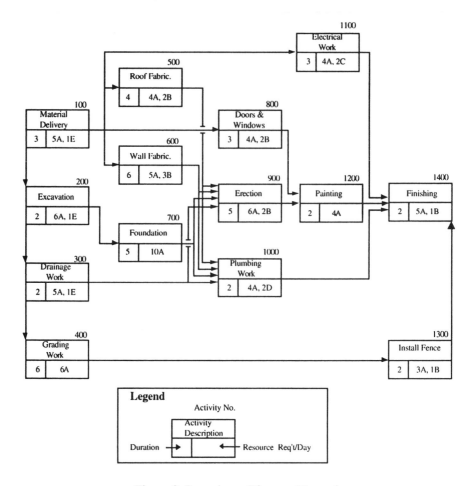

Figure 3 Precedence Diagram Network

82

```
--------------------------------------------------------------------------------------------------------
CIVIL ENGINEERING DEPARTMENT                    PRIMAVERA PROJECT PLANNER

REPORT DATE 23NOV93  RUN NO.   17              PROJECT SCHEDULE REPORT                      START DATE 13MAR93   FIN DATE  9APR93
         21:14
CLASSIC SCHEDULE REPORT - SORT BY ES, TF                                                   DATA DATE  17MAR93  PAGE NO.    1


----- -----  ---- ---- - ---  ----------  ------------------------------------------------  --------  --------  --------  --------  -----
ACTIVITY    ORIG REM                                                                        EARLY     EARLY     LATE      LATE      TOTAL
   ID       DUR  DUR   %  CODE                ACTIVITY DESCRIPTION                           START     FINISH    START     FINISH    FLOAT
----- -----  ---- ---- - ---  ----------  ------------------------------------------------  --------  --------  --------  --------  -----
      100     3    3   0        MATERIAL DELIVERY                                            17MAR93*  19MAR93   17MAR93*  19MAR93      0
      200     2    2   0        EXCAVATION                                                   17MAR93   18MAR93   19MAR93   22MAR93      2
      300     2    2   0        DRAINAGE WORK                                                17MAR93   18MAR93   26MAR93   29MAR93      7
      400     6    6   0        GRADING WORK                                                 17MAR93   24MAR93   31MAR93    7APR93     10
      700     5    5   0        FOUNDATION                                                   19MAR93   25MAR93   23MAR93   29MAR93      2
      600     6    6   0        WALL FABRICATION                                             22MAR93   29MAR93   22MAR93   29MAR93      0
      500     4    4   0        ROOF FABRICATION                                             22MAR93   25MAR93   24MAR93   29MAR93      2
      800     3    3   0        DOORS & WINDOWS                                              22MAR93   24MAR93    1APR93    5APR93      8
     1100     3    3   0        ELECTRICAL WORK                                              22MAR93   24MAR93    5APR93    7APR93     10
     1300     2    2   0        INSTALL FENCE                                                25MAR93   26MAR93    8APR93    9APR93     10
      900     5    5   0        ERECTION                                                     30MAR93    5APR93   30MAR93    5APR93      0
     1000     2    2   0        PLUMBING WORK                                                30MAR93   31MAR93    6APR93    7APR93      5
     1200     2    2   0        PAINTING                                                      6APR93    7APR93    6APR93    7APR93      0
     1400     2    2   0        FINISHING                                                     8APR93    9APR93    8APR93    9APR93      0
```

Figure 4 As-planned Tabular Schedule Report, Precedence Diagram Network

83

.............ACTIVITY DESCRIPTION.............							15	22	29	05	12	19	26	03	10	17
ACTIVITY ID	OD	RD	PCT	CODES	FLOAT	SCHEDULE	MAR	MAR	MAR	APR	APR	APR	APR	MAY	MAY	MAY
							93	93	93	93	93	93	93	93	93	93

EXCAVATION						EARLY	. EE	.	.	.	.	.	.	.	.	.
200	2	2	0		2		. *	.	.	.	.	.	.	.	.	.
DRAINAGE WORK						EARLY	. EE	.	.	.	.	.	.	.	.	.
300	2	2	0		7		. *	.	.	.	.	.	.	.	.	.
MATERIAL DELIVERY						EARLY	. EEE	.	.	.	.	.	.	.	.	.
100	3	3	0		0		. *	.	.	.	.	.	.	.	.	.
GRADING WORK						EARLY	. EEE..EEE	.	.	.	.	.	.	.	.	.
400	6	6	0		10		. *	.	.	.	.	.	.	.	.	.
FOUNDATION						EARLY	. * E..EEEE	.	.	.	.	.	.	.	.	.
700	5	5	0		2		. *	.	.	.	.	.	.	.	.	.
DOORS & WINDOWS						EARLY	. *	EEE	.	.	.	.	.	.	.	.
800	3	3	0		8		. *	.	.	.	.	.	.	.	.	.
ELECTRICAL WORK						EARLY	. *	EEE	.	.	.	.	.	.	.	.
1100	3	3	0		10		. *	.	.	.	.	.	.	.	.	.
ROOF FABRICATION						EARLY	. *	EEEE	.	.	.	.	.	.	.	.
500	4	4	0		2		. *	.	.	.	.	.	.	.	.	.
WALL FABRICATION						EARLY	. *	EEEEE..E	.	.	.	.	.	.	.	.
600	6	6	0		0		. *	.	.	.	.	.	.	.	.	.
INSTALL FENCE						EARLY	. *	. EE	.	.	.	.	.	.	.	.
1300	2	2	0		10		. *	.	.	.	.	.	.	.	.	.
PLUMBING WORK						EARLY	. *		.EE	.	.	.	.	.	.	.
1000	2	2	0		5		. *	.	.	.	.	.	.	.	.	.
ERECTION						EARLY	. *		.EEEE..E	.	.	.	.	.	.	.
900	5	5	0		0		. *	.	.	.	.	.	.	.	.	.
PAINTING						EARLY	. *	.	.	.EE	.	.	.	.	.	.
1200	2	2	0		0		. *	.	.	.	.	.	.	.	.	.
FINISHING						EARLY	. *	.	.	. EE	.	.	.	.	.	.
1400	2	2	0		0		. *	.	.	.	.	.	.	.	.	.

Figure 5 As-planned Bar Chart Schedule Report, Precedence Diagram Network

AT-COMPLETION VARIANCE (ACV) The difference between budget cost and forecasted (estimated) costs at completion for all project work items at any data date (time now).

The at-completion variance (ACV) is one of the C/SCSC elements and represents a performance measurement of whether the project is over or under budget based on the work accomplished to date, which is revised at the current data date. The ACV is calculated using the formula

$$ACV = BAC - EAC$$

(See Budget at completion and Estimate at completion.) Interpretation of the ACV is as follows:

VARIANCE	-	0	+
ACV	Over Budget	On Budget	Under Budget

As shown in the table, if a (–) or (+) value of the ACV is found, the reason should be identified. For example, the following can be possible causes:

1. Change orders
2. Improper EAC or BAC
3. Engineering design changes
4. Actual productivity lower than planned productivity
5. Outright waste greater than normal or estimated waste

To develop the ACV in a project, the following steps are generally followed:

Step 1: Use the original PBS of a project.
Step 2: Use the original cost estimation of the project based on the PBS and baseline schedule.
Step 3: Use the latest updated CPM schedule on the data date (time now) *(see Data date).*
Step 4: Calculate the BAC on the data date (time now).
Step 5: Calculate the EAC on the data date (time now).
Step 6: Calculate the ACV by subtracting the EAC from the BAC. The results of steps 4 to 6 are shown in the monthly cost performance report.
Step 7: Indicate the ACV on the graph using the time-phased and cumulative earned values on the data date (time now).

The preceding seven steps are explained using a residential house for an example.

1. *Step 1:* PBS

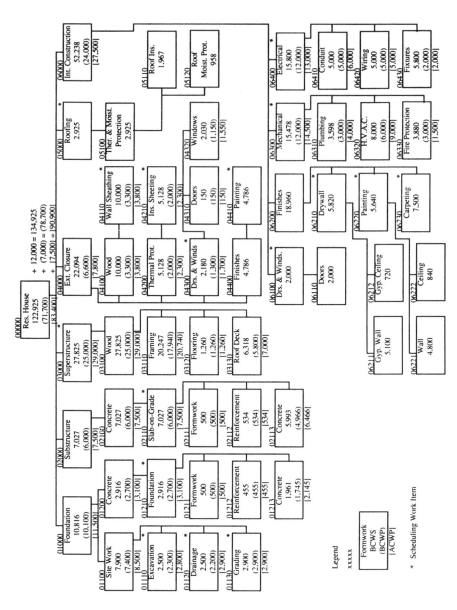

2. *Step 2:* Cost Estimation

DESCRIPTION	U/M	U/P	QTY	TOTAL
01 General Requirements				
Subtotal	L/S	12,000	1	12,000
02 Site Work				
01110 Excavation	C.Y.	1.89	1,323	2,500
01120 Drainage	L.F.	3.11	804	2,500
01130 Grading	C.Y.	1.06	2,736	2,900
Subtotal				7,900
03 Concrete				
01211 Foundation Form	SFCA	2.89	173	500
01212 Foundation Re-bar	Tons	910	0.5	455
01213 Foundation Concrete	C.Y.	103.2	19	1,961
02111 Slab-on-Grade Form	L.F.	1.38	362	500
02112 Slab-on-Grade Re-bar	Tons	890	0.6	534
02113 Slab-on-Grade Concrete	C.Y.	90.8	66	5,993
Subtotal				9,943
06 Wood				
03110 Superstructure Framing	M.B.F.	1,191	17	20,247
03120 Superstructure Flooring	S.F.	0.63	2,000	1,260
03130 Roof Decking	S.F.	5.01	1,261	6,318
04110 Wall Sheathing	S.F.	1.17	8,547	10,000
Subtotal				37,825
07 Thermal & Moisture Protection				
04210 Wall Insulating Sheeting	S.F.	0.6	8,547	5,128
05110 Roof Insulating	S.F.	1.56	1,261	1,967
05120 Roof Moisture Protection	S.F.	0.76	1,261	958
Subtotal				8,053
08 Doors & Windows				
04310 Ext. Doors	Ea.	150	1	150
04320 Ext. Windows	Ea.	203	10	2,030
06110 Int. Doors	Ea.	200	10	2,000
Subtotal				4,180
09 Finishes				
04410 Ext. Painting	S.F.	0.56	8,547	4,786
06211 Int. Gyp. Wallboard	S.F.	0.34	15,000	5,100
06212 Int. Gyp. Ceiling Board	S.F.	0.36	2,000	720
06311 Int. Wall Painting	S.F.	0.32	15,000	4,800
06312 Int. Ceiling Painting	S.F.	0.42	2,000	840
06230 Carpeting	S.Y.	5	1,500	7,500
Subtotal				23,746
15 Mechanical				
06310 Plumbing	L.F.	3.98	904	3,598
06320 H.V.A.C.	L/S	8,000	1	8,000
06330 Fire Protection	Ea.	776	5	3,880
Subtotal				15,478
16 Electrical				
06410 Electrical Conduit	L/S	5,000	1	5,000
06420 Electrical Wiring	L/S	5,000	1	5,000
06430 Electrical Fixtures	L/S	5,800	1	5,800
Subtotal				15,800
GRAND TOTAL				134,925

3. *Step 3:* Updated CPM Network Schedule

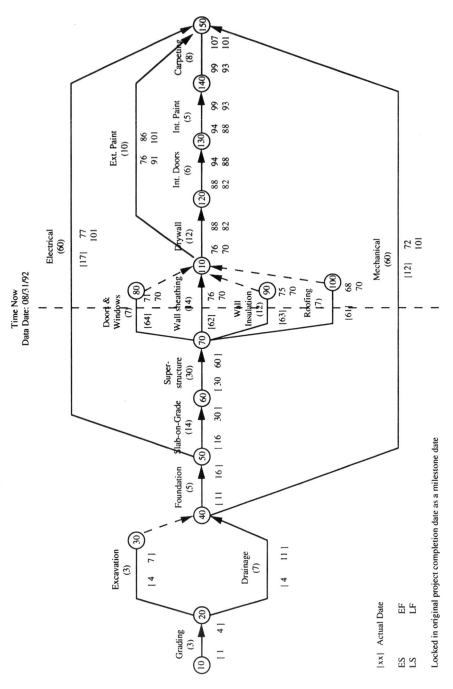

4. *Steps 4 to 6:* Monthly Cost Performance Report

MONTHLY COST PERFORMANCE REPORT: CPM ITEM

PROJECT: RESIDENTIAL HOUSE REPORT DATE: 08/31/92

| ITEM | | CURRENT PERIOD | | | | | | | | | CUMULATIVE TO DATE | | | | | | | | | AT COMPLETION | | |
PBS	CPM	BCWS	BCWP	ACWP	SV	CV	AV	SPI	CPI	PC	BCWS	BCWP	ACWP	SV	CV	AV	SPI	CPI	PC	BAC	EAC	ACV
011 30	10 - 20										2,900	2,900	2,900	0	0	0	1.00	1.00	1.00	2,900	2,900	0
011 10	20 - 30										2,500	2,300	2,800	-200	-500	-300	0.92	0.82	0.92	2,500	2,800	-300
011 20	20 - 40										2,500	2,200	2,800	-300	-600	-300	0.88	0.79	0.88	2,500	2,800	-300
012 10	40 - 50										2,916	2,700	3,100	-216	-400	-184	0.93	0.87	0.93	2,916	3,100	-184
063 00	40 - 150	5,160	5,000	5,200	-160	-200	-40	0.97	0.96	0.32	13,416	12,000	14,500	-1,416	-2,500	-1,084	0.89	0.83	0.78	15,478	16,562	-1,084
021 10	50 - 60										7,027	6,000	7,500	-1,027	-1,500	-473	0.85	0.80	0.85	7,027	7,500	-473
030 00	60 - 70	14,903	13,900	15,000	-1,003	-1,100	-97	0.93	0.93	0.50	27,825	25,000	29,000	-2,825	-4,000	-1,175	0.90	0.86	0.90	27,825	29,000	-1,175
064 00	50 - 150	5,260	5,100	5,300	-160	-200	-40	0.97	0.96	0.32	12,361	12,000	13,000	-361	-1,000	-639	0.97	0.92	0.76	15,800	16,439	-639
043 00	70 - 80	1,555	1,300	1,700	-255	-400	-145	0.84	0.76	0.60	1,555	1,300	1,700	-255	-400	-145	0.84	0.76	0.60	2,180	2,325	-145
042 10	70 - 90	2,135	2,000	2,300	-135	-300	-165	0.94	0.87	0.39	2,135	2,000	2,300	-135	-300	-145	0.94	0.87	0.39	5,128	5,293	-165
041 10	70 - 110	3,570	3,300	3,800	-270	-500	-230	0.92	0.87	0.33	3,570	3,300	3,800	-270	-500	-230	0.92	0.87	0.33	10,000	10,230	-230
050 00	70 - 100																			2,925	2,925	0
062 10	110-120																			5,820	5,820	0
044 10	110-150																			4,786	4,786	0
061 00	120-130																			2,000	2,000	0
062 20	130-140																			5,640	5,640	0
062 30	140-150																			7,500	7,500	0
SUBTOTAL		32,583	30,600	33,300	-1,983	-2,700	-717	0.94	0.92	0.25	78,705	71,700	83,400	-7,005	-11,700	-4,695	0.91	0.86	0.58	122,925	127,620	-4,695
GEN. REQ.		2,400	2,400	2,500		-100	-100	1.00	0.96	0.20	7,200	7,000	7,500	-200	-500	-300	0.97	0.93	0.58	12,000	12,300	-300
TOTAL		34,983	33,000	35,800	-1,983	-2,800	-817	0.94	0.92	0.24	85,905	78,700	90,900	-7,205	-12,200	-4,995	0.92	0.87	0.58	134,925	139,920	-4,995

Note: The project at the data date has -4,995 as the at-completion variance (ACV). It shows that the project is forecast over budget because the ACV is less than 0.

89

5. *Step 7:* ACV

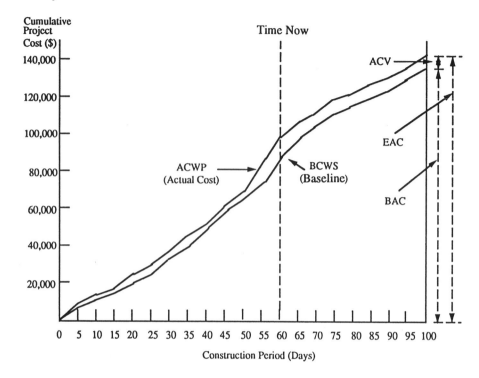

AUDIT (SCHEDULE) A formal or official examination and verification of a schedule by auditors.

The auditing procedures consist of a review of the implemented CPM-oriented information system which provides information with which the managers plan and schedule a project to achieve the contract's expected results. An audit is generally triggered when the following events related to the project occur:

1. The project is behind schedule and the owner requests a review of the network logic.
2. The scope of work is not well defined and many changes have been originated by parties involved with the project.
3. Close to completion, the project is far behind schedule and litigation is inevitable.
4. There is project cancellation or work stoppage due to budgeting constraints, change in scope, or contractor default.
5. Excessive cost overruns have occurred and justifications are needed.
6. Mismanagement is a key issue in publicly financed projects.

The situations mentioned above will generate an audit of the entire project management system. The request for an audit may come from the owner, or the owner's representative (construction manager), but it often comes from the public, which is concerned with the financial consequences. An auditing system should be implemented from project inception and should be used during the project's life time, from adjustment of the planning and scheduling system to changing environment conditions. In auditing a planning and scheduling system, the primary goal is to understand its structure and components:

1. Hardware (computer and peripherals)
2. CPM-oriented software
3. Information and communication channels: the networks, reports, and distribution media/procedures
4. Personnel who are users of the system-generated information

An audit of an implemented planning and scheduling system should take place within a few months after start of the project, and at least once a year. The starting point of the audit are the planning and scheduling specifications, to determine if the performance is meeting the objectives specified. The quality of information provided, accuracy of computation, input data controls, and turnaround time are a few of the items to be audited. The flow of information through the communication channels is audited also, starting with a specific matrix distribution of reports generated. Key users should be interviewed to ascertain if the proper information is received in an adequate format and needed feedback is sent into the system. This denotes the use of computerized reports for management decisions.

AUTHORIZED SCHEDULE The final owner's approval of a network schedule that has been submitted.

Before beginning construction and after it has been begun, a general contractor will prepare various project schedules to show the plan of work. To authorize a schedule, an owner or owner's representative will analyze the submitted schedule to see if it follows the scope of work and is logical. Various types of schedules should be prepared and submitted to the owner during various stages of the project, as follows:

1. *Conceptual schedule (see Conceptual schedule)*. The conceptual schedule is normally submitted with the bidding documents, usually at the owner's request.
2. *Preliminary schedule (see Preliminary schedule)*. Within a specified number of days after notice of the contract award, the contractor, with the owner or owner's representative's approval, submits the preliminary network diagram reflecting the contractor's planned operation during early stages of the project.
3. *Detailed schedule (see Detailed schedule)*. Within a specified number of days after notice of the contract award, the contractor submits for the approval of the owner or owner's representative a network diagram and computerized analysis using the CPM software selected.
4. *As-built schedule to date (see As-built project schedule)*. At each update the status of completed activities (start and completion dates) and actual logic is recorded for future reference and accuracy of analysis.
5. *Updated schedule (see Updated schedule)*. A detailed schedule should be updated on a regular basis as specified in the specifications *(see Specifications)* to reflect the current status of the project. The revised schedule after updating must be approved by the owner or owner's representative.

Review and approval process. Within the number of days specified after receipt of the preliminary and/or completed network diagram(s), the owner or owner's representative will meet with the contractor for a joint review of the proposed plan and schedule. The network diagram, as approved by the owner's representative, called an *as-planned project schedule (see As-planned project schedule),* often constitutes the project schedule until it is updated subsequently in accordance with the requirements in the specifications *(see Specifications)*. During updating, the contractor will describe, on an activity-by-activity basis, all proposed revisions of and adjustments to the network required to reflect the current status of the project. The owner's representative often approves the progress, proposed revisions, and adjustments to the schedule.

B

BACKWARD PASS COMPUTATION Calculation of late finish (LF) and late start (LS) times *(see Late finish date* and *Late start date)* for all uncompleted activities in the network.

In the critical path method (CPM), the backward pass computation determines the latest start and latest finish times for each activity. The activity latest start and latest finish times are the latest times when the activity can start and finish without increasing the project's duration. The calculation is similar to the earliest times *(see Forward pass computation)* but proceeds from right to left or, in other words, from the finish to the start of the project—thus the name *backward pass.*

In a backward pass computation, an activity late finish (LF) is the minimum late start (LS) of the immediately following activities; the activity late start is determined by its late finish minus its duration. The latest allowable finish time for the end network event is set equal to either an arbitrary scheduled completion time or the early finish time computed in the forward pass computation. The following late start and finish times are calculated in the backward pass computation:

1. Late finish time (LF) = minimum late start of immediately following
                              activities
2. Late start time (LS) = late finish time (LF) – activity duration (D)

Figures 1 and 2 show arrow and precedence diagram networks with backward pass computation.

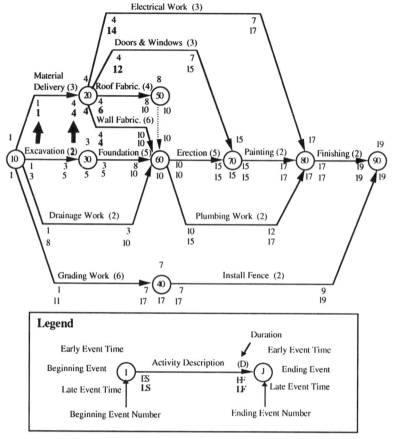

Figure 1 Arrow Diagram Network with Backward Pass Computation

Note:

Activity: Material Delivery 10 – 20

Latest Finish Time = 4 (Minimum Latest Start of the Immediately Following Activities)

Latest Start Time = LF – D = 4 – 3 = 1

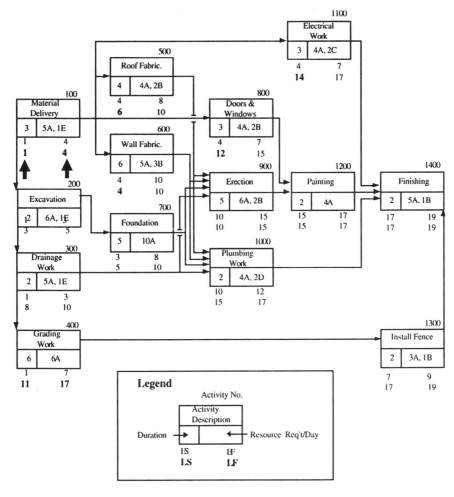

Figure 2 Precedence Diagram Network with Backward Pass Computation

Note:

Activity: Material Delivery 100

Latest Finish Time = 4
(Minimum Latest Start of the Immediately Following Activities)

Latest Start Time = LF – D = 4 – 3 = 1

Figure 2 Precedence Diagram Network with Backward Pass Computation

BAR CHART A graphical presentation of project activities shown in a time-scaled bar line but with no links shown between activities.

A bar chart is a scheduling technique in which an activity duration is drawn to scale on a time base. A bar chart can also be called a Gantt chart since it was developed by Henry L. Gantt. It is one of the most popular and widely used techniques for planning and scheduling activities because the graphical presentation of a bar chart makes it easy to read and understand.

It is recommended that a bar chart schedule be used to present a work schedule to both high and low levels of project personnel because it is at these levels that the need for an easy-to-understand schedule is greatest. Developing a bar chart schedule requires that a scheduler list work activities; then their duration is estimated. During the planning stage, activity start and completion dates must be identified in a calendar date format. After the activities, their durations, and start and completion dates are identified, each activity will be drawn as a horizontal bar in chronological order according to its start date. Vertical lines normally indicate calendar dates. Figure 1 is an example of a bar chart schedule.

Act. ID	Orig. Dur.	Rem. Dur.	Activity Description	March	April
100	3	0	Material Delivery		
200	2	0	Excavation		
300	2	0	Drainage		
400	6	0	Grading Work		
700	5	0	Foundation		
600	6	0	Wall Fabrication		
500	4	0	Roof Fabrication		
800	3	0	Doors and Windows		
1100	3	0	Electrical Work		
1300	2	0	Install Fence		
900	5	0	Erection		
1000	2	0	Plumbing Work		
1200	2	0	Painting		
1400	2	0	Finishing		

Weekend

Figure 1 Bar Chart Schedule

During construction, a bar chart schedule can be used to monitor the progress of a project and its activities by drawing an actual activity date or duration below the as-planned date. Figure 2 illustrates the use of a bar chart in monitoring progress.

Act. ID	Orig. Dur.	Rem. Dur.	Activity Description	March		April	
				17 18 19 20 21 22 23 24 25 26 27 28 29 30 31		1 2 3 4 5 6 7 8 9 10	
100	3	0	Material Delivery				
200	2	0	Excavation				
300	2	0	Drainage				
400	6	0	Grading Work				
700	5	0	Foundation				
600	6	0	Wall Fabrication				
500	4	0	Roof Fabrication				
800	3	0	Doors and Windows				
1100	3	0	Electrical Work				
1300	2	0	Install Fence				
900	5	0	Erection				
1000	2	0	Plumbing Work				
1200	2	0	Painting				
1400	2	0	Finishing				

Weekend
As-planned Activity Duration
Actaul Activity Duration

Figure 2 Updated Bar Chart Schedule

BASELINE SCHEDULE *(see As-planned project schedule)*

BCWP *(see Budgeted cost of work performed)*

BCWS *(see Budgeted cost of work schedule)*

BEGINNING EVENT (PRECEDING EVENT) An event that signifies the beginning of an activity or activities in ADM scheduling.

The beginning event can represent the joint initiation of more than one activity and can also be the ending event *(see Ending event)* for one or more preceding activities. All arrow activities commencing at the beginning event can be started only when all activities ending at that event are completed. For network computation, all activities leading out of the same beginning event will have the same early start date unless contract dates are assigned to any particular activity *(see Contract date)*. Figure 1 shows an arrow diagram network indicating beginning events.

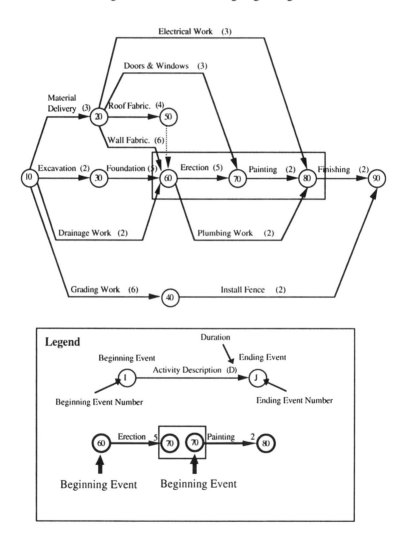

Figure 1 Arrow Diagram Network Indicating Beginning Events

BEGINNING NETWORK EVENT(S) An event in ADM scheduling that signifies the beginning of a network. No activity precedes that event.

A beginning network event represents the starting point of the project in an arrow diagram network. There are two types of the beginning network events: (1) one beginning network event, and (2) multiple beginning network events.

1. *One beginning network event.* This type of beginning network event is that most commonly used in network construction. One beginning event indicates that every project activity must be initiated by events logically dependent—directly or indirectly—on the starting event of the project. All beginning activities will have the same starting event time. An example of an arrow diagram network with one beginning network event is shown in Figure 1.

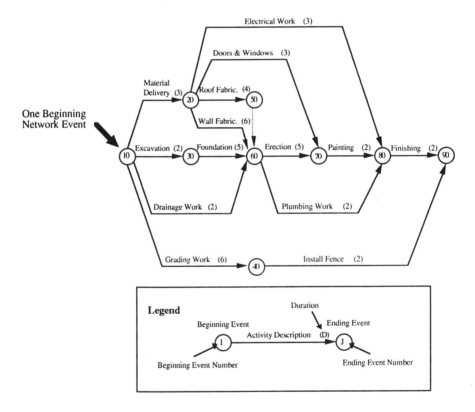

Figure 1 Arrow Diagram Network with One Beginning Network Event

2. *Multiple beginning network events.* Multiple beginning events are used when a project has more than one starting point. In this case, network computations are

inconvenient or incorrect if all beginning events are artificially brought together in a single initial event. Multiple beginning events are suitable for long-range projects with different subprojects since some subprojects may have no need to be started at the beginning of the project. For example, in the development of multicomplex office buildings, the main building may be started first; then other buildings may be started sometime after the start of the main building. An example of an arrow diagram network with multiple beginning network events is shown in Figure 2.

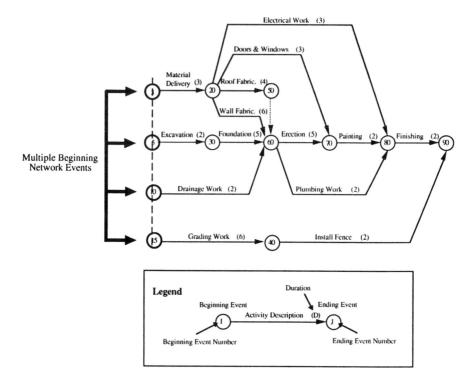

Figure 2 Arrow Diagram Network with Multiple Beginning Network Events

BUDGET AT COMPLETION (BAC) The total baseline cost (BCWS) of a project or work package. Budget at completion (BAC) is one of the C/SCSC elements and represents the total baseline of the performance measurement with its S curve *(see Cost and schedule control system criteria)*. To develop the BAC in a project, the same steps are followed as for the BCWS, and finally the total BCWS is indicated.

To develop the BAC in a project, the same steps are followed as for the BCWS and finally, the total BCWS is indicated.

Step 1: Develop the project breakdown structure (PBS) of a project *(see Project breakdown structure)*.
Step 2: Prepare the cost estimation for each work item based on the PBS.
Step 3: Acquire the basic CPM data by using the PBS and the cost estimation. The basic CPM data sheet format is used.
Step 4: Schedule each work item in the basic CPM data sheet, and develop the CPM network schedule.
Step 5: Calculate the cumulative and time-phased budget cost by adding up the budget cost of each work item in a given period by period, based on an early start CPM schedule.
Step 6: Develop the BCWS curve on the graph by using the cumulative and time-phased budget costs.
Step 7: Indicate the total amount at the endpoint of the construction period.

The preceding seven steps are explained using a residential house as an example.

1. *Step 1:* PBS

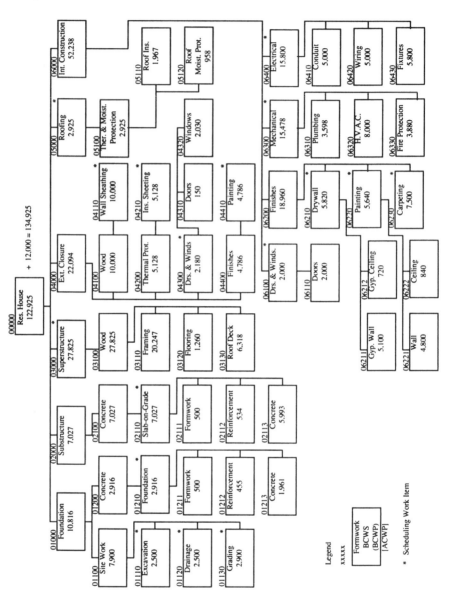

2. *Step 2:* Cost Estimation

DESCRIPTION	U/M	U/P	Q'TY	TOTAL
01 General Requirements				
Subtotal	L/S	12,000	1	12,000
02 Site Work				
01110 Excavation	C.Y.	1.89	1,323	2,500
01120 Drainage	L.F.	3.11	804	2,500
01130 Grading	C.Y.	1.06	2,736	2,900
Subtotal				7,900
03 Concrete				
01211 Foundation Form	SFCA	2.89	173	500
01212 Foundation Re-bar	Tons	910	0.5	455
01213 Foundation Concrete	C.Y.	103.2	19	1,961
02111 Slab-on-Grade Form	L.F.	1.38	362	500
02112 Slab-on-Grade Re-bar	Tons	890	0.6	534
02113 Slab-on-Grade Concrete	C.Y.	90.8	66	5,993
Subtotal				9,943
06 Wood				
03110 Superstructure Framing	M.B.F.	1,191	17	20,247
03120 Superstructure Flooring	S.F.	0.63	2,000	1,260
03130 Roof Decking	S.F.	5.01	1,261	6,318
04110 Wall Sheathing	S.F.	1.17	8,547	10,000
Subtotal				37,825
07 Thermal & Moisture Protection				
04210 Wall Insulating Sheeting	S.F.	0.6	8,547	5,128
05110 Roof Insulating	S.F.	1.56	1,261	1,967
05120 Roof Moisture Protection	S.F.	0.76	1,261	958
Subtotal				8,053
08 Doors & Windows				
04310 Ext. Doors	Ea.	150	1	150
04320 Ext. Windows	Ea.	203	10	2,030
06110 Int. Doors	Ea.	200	10	2,000
Subtotal				4,180
09 Finishes				
04410 Ext. Painting	S.F.	0.56	8,547	4,786
06211 Int. Gyp. Wallboard	S.F.	0.34	15,000	5,100
06212 Int. Gyp. Ceiling Board	S.F.	0.36	2,000	720
06311 Int. Wall Painting	S.F.	0.32	15,000	4,800
06312 Int. Ceiling Painting	S.F.	0.42	2,000	840
06230 Carpeting	S.Y.	5	1,500	7,500
Subtotal				23,746
15 Mechanical				
06310 Plumbing	L.F.	3.98	904	3,598
06320 H.V.A.C.	L/S	8,000	1	8,000
06330 Fire Protection	Ea.	776	5	3,880
Subtotal				15,478
16 Electrical				
06410 Electrical Conduit	L/S	5,000	1	5,000
06420 Electrical Wiring	L/S	5,000	1	5,000
06430 Electrical Fixtures	L/S	5,800	1	5,800
Subtotal				15,800
GRAND TOTAL				134,925

3. *Step 3:* CPM Basic Data Sheet

PBS NO.		WORK ITEMS	DUR.	BUDGET	PRECEDENCES
01130	10 - 20	Grading	3	2,900	-
01110	20 - 30	Excavation	3	2,500	10 - 20
01120	20 - 40	Drainage	5	2,500	10 - 20
-	30 - 40	Dummy	0	0	20 - 30
01210	40 - 50	Foundation	5	2,916	20 - 40, 30 - 40
06300	40 - 150	Mechanical	60	15,478	20 - 40, 30 - 40
02110	50 - 60	Slab-on-Grade	14	7,027	40 - 50
03000	60 - 70	Superstructure	28	27,825	50 - 60
06400	50 - 150	Electrical	60	15,800	40 - 60
04300	70 - 80	Doors & Windows	7	2,180	60 - 70
04210	70 - 90	Wall Insulation	12	5,128	60 - 70
04110	70 - 110	Wall Sheathing	14	10,000	60 - 70
-	80 - 110	Dummy	0	0	70 - 80
-	90 - 110	Dummy	0	0	70 - 90
-	100 - 110	Dummy	0	0	70 - 100
05000	70 - 100	Roofing	7	2,925	60 - 70
06210	110 - 120	Drywall	12	5,820	80-110, 70-110, 90-110, 100-110
04410	110 - 150	Ext. Paint	10	4,786	80-110, 70-110, 90-110, 110-110
06100	120 - 130	Int. Doors	6	2,000	110 - 120
06220	130 - 140	Int. Paint	5	5,640	120 - 130
06230	140 - 150	Carpeting	8	7,500	130 - 140
		Total		122,925	

4. *Step 4:* CPM Network Schedule with Budget Cost Allocated to Activities (Assume uniform cost over activity duration)

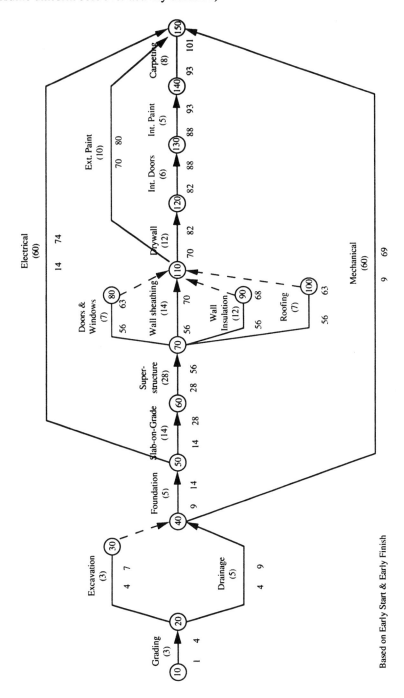

5. *Step 5:* Time-Phased and Cumulative Budget Cost

WORK	TIME																				TOTAL
ITEMS	5	10	15	20	25	30	35	40	45	50	55	60	65	70	75	80	85	90	95	100	
10 - 20	2,900																				2,900
20 - 30	1,667	833																			2,500
20 - 40	1,000	1,500																			2,500
40 - 50		1,166	1,750																		2,916
40 - 150		516	1,290	1,290	1,290	1,290	1,290	1,290	1,290	1,290	1,290	1,290	1,290	772							15,478
50 - 60			1,004	2,510	2,510	1,003															7,027
60 - 70						2,982	4,970	4,970	4,970	4,970	4,963										27,825
50 - 150			526	1,315	1,315	1,315	1,315	1,315	1,315	1,315	1,315	1,315	1,315	1,315	809						15,800
70 - 80												1,555	625								2,180
70 - 90												2,135	2,135	858							5,128
70 - 110												3,570	3,570	2,860							10,000
70 - 100												2,090	835								2,925
110 - 120														485	2,425	2,425	485				5,820
110 - 150														479	2,395	1,912					4,786
120 - 130																	1,332	668			2,000
130 - 140																		3,384	2,256		5,640
140 - 150																			2,813	4,687	7,500
Subtotal	5,567	4,015	4,570	5,115	5,115	6,590	7,575	7,575	7,575	7,575	7,568	11,955	9,770	6,769	5,629	4,337	1,817	4,052	5,069	4,687	122,925
Gen. Req.	600	600	600	600	600	600	600	600	600	600	600	600	600	600	600	600	600	600	600	600	12,000
Total	6,167	4,615	5,170	5,715	5,715	7,190	8,175	8,175	8,175	8,175	8,168	12,555	10,370	7,369	6,229	4,937	2,417	4,652	5,669	5,287	134,925
Cumulat.	6,167	10,782	15,952	21,667	27,382	34,572	42,747	50,922	59,097	67,272	75,440	87,995	98,365	105,734	111,963	116,900	119,317	123,969	129,638	134,925	134,925

6. *Steps 6 and 7:* BCWS Curve and BAC

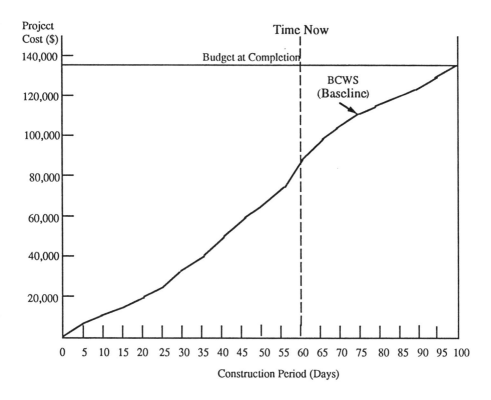

BUDGETED COST OF WORK PERFORMED (BCWP) (VALUE OF WORK; EARNED VALUE) The planned cost for completed work items, including that part of work items in progress at a given data date.

The budgeted cost of work performed (BCWP) is one element of the C/SCSC and represents the earned value of the performance measurement. The BCWP is acquired by multiplying the quantities in place by the budgeted unit price of each work item *(see Cost and schedule control system criteria).* To develop the BCWP in a project, the following steps are generally followed:

Step 1: Use the original PBS of a project.
Step 2: Use the original cost estimation of the project based on the PBS.
Step 3: Use the latest updated CPM schedule by the data date *(see Data date).*
Step 4: Measure the work completed at the activity level, and calculate the BCWP by multiplying the actual quantities in place by the budgeted (original estimate) unit price of each work item by the data date. This measurement and calculation are shown in the monthly cost performance report (manual or computerized).

Step 5: Develop the BCWP curve on the graph using the cumulative and time-phased earned values by the data date (actual quantities multiplied by original estimated unit price).

The preceding five steps are explained using a residential house as an example.

1. *Step 1:* PBS

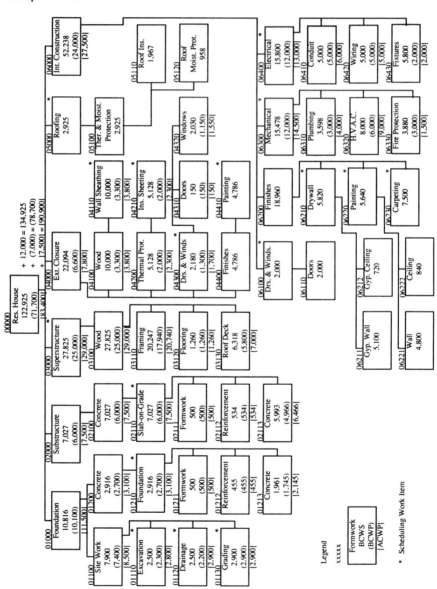

2. *Step 2:* Cost Estimation

DESCRIPTION	U/M	U/P	Q'TY	TOTAL
01 General Requirements				
Subtotal	L/S	12,000	1	12,000
02 Site Work				
01110 Excavation	C.Y.	1.89	1,323	2,500
01120 Drainage	L.F.	3.11	804	2,500
01130 Grading	C.Y.	1.06	2,736	2,900
Subtotal				7,900
03 Concrete				
01211 Foundation Form	SFCA	2.89	173	500
01212 Foundation Re-bar	Tons	910	0.5	455
01213 Foundation Concrete	C.Y.	103.2	19	1,961
02111 Slab-on-Grade Form	L.F.	1.38	362	500
02112 Slab-on-Grade Re-bar	Tons	890	0.6	534
02113 Slab-on-Grade Concrete	C.Y.	90.8	66	5,993
Subtotal				9,943
06 Wood				
03110 Superstructure Framing	M.B.F.	1,191	17	20,247
03120 Superstructure Flooring	S.F.	0.63	2,000	1,260
03130 Roof Decking	S.F.	5.01	1,261	6,318
04110 Wall Sheathing	S.F.	1.17	8,547	10,000
Subtotal				37,825
07 Thermal & Moisture Protection				
04210 Wall Insulating Sheeting	S.F.	0.6	8,547	5,128
05110 Roof Insulating	S.F.	1.56	1,261	1,967
05120 Roof Moisture Protection	S.F.	0.76	1,261	958
Subtotal				8,053
08 Doors & Windows				
04310 Ext. Doors	Ea.	150	1	150
04320 Ext. Windows	Ea.	203	10	2,030
06110 Int. Doors	Ea.	200	10	2,000
Subtotal				4,180
09 Finishes				
04410 Ext. Painting	S.F.	0.56	8,547	4,786
06211 Int. Gyp. Wallboard	S.F.	0.34	15,000	5,100
06212 Int. Gyp. Ceiling Board	S.F.	0.36	2,000	720
06311 Int. Wall Painting	S.F.	0.32	15,000	4,800
06312 Int. Ceiling Painting	S.F.	0.42	2,000	840
06230 Carpeting	S.Y.	5	1.500	7,500
Subtotal				23,746
15 Mechanical				
06310 Plumbing	L.F.	3.98	904	3,598
06320 H.V.A.C.	L/S	8,000	1	8,000
06330 Fire Protection	Ea.	776	5	3,880
Subtotal				15,478
16 Electrical				
06410 Electrical Conduit	L/S	5,000	1	5,000
06420 Electrical Wiring	L/S	5,000	1	5,000
06430 Electrical Fixtures	L/S	5,800	1	5,800
Subtotal				15,800
GRAND TOTAL				134,925

3. *Step 3:* Updated CPM Network Schedule

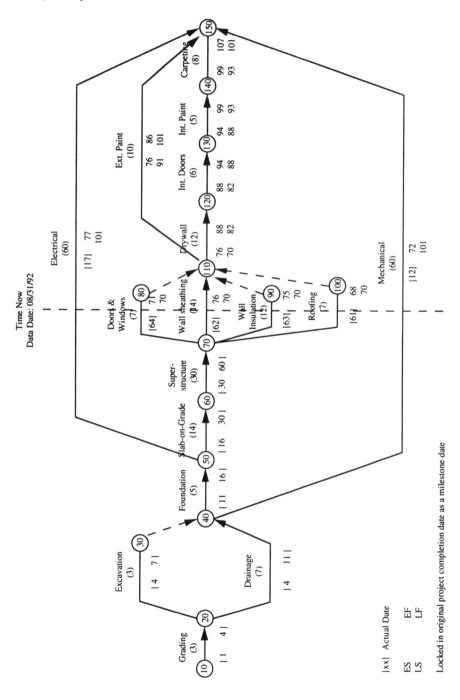

4. *Step 4:* Monthly Cost Performance Report

MONTHLY COST PERFORMANCE REPORT: CPM ITEM

PROJECT: RESIDENTIAL HOUSE   REPORT DATE: 08/31/92

| ITEM | | CURRENT PERIOD | | | | | | | | | CUMULATIVE TO DATE | | | | | | | | | AT COMPLETION | | |
| | | BUDGET COST | | ACWP | VARIANCE | | | INDEX | | PC | BUDGET COST | | ACWP | VARIANCE | | | INDEX | | PC | BAC | EAC | ACV |
PBS	CPM	BCWS	BCWP	ACWP	SV	CV	AV	SPI	CPI	PC	BCWS	BCWP	ACWP	SV	CV	AV	SPI	CPI	PC	BAC	EAC	ACV
01130	10 - 20										2.900	2.900	2.900	0	0	0	1.00	1.00	1.00	2.900	2.900	
01110	20 - 30										2.500	2.300	2.800	-200	-500	-300	0.92	0.82	0.92	2.500	2.800	-300
01120	20 - 40										2.500	2.200	2.800	-300	-600	-300	0.88	0.79	0.88	2.500	2.800	-300
01210	40 - 50										2.916	2.700	3.100	-216	-400	-184	0.93	0.87	0.93	2.916	3.100	-184
06300	40 - 150	5.160	5.000	5.200	-160	-200	-40	0.97	0.96	0.32	13.416	12.000	14.500	-1.416	-2.500	-1.084	0.89	0.83	0.78	15.478	16.562	-1.084
02110	50 - 60										7.027	6.000	7.500	-1.027	-1.500	-473	0.85	0.80	0.85	7.027	7.500	-473
03000	60 - 70	14.903	13.900	15.000	-1.003	-1.100	-97	0.93	0.93	0.50	27.825	25.000	29.000	-2.825	-4.000	-1.175	0.90	0.86	0.90	27.825	29.000	-1.175
06400	50 - 150	5.260	5.100	5.300	-160	-200	-40	0.97	0.96	0.32	12.361	12.000	13.000	-361	-1.000	-639	0.97	0.92	0.76	15.800	16.439	-639
04300	70 - 80	1.555	1.300	1.700	-255	-400	-145	0.84	0.76	0.60	1.555	1.300	1.700	-255	-400	-145	0.84	0.76	0.60	2.180	2.325	-145
04210	70 - 90	2.135	2.000	2.300	-135	-300	-165	0.94	0.87	0.39	2.135	2.000	2.300	-135	-300	-165	0.94	0.87	0.39	5.128	5.293	-165
04110	70 - 110	3.570	3.300	3.800	-270	-500	-230	0.92	0.87	0.33	3.570	3.300	3.800	-270	-500	-230	0.92	0.87	0.33	10.000	10.230	-230
05000	70 - 100																			2.925	2.925	0
06210	110 - 120																			5.820	5.820	0
04410	110 - 150																			4.786	4.786	0
06100	120 - 130																			2.000	2.000	0
06220	130 - 140																			5.640	5.640	0
06230	140 - 150																			7.500	7.500	0
SUBTOTAL		32.583	30.600	33.300	-1.983	-2.700	-717	0.94	0.92	0.25	78.705	71.700	83.400	-7.005	-11.700	-4.695	0.91	0.86	0.58	122.925	127.620	-4.695
GEN. REQ.		2.400	2.400	2.500		-100	-100	1.00	0.96	0.20	7.200	7.000	7.500	-200	-500	-300	0.97	0.93	0.58	12.000	12.300	-300
TOTAL		34.983	33.000	35.800	-1.983	-2.800	-817	0.94	0.92	0.24	85.905	78.700	90.900	-7.205	-12.200	-4.995	0.92	0.87	0.58	134.925	139.920	-4.995

Note: The project at the data date has -817 in the current period and -4,995 cumulatively as the accounting variance (AV). It shows that the project is over budget because the AV is less than 0.
The same comment is valid for all work items except for item 01130, which is on schedule (AV=0).

111

5. *Step 5:* BCWP (Earned Value or Value of Work)

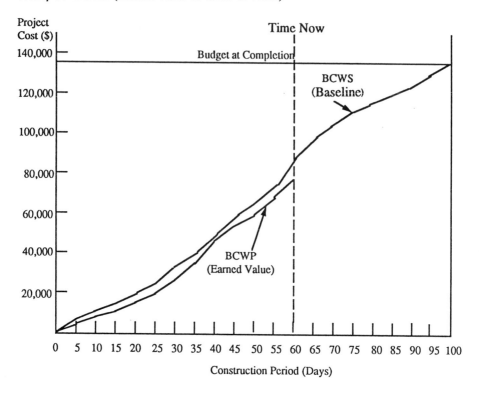

BUDGETED COST OF WORK SCHEDULE (BCWS) (BASELINE) The planned costs for all work items as originally scheduled at a given data date.

The budgeted cost of work schedule (BCWS) is one element of the C/SCSC and represents the baseline of the performance measurement with its S curve *(see Cost and schedule control system criteria).* To develop the BCWS in a project, the following steps are generally followed.

Step 1: Develop the project breakdown structure (PBS) of a project *(see Project breakdown structure).*

Step 2: Prepare the cost estimation for each work item based on the PBS.

Step 3: Acquire the CPM basic data by using the PBS and the cost estimation. The basic CPM data sheet format is used.

Step 4: Schedule each work item in the CPM basic data sheet, and develop the CPM network schedule.

Step 5: Calculate the cumulative and time-phased budgeted costs by adding up the budget cost of each work item period by period, based on an early start CPM schedule.

Step 6: Develop the BCWS curve on the graph by using the cumulative and time-phased budgeted costs.

The preceding six steps are explained using a residential house as an example.

1. *Step 1:* PBS

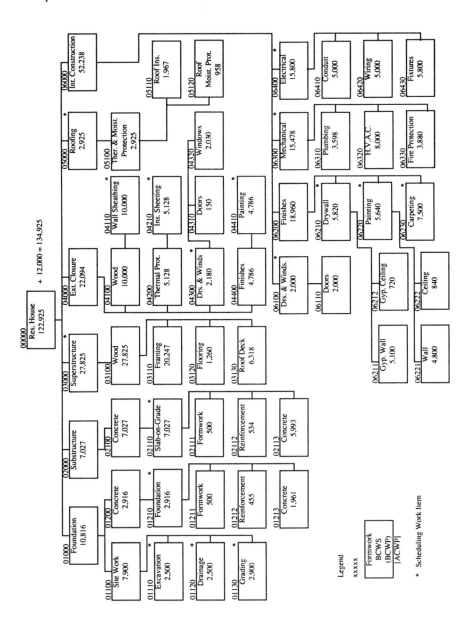

2. *Step 2:* Cost Estimation

DESCRIPTION	U/M	U/P	QTY	TOTAL
01 General Requirements				
Subtotal	L/S	12,000	1	12,000
02 Site Work				
01110 Excavation	C.Y.	1.89	1,323	2,500
01120 Drainage	L.F.	3.11	804	2,500
01130 Grading	C.Y.	1.06	2,736	2,900
Subtotal				7,900
03 Concrete				
01211 Foundation Form	SFCA	2.89	173	500
01212 Foundation Re-bar	Tons	910	0.5	455
01213 Foundation Concrete	C.Y.	103.2	19	1,961
02111 Slab-on-Grade Form	L.F.	1.38	362	500
02112 Slab-on-Grade Re-bar	Tons	890	0.6	534
02113 Slab-on-Grade Concrete	C.Y.	90.8	66	5,993
Subtotal				9,943
06 Wood				
03110 Superstructure Framing	M.B.F.	1,191	17	20,247
03120 Superstructure Flooring	S.F.	0.63	2,000	1,260
03130 Roof Decking	S.F.	5.01	1,261	6,318
04110 Wall Sheathing	S.F.	1.17	8,547	10,000
Subtotal				37,825
07 Thermal & Moisture Protection				
04210 Wall Insulating Sheeting	S.F.	0.6	8,547	5,128
05110 Roof Insulating	S.F.	1.56	1,261	1,967
05120 Roof Moisture Protection	S.F.	0.76	1,261	958
Subtotal				8,053
08 Doors & Windows				
04310 Ext. Doors	Ea.	150	1	150
04320 Ext. Windows	Ea.	203	10	2,030
06110 Int. Doors	Ea.	200	10	2,000
Subtotal				4,180
09 Finishes				
04410 Ext. Painting	S.F.	0.56	8,547	4,786
06211 Int. Gyp. Wallboard	S.F.	0.34	15,000	5,100
06212 Int. Gyp. Ceiling Board	S.F.	0.36	2,000	720
06311 Int. Wall Painting	S.F.	0.32	15,000	4,800
06312 Int. Ceiling Painting	S.F.	0.42	2,000	840
06230 Carpeting	S.Y.	5	1,500	7,500
Subtotal				23,746
15 Mechanical				
06310 Plumbing	L.F.	3.98	904	3,598
06320 H.V.A.C.	L/S	8,000	1	8,000
06330 Fire Protection	Ea.	776	5	3,880
Subtotal				15,478
16 Electrical				
06410 Electrical Conduit	L/S	5,000	1	5,000
06420 Electrical Wiring	L/S	5,000	1	5,000
06430 Electrical Fixtures	L/S	5,800	1	5,800
Subtotal				15,800
GRAND TOTAL				134,925

3. *Step 3:* CPM Basic Data Sheet

PBS NO.		WORK ITEMS	DUR.	BUDGET	PRECEDENCES
01130	10 - 20	Grading	3	2,900	-
01110	20 - 30	Excavation	3	2,500	10 - 20
01120	20 - 40	Drainage	5	2,500	10 - 20
-	30 - 40	Dummy	0	0	20 - 30
01210	40 - 50	Foundation	5	2,916	20 - 40, 30 - 40
06300	40 - 150	Mechanical	60	15,478	20 - 40, 30 - 40
02110	50 - 60	Slab-on-Grade	14	7,027	40 - 50
03000	60 - 70	Superstructure	28	27,825	50 - 60
06400	50 - 150	Electrical	60	15,800	40 - 60
04300	70 - 80	Doors & Windows	7	2,180	60 - 70
04210	70 - 90	Wall Insulation	12	5,128	60 - 70
04110	70 - 110	Wall Sheathing	14	10,000	60 - 70
-	80 - 110	Dummy	0	0	70 - 80
-	90 - 110	Dummy	0	0	70 - 90
-	100 - 110	Dummy	0	0	70 - 100
05000	70 - 100	Roofing	7	2,925	60 - 70
06210	110 - 120	Drywall	12	5,820	80-110, 70-110, 90-110, 100-110
04410	110 - 150	Ext. Paint	10	4,786	80-110, 70-110, 90-110, 110-110
06100	120 - 130	Int. Doors	6	2,000	110 - 120
06220	130 - 140	Int. Paint	5	5,640	120 - 130
06230	140 - 150	Carpeting	8	7,500	130 - 140
		Total		122,925	

4. *Step 4:* CPM Network Schedule

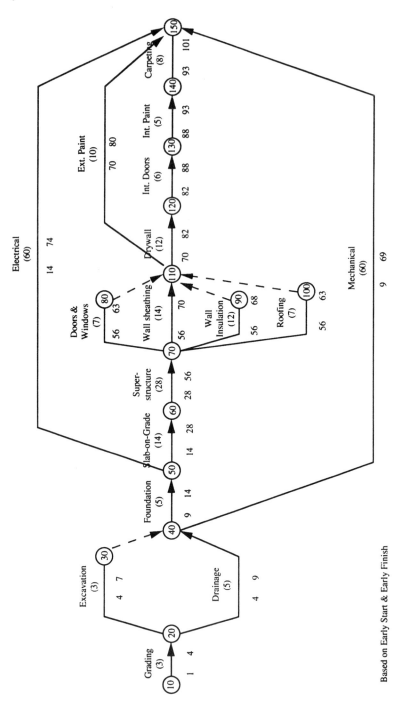

Based on Early Start & Early Finish

116

5. *Step 5:* Cumulative and Time-Phased Budget Costs

WORK ITEMS	TIME																				TOTAL
	5	10	15	20	25	30	35	40	45	50	55	60	65	70	75	80	85	90	95	100	
10 - 20	2,900																				2,900
20 - 30	1,667	833																			2,500
20 - 40	1,000	1,500																			2,500
40 - 50		1,166	1,750																		2,916
40 - 150		516	1,290	1,290	1,290	1,290	1,290	1,290	1,290	1,290	1,290	1,290	1,290	772							15,478
50 - 60			1,004	2,510	2,510	1,003															7,027
60 - 70						2,982	4,970	4,970	4,970	4,970	4,963										27,825
50 - 150			526	1,315	1,315	1,315	1,315	1,315	1,315	1,315	1,315	1,315	1,315	1,315	809						15,800
70 - 80												1,555	625								2,180
70 - 90												2,135	2,135	858							5,128
70 - 110												3,570	3,570	2,860							10,000
70 - 100												2,090	835								2,925
110 - 120														485	2,425	2,425	485				5,820
110 - 150														479	2,395	1,912					4,786
120 - 130																	1,332	668			2,000
130 - 140																		3,384	2,256		5,640
140 - 150																			2,813	4,687	7,500
Subtotal	5,567	4,015	4,570	5,115	5,115	6,590	7,575	7,575	7,575	7,575	7,568	11,955	9,770	6,769	5,629	4,337	1,817	4,052	5,069	4,687	122,925
Gen. Req.	600	600	600	600	600	600	600	600	600	600	600	600	600	600	600	600	600	600	600	600	12,000
Total	6,167	4,615	5,170	5,715	5,715	7,190	8,175	8,175	8,175	8,175	8,168	12,555	10,370	7,369	6,229	4,937	2,417	4,652	5,669	5,287	134,925
Cumulat.	6,167	10,782	15,952	21,667	27,382	34,572	42,747	50,922	59,097	67,272	75,440	87,995	98,365	105,734	111,963	116,900	119,317	123,969	129,638	134,925	134,925

117

6. *Step 6:* BCWS (Baseline, S Curve)

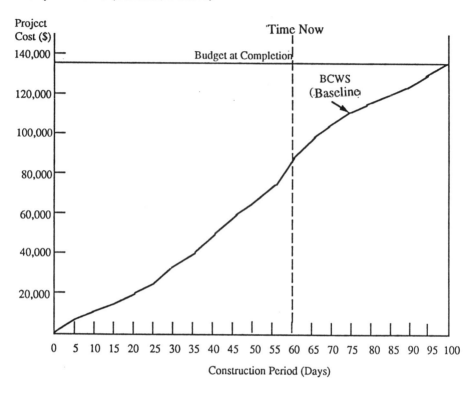

BURDEN (OVERHEAD) The cost of all other factors contributing to the completion of any activity or project (not direct costs) *(see Activity indirect costs)*.

Cost structure for a construction project is established as shown in Figure 1. In the cost structure, total project costs are divided into four categories: direct cost, burden, contingency, and profit. Burden is classified into two subsets: job overhead and general overhead.

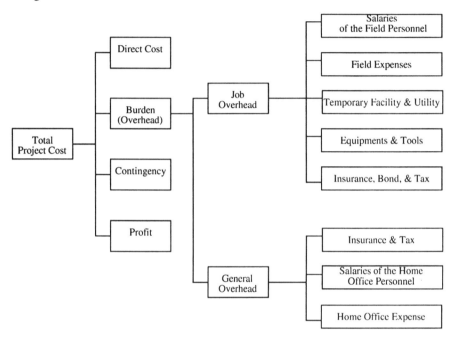

Figure 1 Cost Structure for Construction Project

1. *Job overhead.* Job overhead includes those items that are related to a particular project and is composed of salaries (field personnel), field expense, temporary facilities and utilities, equipment and tools, insurance, bonds, and taxes.

1.1 *Salaries of field personnel.* The following salaries are examples of those included in this category: project manager, job superintendent, assistant job superintendent, job engineer, timekeeper/clerk, secretary, watchman, general-purpose labor, and general-purpose carpenter.

1.2 *Field expense.* The following expenses are examples of those included in this category: living expense, permit, job cleanup, clean glass, trash chutes, rubbish removal, chemical toilet, first-aid supplies, travel expenses, job telephone, light bill, water bill, and office supplies.

1.3 *Temporary facilities and utilities.* The following facilities are examples of those included in this category: temporary job fences, temporary roads, temporary stor-

119

ages, office trailers, job tool house, camps, barriers and enclosures, and temporary shops.

1.4 *Equipment and tools.* The following equipment and tools are examples of those included in this category: small tools, crane rental, tower hoist, air compressor, welding machine, generators, pickup truck, pumps, fork lift, and scaffolding.

1.5 *Insurance, bonds, and taxes.* The following insurance, bonds, and taxes are examples of those included in this category: employer's FICA (Federal Insurance Contributions Act) expense, federal unemployment insurance (FUI), state unemployment insurance (SUI), worker's compensation insurance (WCI), employer's liability insurance (EL), public damage insurance (PD), builder's risk insurance, performance bond, subcontract bond, and material sales tax.

2. *General overhead.* General overhead includes items related to a home office and consists of insurance, taxes, salaries (home office personnel), and home office expense.

2.1 *Insurance and taxes.* Insurance and taxes, which are related to the home office, are included in this category.

2.2 *Salaries of home office personnel.* The following salaries are included in this category: executives, lawyer, accountant, and secretary.

2.3 *Home office expense.* The following expenses are included in this category: home office rent, postage, publications, public relations, advertising, training, travel, and subscriptions.

Generally, burdens are a function of time. For example, as the project duration increases, burdens tend to increase.

C

CALENDAR RANGE (CALENDAR SPAN) The maximum number of time units included in any given or defined calendar. The calendar start date is time unit number one. The calendar range is usually expressed in years.

A calendar range is a time span during which a project can be scheduled. Planning and scheduling software normally has a large enough time range for any project. For example, Primavera Project Planner 5.0 accepts dates only from 1972 to 2060 (88 years). A project schedule must be within a calendar range, as shown in Figure 1.

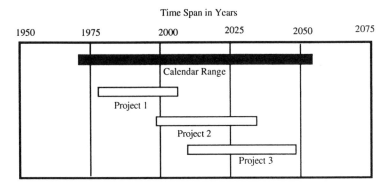

Figure 1 Project Schedule in Correlation with Calendar Range

CALENDAR(S) The time frames used in scheduling project activities. The calendar identifies holidays, working days, and the length of the working day in time units.

Calendars are used for computerized project time computations after the network logic has been developed and activities' durations have been assigned in order to represent project and activity times in terms of calendar dates. It is done by matching to a project calendar the computed time units with the specified starting calendar date of the project. Most CPM-oriented software allows a calendar file to be created with a multitude of calendars. A different number of working days and holidays can be assigned to different calendars. For working hours, some CPM-oriented software allows working hours to be assigned directly in each calendar, and some requires a daily calendar *(see Daily calendar)* with specified working hours to be created and assigned to a calendar.

By creating more than one calendar for a project, multiple calendars can be utilized against a project with a variable workweek for different activities and reasons (e.g., some activities may be on a 5-day week, some on a 7-day week). Also,

different working hours can be assigned to different calendars or activities, such as some activities on a 7:00 am-4:00 pm schedule and some on a 8:00 am-5:00 pm schedule. The starting day for each week may vary (e.g., the starting day is Saturday in the Middle East). The use of multiple calendars is necessary for large-scale projects requiring that some components be done at different countries or locations or that various trades work at different days or times.

CALENDAR START DATE The date assigned to the first time unit of the calendar being used.

Project calendars define a time span during which a project can be scheduled. A calendar start date indicates the earliest time when a project activity can begin. Normally, planning and scheduling software provides a default calendar start date for all calendars defined within a project. In Primavera Project Planner 5.0, the default calendar start date is 1 year, 4 years, and 12 years, where time units are in days, weeks, and months, respectively. A calendar start date, however, can be modified by the user, but this date cannot be later than the project start date. The sample of a calendar start date represented in calendar information is shown in Figure 1.

```
                      GLOBAL CALENDAR INFORMATION
═══════════════════════════════════════════════════════════════════════

       Planning Unit: D = Day                    Week starts on: MON
     Cal. Start:09MAR92  Start date:13MAR93  Data date:17MAR93  Finish date:

───────────────────────────────────────────────────────────────────────

   If a holiday occurs on a weekend, make the nearest workday a holiday (Y/N)? Y

───────────────────────────────────────────────────────────────────────

         Calendar ID specified         |         Calendar ID available
     1                                  |    23456789ABCDEFGHJKLMNPQRSTUXYZ

───────────────────────────────────────────────────────────────────────

   Nonwork periods          Page:  1 | Exceptions               Page   1
                                      |
        Start            End          |    Start             End
     ~~~~~~~~~~~~~~~  ~~~~~~~~~~~~~~~  |  ~~~~~~~~~~~~~~~  ~~~~~~~~~~~~~~~
                                      |
                                      |
                                      |
                                      |
                                      |
                                      |
         Scroll using Home, End       |     Scroll using PgUp, PgDn
═══════════════════════════════════════════════════════════════════════

   Commands: Add Delete Edit Help More Next Print Return Transfer View Window
   Windows : Calendar Global List
```

Figure 1 Calendar Information (Primavera Project Planner 5.0)

122

CALENDAR UNIT (TIME UNIT) The smallest time unit of the calendar, used in estimating activity duration. This unit is generally minutes, hours, shifts, days (the most common time unit in a calendar), weeks, or months.

Various types of time units, such as minutes, hours, shifts, days, or weeks, can be assigned to work activities, depending on work types. The use of a calendar unit brings all activity durations to a common denominator in terms of a normal working time unit. Once the type of time unit is specified, it must be used consistently for all activities in the project. In other words, different time units for each activity in a project are not allowed.

The calendar unit is related with a project calendar *(see Calendar)* and a daily calendar *(see Daily calendar)*. When minutes or hours are used as activity time units, the working hours in a working day will be indicated in the daily calendar, and the number of working days in a week will be specified in the calendar. After the activity time units are assigned, the CPM calculation will convert project or activity time units into a specified calendar unit.

CHANGE ORDER A contract document authorizing or directing a project scope change or correction.

A contract change order is used when a change is made in the original contract that will affect the scope of work. The change order is authorized by the owner and is often initiated by the contractor, subcontractor, or architect/engineer.

The change order form is completed by the contractor/subcontractor to record the estimated cost of the change. The form is then submitted to the owner or owner's agent (A/E or construction manager) for approval. The work cannot proceed without written approval.

A sample of a contract change order is shown in Figures 1 and 2. Sometimes the change order is preceded by an initiator change order request, which describes proposed changes and their justification and issuance of a change order by the proper authorities. Usually, this document is prepared by any principal involved in the project and then the owner is advised. A sample of an initiator change order request is shown in Figure 3.

For documenting the impact on a CPM schedule of a change order, it is recommended that a timely project change order summary be prepared. Figure 4 shows a sample form to be used with the CPM procedure diagramming technique, and Figure 5 shows a similar form to be used with the CPM arrow diagram method.

**CONTRACT
CHANGE ORDER**

FROM:

J. D. JONES ARCH.

JONES AND ASSOCIATES

SAN ANTONIO , TEXAS

TO:

L. M. PARKER - PROJECT MANAGER.

TEXAS BUILDERS CO.

DALLAS , TEXAS

CHANGE ORDER NO.	93 - 10 - 7
DATE	OCT 15, 1993.
PROJECT	TEXAS BANK - INTERIOR RENOVATION
LOCATION	AUSTIN , TEXAS
JOB NO.	87

ORIGINAL CONTRACT AMOUNT	$	2 8 5 4 0 0 °°
TOTAL PREVIOUS CONTRACT CHANGES		7 2 0 0 °°
TOTAL BEFORE THIS CHANGE ORDER		2 9 2 6 0 0 °°
AMOUNT OF THIS CHANGE ORDER		1 2 0 0 0 °°
REVISED CONTRACT TO DATE		3 0 4 6 0 0 °°

Gentlemen:

This CHANGE ORDER includes all Material, Labor and Equipment necessary to complete the following work and to adjust the total contract as indicated;

☐ the work below to be paid for at actual cost of Labor, Materials and Equipment plus _____ percent (_____%)

☒ the work below to be completed for the sum of _____ $ 12,000. °°

twelve thousands dollars — _____ dollars ($ 12,000.°°)

REMOVE AND DISPOSE THE EXISTING CARPET AND PADING LOCATED
IN THE BANK MAIN LOBBY

INSTALL NEW WOOD PLANK FLOORING ON WOOD SUBFLOOR IN THE
MAIN LOBBY AS INDICATED IN THE ATTACHED DRAWINGS

THE PROJECT COMPLETION DATE IS EXTENDED WITH 6 CALENDAR
DAYS .

CHANGES APPROVED

The work covered by this order shall be performed under the same Terms and
Conditions as that included in the original contract unless stated otherwise above.

By L .M. PARKER

PROJECT MANAGER

By_____

Signed J. D. Jones

By_____

Figure 1 Contract Change Order

CHANGE ORDER

PROJECT TITLE___TEXAS BANK - INTERIOR RENOVATION___1/12/92___

PROJECT NO.___87___ CONTRACT NO.___92-020___ CONTRACT DATE _____

CONTRACTOR___TEXAS BUILDERS CO.___

The following changes are hereby made to the Contract Documents:

Remove existing carpet and padding in the main lobby
install wood plank flooring on wood subfloor as
specified by A/E

Justification:

OWNER REQUEST

CHANGE TO CONTRACT PRICE

Original Contract Price: $___285,400___

Current contract price, as adjusted by previous change orders: $___292,600.00___

The Contract Price due to this Change Order will be [(increased)] [decreased] by $___12,000.00___

The new Contract Price due to this Change Order will be: $___304,600.00___

CHANGE TO CONTRACT TIME

The Contract Time will be [increased] [decreased] by ___6___ calendar days.

The date for completion of all work under the contract will be ___12/10/1993___

Approvals Required:

To be effective, this order must be approved by the Owner if it changes the scope or objective of the project, or as may otherwise be required under the terms of the Supplementary General Conditions of the Contract.

Requested by ___C. T. SMITH___ date OCT 5, 1993

Recommended by ___J. D. JONES   AIA___ date OCT 15, 1993

Ordered by ___J. D. JONES AIA___ date OCT 15, 1993

Accepted by___L. M. PARKER___ date___OCT 19, 1993___

Wiley-Fisk Form 15-3

Figure 2 Contract Change Order

INITIATOR CHANGE ORDER REQUEST

Project Title _____ TEXAS BANK - INTERIOR RENOVATION _____

Project No. _____ 87 _____ Contract No. _____ 92-020 _____ Contract Date _____ 1/12/92 _____

Contractor _____ TEXAS BUILDERS CO. _____

Proposed By: _____ J.D. JONES - ARCH. _____ Date _ OCT 15, 93 _ ☐ Owner
(Name)   ☒ Arch. & Engr.

Submitted By: _____ L.M. PARKER (PROJ. MNG) _____ Date _ OCT 16, 1993 _ ☐ Construction
(Name)

Actual job conditions in area of proposed change:

CARPET 8 YEARS OLD

Change order justification:

WORN OUT DO TO TRAFFIC & MANY SPOTS

Contractor authorized to proceed with this change ☒ YES ☐ NO on _____

Other contracts involved are as follows (List Contracts by No.): _____ Is Dwg. Req.? ☒ NO ☐ YES _____
(Sheet No.)

Description of Work to be Performed:

REMOVE AND DISPOSE THE EXISTING CARPET AND PADD
LOCATED IN THE BANK MAIN LOBBY
INSTALL WOOD PLANK FLOORING ON WOOD SUBFLOOR AS
INDICATED IN THE ATTACHED DRAWING.

| Estimated effect on costs of: A&E | NONE | Inspection: $ | NONE |

Approved _____ J.D. JONES _____ OCT 15, 1993 _____
(Date)

Wiley-Fisk Form 16-2

Figure 3 Initiator Contract Change Order

126

PROJECT CHANGE ORDER SUMMARY

CHANGE ORDER NO: 93-10-7

PROJECT CHANGE NO: 27

A.

ORIGINATOR	DATES				COST		
	NOTICE OF CHANGE	SUBMITTED	NOTICE TO PROCEED	COMPLETION	SUBMITTED	APPROVED	PENDING
A/E	OCT 5,93	OCT 15,93	OCT 19,93		12,000°°	12,000°°	—

ORIGINATORS: CLIENT, (ARCHITECT/ENGINEER,) CLIENT/SUBCONTRACTORS, GENERAL CONTRACTORS, GENERAL CONTRACTOR ⇌ SUBCONTRACTORS, OTHERS___

B. PROJECT CHANGE DESCRIPTION:

REMOVE EXISTING CARPET AND PADDING FROM THE MAIN LOBBY
INSTALL WOOD PLANK FLOORING ON WOOD SUBFLOORING AS
SPECIFIED

C. IMPACTED ACTIVITIES:

NOTICE TO PROCEED
OCT 19, 1993 ___27₁

PC# __27___
PREPARATORY TIME
4 | TB

PC# __27____
CONSTRUCTION TIME
2 | TB

___27₂

IMPACTED ACTIVITIES
FS 350
FS 380
FS 400

DURATION IN WORKING DAYS   RESPONSIBILITY CODE

D. PREPARATORY TIME & COST ANALYSIS

REMOVE AND DISPOSE EXISTING CARPET AND PADDING	1 DAY
PROCURE WOOD PLANK FLOORING, AND SUBFLOORING FROM THE LOCAL SUPPLIER	3 days

E CONSTRUCTION TIME & COST ANALYSIS

SUBFLOORING INSTALLATION --	1 DAY
PLANK FLOORING INSTALLATION	1 DAY

PREPARED BY: L. M. Parker   DATE: OCT 25, 1993

Figure 4 Project Change Order Summary for Precedence Diagramming

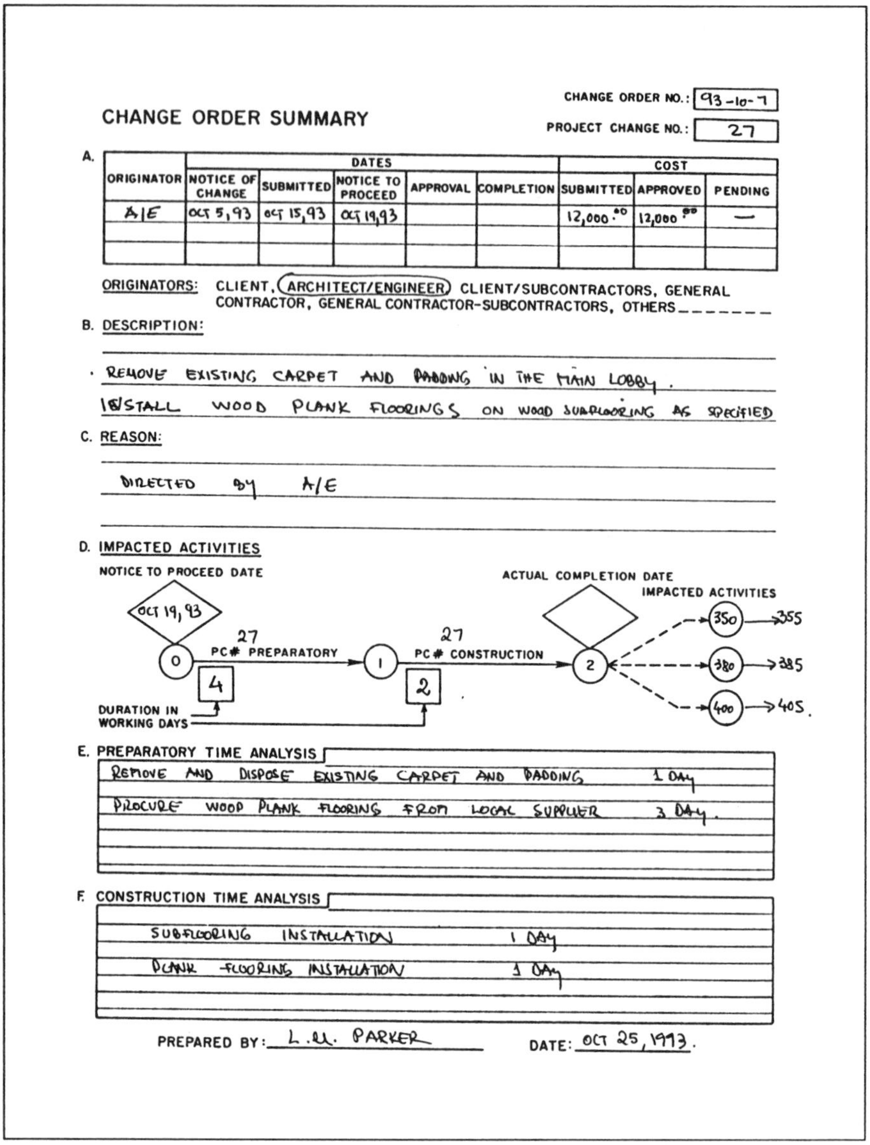

Figure 5 Project Change Order Summary for Arrow Diagramming

CHANGE ORDER, NOTICE OF CHANGE The date when the contractor or subcontractor has been informed verbally or in writing regarding a project change.

The notice of change date is important for documenting the change order impact on the approved baseline CPM diagram (precedence or arrow diagram) schedule.

128

This date is treated as a "not earlier than" date associated with a new activity to be added to the baseline schedule (see Baseline schedule). Figure 1 shows a project change order summary form and its impact on the new activity project change number preparatory time, which includes the time needed to estimate the proposed change, to submit it for approval, and to be informed (in writing) of the date to proceed. Figure 1 indicates that a contractor has been informed about the change verbally by J. D. Jones on October 5, 1993.

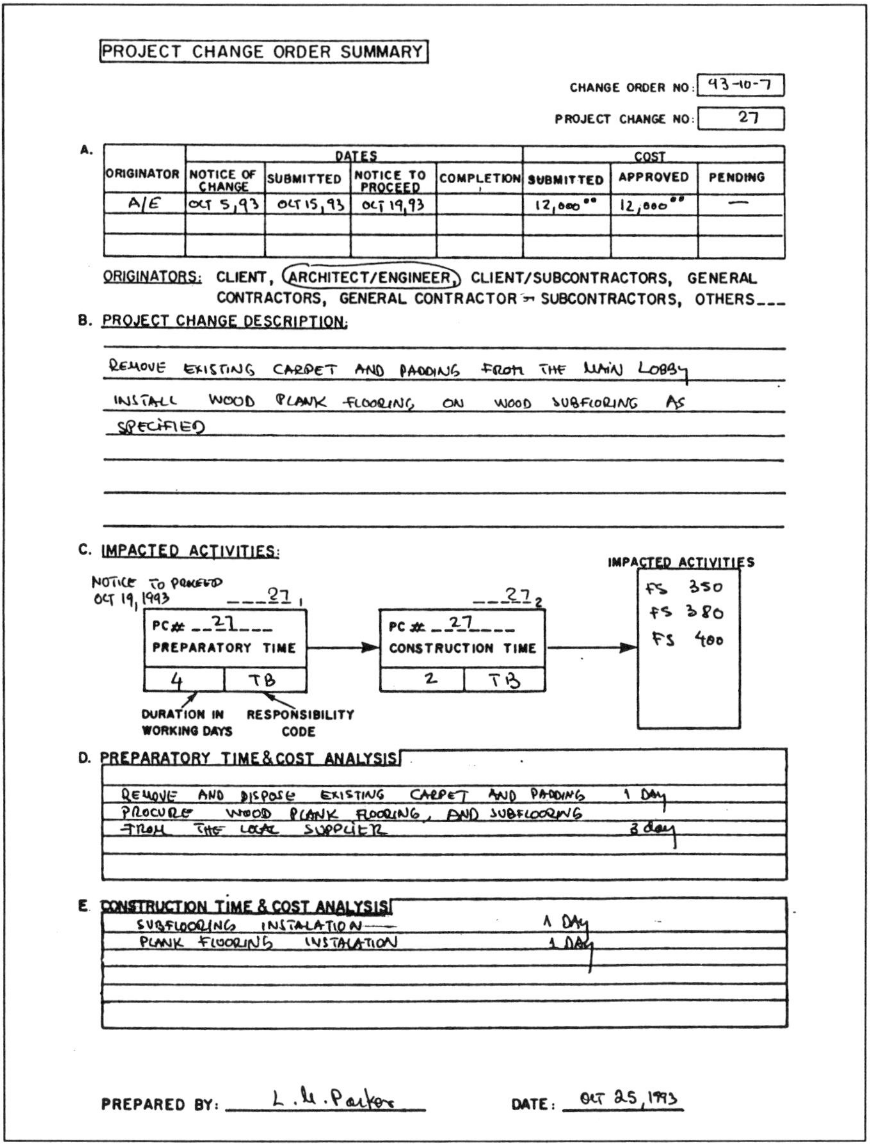

Figure 1 Notice of Change on Project Change Order Summary

CHANGE ORDER, NOTICE TO PROCEED The date when the contractor or subcontractor is authorized in writing to implement a project change order.

The notice to proceed date is considered a "not earlier than" date in documenting the impact of the change order on an approved baseline schedule. After the approved date, the estimated time for change order implementation will be considered in the network computation. Each change order considered to have an impact on the project approved schedule will be represented by two new activities.

The first activity, reflecting the preparatory time for implementing the change, is treated as a start activity with an imposed date not earlier than the notice to proceed date *(see Change order, preparatory duration)*. The second activity, construction and implementation time, is preceded by the preparatory time activity and is succeeded by the appropriate activity from the existing CPM baseline schedule. Figures 1 and 2 show two project change order summaries in the precedence and arrow techniques, indicating the notice to proceed date, October 19, 1993.

130

```
┌─────────────────────────────┐
│ PROJECT CHANGE ORDER SUMMARY │
└─────────────────────────────┘
```

CHANGE ORDER NO: 93-10-7

PROJECT CHANGE NO: 27

A.

ORIGINATOR	DATES				COST		
	NOTICE OF CHANGE	SUBMITTED	NOTICE TO PROCEED	COMPLETION	SUBMITTED	APPROVED	PENDING
A/E	OCT 5,93	OCT 15,93	OCT 19,93		12,000°°	12,000°°	—

ORIGINATORS: CLIENT, (ARCHITECT/ENGINEER), CLIENT/SUBCONTRACTORS, GENERAL CONTRACTORS, GENERAL CONTRACTOR → SUBCONTRACTORS, OTHERS___

B. PROJECT CHANGE DESCRIPTION:

REMOVE EXISTING CARPET AND PADDING FROM THE MAIN LOBBY

INSTALL WOOD PLANK FLOORING ON WOOD SUBFLOORING AS SPECIFIED

C. IMPACTED ACTIVITIES:

NOTICE TO PROCEED
OCT 19, 1993

```
            ___27₁                        ___27₂

┌─────────────────┐         ┌─────────────────┐         IMPACTED ACTIVITIES
│ PC# __21___     │         │ PC# _27____     │         FS  350
│ PREPARATORY TIME│ ──────▶ │ CONSTRUCTION TIME│ ──────▶ FS  380
├────────┬────────┤         ├────────┬────────┤         FS  400
│   4    │   TB   │         │   2    │   TB   │
└────────┴────────┘         └────────┴────────┘
 DURATION IN  RESPONSIBILITY
 WORKING DAYS    CODE
```

D. PREPARATORY TIME & COST ANALYSIS:

REMOVE AND DISPOSE EXISTING CARPET AND PADDING	1 DAY
PROCURE WOOD PLANK FLOORING, AND SUBFLOORING FROM THE LOCAL SUPPLIER	3 day

E. CONSTRUCTION TIME & COST ANALYSIS:

SUBFLOORING INSTALATION	1 DAY
PLANK FLOORING INSTALATION	1 DAY

PREPARED BY: _L. M. Parker_   DATE: _OCT 25, 1993_

Figure 1   Notice to Proceed on Project Change Order Summary
for Precedence Diagram Technique

CHANGE ORDER SUMMARY

CHANGE ORDER NO.: 93-10-7
PROJECT CHANGE NO.: 27

A.

ORIGINATOR	DATES					COST		
	NOTICE OF CHANGE	SUBMITTED	NOTICE TO PROCEED	APPROVAL	COMPLETION	SUBMITTED	APPROVED	PENDING
A/E	OCT 5, 93	OCT 15, 93	OCT 19, 93			12,000.00	12,000.00	—

ORIGINATORS: CLIENT, (ARCHITECT/ENGINEER) CLIENT/SUBCONTRACTORS, GENERAL CONTRACTOR, GENERAL CONTRACTOR-SUBCONTRACTORS, OTHERS _ _ _ _ _ _ _

B. DESCRIPTION:

REMOVE EXISTING CARPET AND PADDING IN THE MAIN LOBBY.
INSTALL WOOD PLANK FLOORINGS ON WOOD SUBFLOORING AS SPECIFIED

C. REASON:

DIRECTED BY A/E

D. IMPACTED ACTIVITIES

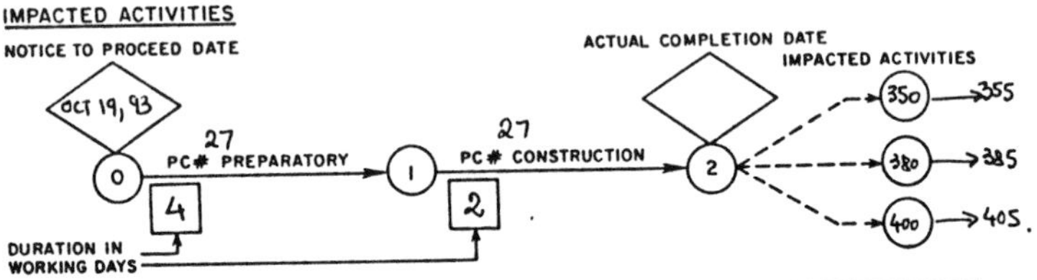

NOTICE TO PROCEED DATE

ACTUAL COMPLETION DATE
IMPACTED ACTIVITIES

OCT 19, 93

27
PC# PREPARATORY

27
PC# CONSTRUCTION

4

2

DURATION IN WORKING DAYS

350 → 355
380 → 385
400 → 405

E. PREPARATORY TIME ANALYSIS

REMOVE AND DISPOSE EXISTING CARPET AND PADDING	1 DAY
PROCURE WOOD PLANK FLOORING FROM LOCAL SUPPLIER	3 DAY

F. CONSTRUCTION TIME ANALYSIS

SUBFLOORING INSTALLATION	1 DAY
PLANK FLOORING INSTALLATION	1 DAY

PREPARED BY: L. U. PARKER   DATE: OCT 25, 1993.

Figure 2 Notice to Proceed on Project Change Order Summary for Arrow Diagram Technique

132

CHANGE ORDER, PREPARATORY DURATION (PREPARATORY TIME)

The estimated (documented) time to procure materials and parts, fix equipment, and mobilize the labor force and equipment required to start construction of a change order.

When a scheduler prepares project changes to the baseline schedule, each change needs to be represented (incorporated) in the CPM model as two new activities. One represents the project change preparatory duration as preparing a purchase order, mailing, fabrication, or procurement, and delivery of specific equipment, component, or materials requested by the project change. In addition to the materials or fixed equipment procurement, obtaining the skilled labor or specific construction equipment or specialized tool required is also a part of the preparatory time.

The second activity to model the change is related to implementation itself. All the needed components, resources existing on the job, and the installation time based on estimated productivity or production time need to be documented. Figure 1 shows the typical model used for incorporating the project change into the project CPM model (precedence technique).

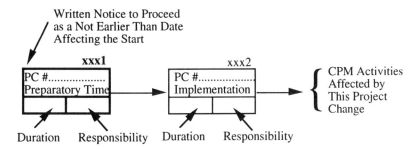

Figure 1 Project Change Number Preparatory Time Treated as New Start Activity with Imposed Start Date Not Earlier Than Written Notice to Proceed

CHANGE ORDER IMPLEMENTATION (CONSTRUCTION) DURATION

The length of time from the start to the finish of the implementation (construction) of a directed (approved) project change order, including collateral impacts.

Assume that all materials, trades, equipment, and specialized tools are on the project site. The time needed to implement the change to perform the work under normal conditions (productivity and environmental conditions) will be estimated in the time units selected for the project.

Estimated preparatory time, *(see Change order, preparatory duration)* and implementation time are of primary importance in timely documentation of the change order on an approved project schedule. The project manager plays an important role in documenting the implementation time and with the help of the scheduling engi-

133

neer in identifying activities that will be affected. The impact, if any, will be on the start or completion of succeeding activities identified from the CPM model. Figure 1 is a project change order summary indicating implementation time documentation and activities affected by the change using a precedence technique.

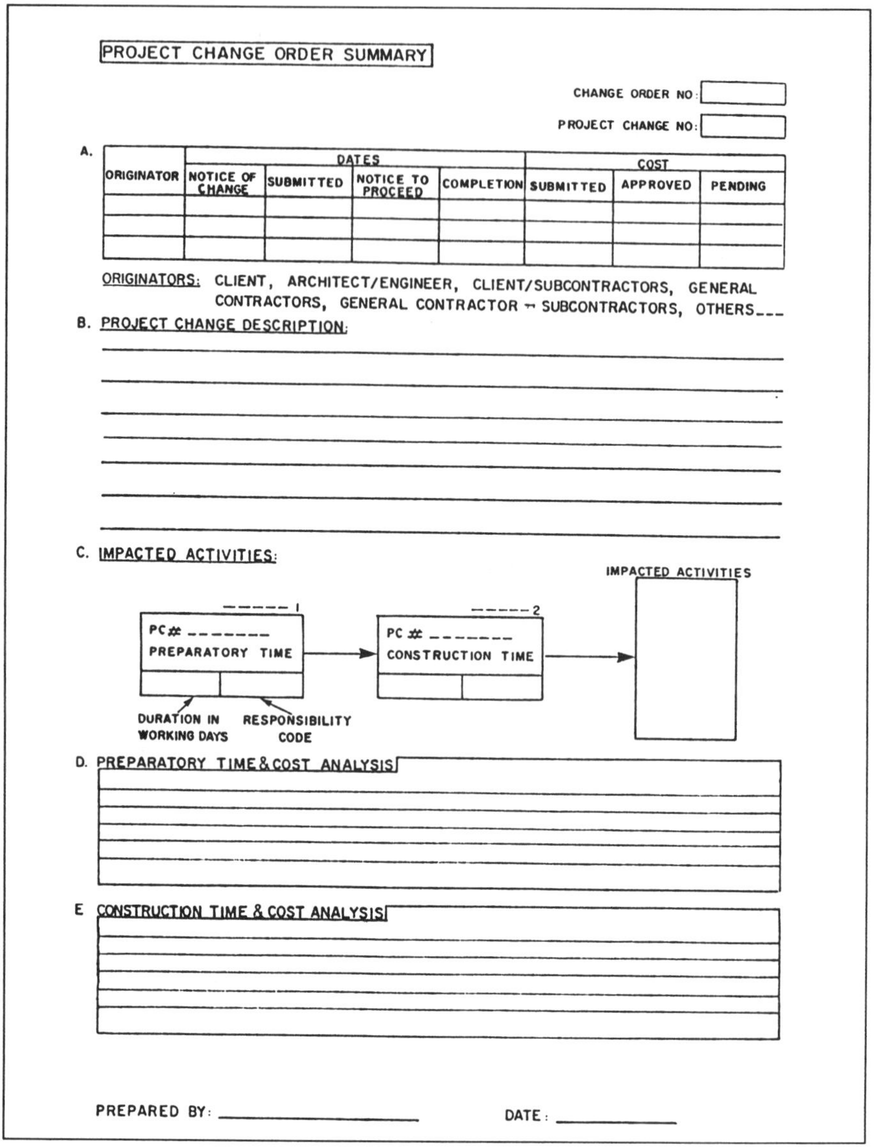

Figure 1 Project Change Order Summary

CHANGE ORDER SUBMITTED DATE The date when the contractor or sub-contractor completed pricing of a directed project change and submitted it in writing to the client's representative for approval.

From the notice of change, a contractor or subcontractor affected by the change(s) needs to prepare cost estimates. The change(s) may also affect both the estimated time and final project completion date. The date when documentation related to a requested project change is submitted to the architect/engineer or owner for approval is important because most contracts specify clearly the time needed for the owner or owner representative to review, approve, or request a resubmittal. The change order submitted date for each change order or project change will be recorded in the change order log for future reference and for monitoring the project change order status.

CHANGE REQUEST (REQUEST FOR PROPOSAL) A document originating from any member of the project team proposing a change or clarification of the contract documents, or requesting an estimate of the cost or schedule impact of a change to the contract work.

CHARGE NUMBER A numerical identification of the costs or labor hours charged to a project work package.

The charge number is one component of the account code structure *(see Account code structure)*. In the structure, a number is assigned to each work package item and called the charge number. The charge number is used both for cost estimation and for collection of actual costs charged to each work package item. The charge number should be tied into the company's existing accounting system. For example, the account code structure for a commercial building shown in Figure 1 shows use of the charge number.

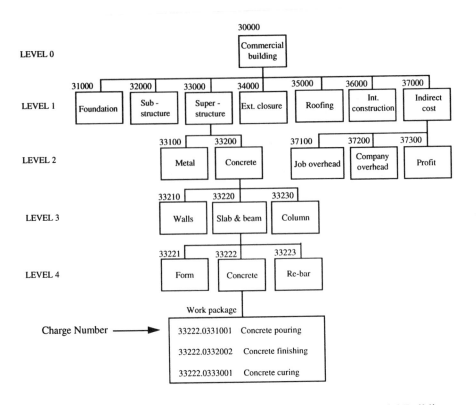

LEVEL 0 — 30000 Commercial building

LEVEL 1 — 31000 Foundation | 32000 Sub-structure | 33000 Super-structure | 34000 Ext. closure | 35000 Roofing | 36000 Int. construction | 37000 Indirect cost

LEVEL 2 — 33100 Metal | 33200 Concrete | 37100 Job overhead | 37200 Company overhead | 37300 Profit

LEVEL 3 — 33210 Walls | 33220 Slab & beam | 33230 Column

LEVEL 4 — 33221 Form | 33222 Concrete | 33223 Re-bar

Work package

Charge Number → 33222.0331001 Concrete pouring
33222.0332002 Concrete finishing
33222.0333001 Concrete curing

Figure 1 Account Code Structure and Charge Number for Commercial Building

When referring to the "concrete" work item in the account code structure, the charge numbers 33222.0331001, 33222.0332002, and 33222.0333001 for "concrete pouring," "concrete finishing," and "concrete curing," respectively, are used for the collection of actual costs. For example, actual labor cost related to "concrete finishing" should be charged to account number 33222.0332002. In addition to the labor cost, material, equipment, and subcontract costs for "concrete finishing" should be charged to the same account number, 33222.0332002.

COLLATERAL IMPACTS Impacts caused by a delay event which are incidental to the primary impact of the delay event. For instance, by unforeseen site conditions, the owner may delay critical path foundation concrete placement from the fall of the year into midwinter. The primary impact of the delay is the day-for-day movement of the critical concrete work into winter; however, the collateral impact of the delay is the longer-than-planned duration required to actually place the concrete during winter conditions.

136

COLOR GRAPH A monitoring technique for repetitive-type small projects which shows the progress of work as in-progress or completed activities. It is adequate for overall loose control of a project.

The color graph shows the work progress for a project or repetitive projects at a given data date or time now *(see Data date)* but cannot indicate whether the work performed is ahead or behind schedule. It is only time oriented and cannot capture the profit or loss for project activities monitored using this technique.

An example of a color graph for monitoring residential housing construction for units with similar floor plans is presented in Figure 1. The color graph must be updated regularly and the status of various activities related to a given data date. In other words, it is like a picture taken of a project, where all activities can be seen.

WINDMILL RUN ESTATE, AUSTIN, TEXAS
(80 Houses)   DATA DATE: April 30, 1993

House Activities	House 1	House 2	House 3	House 4	House 5	House N
Survey and Corner Stakes	●	●	●	●	◕	
Rough-in Sewer & Water Line	●	●	●	◕	◕	
Concrete Slab	●	●	●	◕	◕	
Wood Framing	●	●	◕	◕		
Exterior Doors & Windows	●	●	◕	◕		
Roof Deck & Felt	●	●	◕	◕		
Rough-in Plumbing	●	●	◕			
Rough-in Electrical	●	◕	◕			
A/C Ductwork	◕	◕				

Activities Listed in Logical Order
as Much as Possible

Indicate Status at the Data Date

Legend: ● Completed Activity
         ◕ In-progress Activity

Figure 1 Color Graph Representing Residential Housing Construction Progress

COMPLETED ACTIVITY An activity in the network that has been finished, indicating actual start and finish dates.

Completed activities are identified during the updating process *(see Monitoring)*. Actual start and finish dates are identified in the network to keep the network current. A completed activity normally needs no attention during the course of con-

struction unless disputes have arisen or crucial lessons learned from past events are considered.

The completed activity can be marked directly on logic drawings, bar charts, or computerized reports. There are three ways of considering completed activities within a network analysis of a partly completed project:

1. All completed activity durations are altered to zero, and all partially completed activity durations *(see In-progress activity)* are changed to their estimated remaining duration.
2. All completed activities are dropped from the network, and uncompleted activities left without predecessors by this means are set to be start activities at the present date *(see Data date)*.
3. All completed activities are maintained with actual start and completion dates, and partially completed activities with their remaining duration; computation is performed from the original start date.

Figures 1 and 2 show updated arrow and precedence diagram networks with completed activities. Figures 3 and 4 show tabular and bar chart schedule reports using the precedence diagram method.

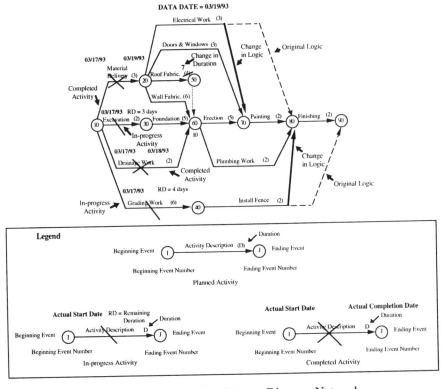

Figure 1 Partly Completed Arrow Diagram Network

138

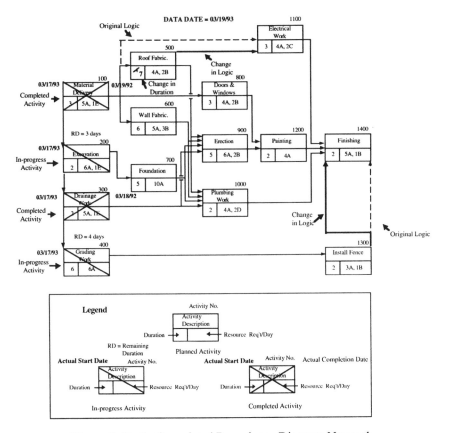

Figure 2 Partly Completed Precedence Diagram Network

```
-------------------------------------------------------------------------------------------
CIVIL ENGINEERING DEPARTMENT              PRIMAVERA PROJECT PLANNER

REPORT DATE 23NOV93 RUN NO.    2          PROJECT SCHEDULE REPORT              START DATE 13MAR93  FIN DATE 12APR93
                    21:37
CLASSIC SCHEDULE REPORT - SORT BY ES, TF                                      DATA DATE  19MAR93  PAGE NO.    1
```

ACTIVITY ID	ORIG DUR	REM DUR	%	CODE	ACTIVITY DESCRIPTION	EARLY START	EARLY FINISH	LATE START	LATE FINISH	TOTAL FLOAT
100	3	0	100		MATERIAL DELIVERY	17MAR93A	19MAR93A			
300	2	0	100		DRAINAGE WORK	17MAR93A	18MAR93A			
200	2	3	0		EXCAVATION	17MAR93A	23MAR93		23MAR93	0
400	6	4	33		GRADING WORK	17MAR93A	24MAR93		8APR93	11
500	7	7	0		ROOF FABRICATION	19MAR93	29MAR93	22MAR93	30MAR93	1
600	6	6	0		WALL FABRICATION	19MAR93	26MAR93	23MAR93	30MAR93	2
800	3	3	0		DOORS & WINDOWS	19MAR93	23MAR93	2APR93	6APR93	10
1100	3	3	0		ELECTRICAL WORK	19MAR93	23MAR93	6APR93	8APR93	12
700	5	5	0		FOUNDATION	24MAR93	30MAR93	24MAR93	30MAR93	0
1300	2	2	0		INSTALL FENCE	25MAR93	26MAR93	9APR93	12APR93	11
900	5	5	0		ERECTION	31MAR93	6APR93	31MAR93	6APR93	0
1000	2	2	0		PLUMBING WORK	31MAR93	1APR93	7APR93	8APR93	5
1200	2	2	0		PAINTING	7APR93	8APR93	7APR93	8APR93	0
1400	2	2	0		FINISHING	9APR93	12APR93	9APR93	12APR93	0

Figure 3 Updated Tabular Schedule Report, Precedence Diagramming Method

CIVIL ENGINEERING DEPARTMENT   PRIMAVERA PROJECT PLANNER

REPORT DATE 23NOV93 RUN NO.   3   PROJECT SCHEDULE REPORT   START DATE 13MAR93 FIN DATE 12APR93
         21:45
BAR CHART BY ES, EF, TP   DATA DATE 19MAR93   PAGE NO.   1

                                                                                    DAILY-TIME PER.   1

...........ACTIVITY DESCRIPTION.............						15	22	29	05	12	19	26	03	10	17
ACTIVITY ID	OD	RD	PCT	CODES	FLOAT	SCHEDULE	MAR	MAR	MAR	APR	APR	APR	APR	MAY	MAY	MAY
							93	93	93	93	93	93	93	93	93	93

DRAINAGE WORK   CURRENT   . AA* .   .   .   .
        300   2   0 100   PLANNED   . EE* .   .   .   .

MATERIAL DELIVERY   CURRENT   . AA* .   .   .   .
        100   3   0 100   PLANNED   . EEE .   .   .   .

EXCAVATION   CURRENT   . AAE..EE   .   .   .   .
        200   2   3   0   0   PLANNED   . EE* .   .   .   .

GRADING WORK   CURRENT   . AAE..EEE   .   .   .   .
        400   6   4 33   11   PLANNED   . EEE..EEE   .   .   .   .

DOORS & WINDOWS   CURRENT   . E..EE   .   .   .   .
        800   3   3   0   10   PLANNED   . * EEE   .   .   .   .

ELECTRICAL WORK   CURRENT   . E..EE   .   .   .   .
       1100   3   3   0   12   PLANNED   . * EEE   .   .   .   .

WALL FABRICATION   CURRENT   . E..EEEEE .   .   .   .
        600   6   6   0   2   PLANNED   . * EEEEE..E .   .   .   .

ROOF FABRICATION   CURRENT   . E..EEEEE..E .   .   .   .
        500   7   7   0   1   PLANNED   . * EEEE   .   .   .   .

FOUNDATION   CURRENT   . *   . EEE..EE .   .   .   .
        700   5   5   0   0   PLANNED   . E..EEEE   .   .   .   .

INSTALL FENCE   CURRENT   . *   . EE .   .   .   .
       1300   2   2   0   11   PLANNED   . *   . EE .   .   .   .

PLUMBING WORK   CURRENT   . *   .   . EE .   .   .   .
       1000   2   2   0   5   PLANNED   . *   .   .EE   .   .   .   .

ERECTION   CURRENT   . *   .   . EEE..EE .   .   .   .   .   .
        900   5   5   0   0   PLANNED   . *   .   .EEEE..E   .   .   .   .   .   .

PAINTING   CURRENT   . *   .   .   . EE .   .   .   .   .   .
       1200   2   2   0   0   PLANNED   . *   .   .   .EE   .   .   .   .   .   .

FINISHING   CURRENT   . *   .   .   . E..E .   .   .   .   .
       1400   2   2   0   0   PLANNED   . *   .   .   . EE .   .   .   .   .   .

Figure 4 Updated Bar Chart Schedule Report, Precedence Diagramming Method

140

CONCEPTUAL SCHEDULE (PROPOSAL SCHEDULE) A conceptual schedule used primarily to give a general idea of the project scope and overall project duration.

A conceptual schedule is a condensed schedule submitted with a bid with or without an owner request. It is recommended that the owner or owner's representative require prospective bidders to submit a visualization of the project development/ implementation associated with bidding costs. For large and complex projects, a conceptual schedule will normally be requested when the bid is submitted. The project breakdown structure (PBS) *(see Project breakdown structure)* simplifies this schedule by organizing the work so as to minimize the number of activities. Figure 1 shows the correlation of a PBS level to a conceptual schedule.

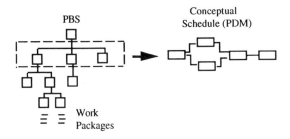

Figure 1 Correlation of PBS Level to Conceptual Schedule

The conceptual schedule contains the duration estimates of the project and major activities. Once the bid is awarded, this conceptual schedule may be expanded to show each activity identified in more detail at a lower PBS level, as shown in Figure 2. The conceptual schedule can be represented by a network diagram (*see Network)* or bar chart *(see Bar chart)*, unless specified in the contract. Figure 3 shows the time span for a conceptual schedule related to other schedules.

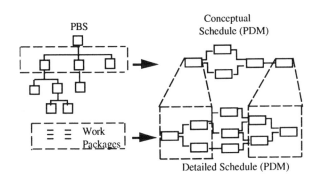

Figure 2 Correlation of PBS Levels to Conceptual and Detailed Schedules

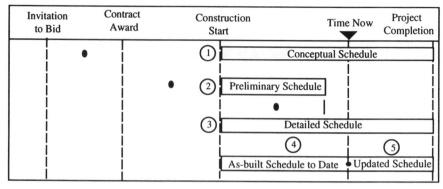

Invitation to Bid	Contract Award	Construction Start	Time Now ▼	Project Completion

● Submitted Date

Figure 3 Time Span of Project Schedules

CONCURRENT DELAY *(see Delay, concurrent)*

CONSTRAINT (RESTRAINT) An imposed condition affecting the start or completion of a project activity.

Constraints define the sequence and relationship of all activities in a project. Although many of them may proceed concurrently, certain activities must be completed in a sequence or chain. By assigning constraints, each activity is proposed to represent the overall project schedule, determining the necessary sequences. The constraints should be set by an experienced planner or scheduler who can identify the relationship between activities as realistically as possible; otherwise, the project schedule will be difficult to follow, resulting in poor project execution. Three types of constraints can be identified in the network:

1. *Technological constraints.* This type of constraint is the first and most obvious constraint to be identified. Technological constraints are unavoidable unless the technology is changed. Activities' technological constraints appear as soon as each activity in the project is subjected to the following questions:

- What activities must precede this activity?
- What activities can be done concurrently with this activity?
- What activities must follow this activity?

By answering these questions, each activity and its relationship with others will be determined. For example, the concrete slab cannot be poured until the slab formwork, reinforcing bars, and embedded conduits are installed. Figures 1 and 2 show portions of arrow and precedence diagram networks with technological constraints.

142

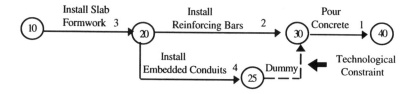

Figure 1 Technological Constraints in Arrow Diagram Network

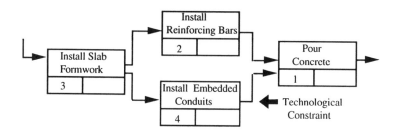

Figure 2 Technological Constraints in Precedence Diagram Network

2. Managerial constraints. Management constraints normally are related to resources such as materials, equipment, and crews. They occur when it is essential to
reschedule because resources for certain operations cannot be made available or
management feels that operations could be better sequenced if activities are delayed
from an early start. For example, due to the limitations of the concrete crews, the
concrete slab can be poured after the slab formwork is set, reinforcing bars and embedded conduits are installed, and the concrete elevator shafts are poured. Sequencing the concrete pour is a managerial decision. Figures 3 and 4 show portions of
arrow and precedence diagram networks with managerial constraints.

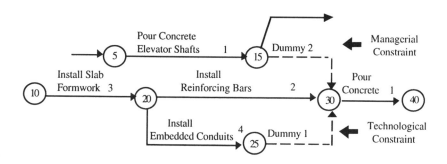

Figure 3 Managerial Constraints in Arrow Diagram Network

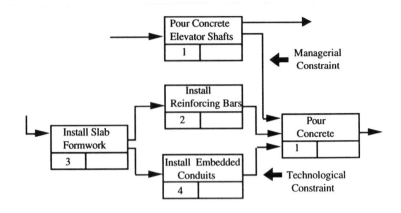

Figure 4 Managerial Constraints in Precedence Diagram Network

3. *External Constraints.* These constraints are difficult to identify when a network is being developed. Typical constraints are safety and regulatory (federal, state, or community). For example, after OSHA inspection, the concrete slab cannot be poured because proper protection (guardrails) is not provided. Therefore, the concrete slab cannot be poured until the guardrails, slab formwork, reinforcing bars, and embedded conduits are installed, and the concrete elevator shafts are poured. Figures 5 and 6 show portions of arrow and precedence diagram networks with external constraints.

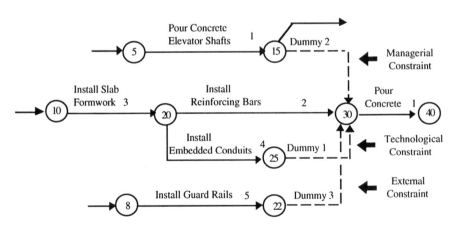

Figure 5 External Constraints in Arrow Diagram Network

144

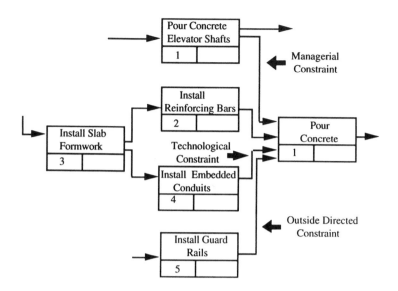

Figure 6 External Constraints in Precedence Diagram Network

CONSTRUCTIVE ACCELERATION *(see Acceleration)* A unilateral acceleration that takes place when an owner refuses to grant a time extension to a contractor
who has otherwise submitted a proper request for an entitlement to an extended performance period due to excusable delay, and by the owner's refusal to grant the time
extension, forces the contractor to complete the work scope within the currently recognized contract duration.

CONTRACT DATE(S) (SCHEDULED, PLUG, OR IMPOSED DATES) Any
date specified in the contract or imposed on any project activity or event that affects
activity and project time computation.

Contractually, the owner may impose detailed dates on various systems or subsystems that will have an effect on the start and finish dates of any major event or
activity. The most commonly applied schedule dates are those of the final completion of the project, but any intermediate activity may be treated in the same way
when the owner feels that these intermediate activities are critical for project completion. The use of contract dates allows the owner to exert control over a project schedule and may have the effect of producing negative floats *(see Activity total float* and
Free float) on activities. As a result, assigning too many imposed dates without
careful consideration can cause a general contractor to be faced with difficult problems in optimizing resources. When a scheduled or imposed date is assigned to a
specific activity, it will affect only that activity, but when it is assigned to an event, it

will affect all preceding and succeeding activities of that event.

It is recommended that care be taken when schedule dates are imposed on a project. The three types of contract dates described below are applicable. Selection of all these milestones has a tremendous impact on a project schedule. Figure 1 shows various types of imposed milestone dates.

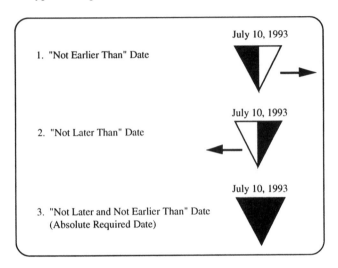

Figure 1 Various Types of Imposed Milestone Dates

1. *"Not earlier than" date:* an imposed date that places a limit on an event, activity, or milestone *(see Event, Activity, and Milestone)* so that it cannot take place earlier than a specified date, but which does not restrain it from occurring later than the specified date. The impact of a "not earlier than" date is shown in Figure 2.

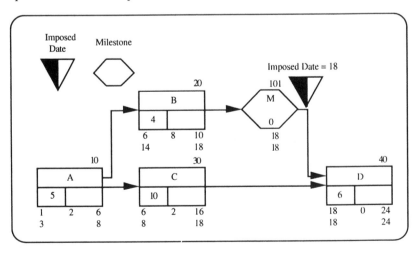

Figure 2 "Not Earlier Than" Date Impact on Project Schedule

146

Assume that the precedence technique for project scheduling has been selected. In this case a "not earlier than" date which is earlier than the computed early start of a specified milestone does not affect either preceding or succeeding activities. If the "not earlier than" date is later than the computed early start of the succeeding activity as shown in the example, this imposed date will be considered for forward pass computation *(see Forward pass computation)* and will reduce the total float of all succeeding activities and milestones.

2. *"Not later than" date:* an imposed date that places a limit on an event, activity, or milestone *(see Event, Activity, and Milestone)* so that it cannot take place later than a specified date, but which does not restrain it from falling earlier than the specified date. The impact of a "not later than" date on a CPM network schedule is shown in Figure 3.

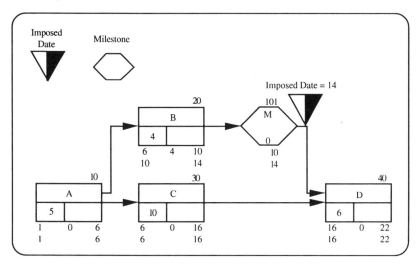

Figure 3 "Not Later Than" Date Impact on Project Schedule

A "not later than" date that is greater than the late computed finish date as shown in this example does not affect computed dates for any milestones or activities in the network that follow the activity affected by the imposed date. When a "not later than" date is earlier than the computed late finish date of the activity affected, this date will be considered during the backward pass computation *(see Backward pass computation)*. As a result, the total float of all preceding activities and milestones will be reduced by the difference between the late finish date and the imposed date.

3. *"Not later and not earlier than" date (absolute required date):* A situation that generates a major impact in scheduling, in which the owner specifies an "absolute" date, as shown in Figure 4. In this example, the milestone *(see Milestone)* is affected by an absolute date for an activity to take place on the tenth calendar day. The early and late times, in this case, are assigned to day 10; no float exists and all

preceding and succeeding activities are affected by this date. A temporary case arises when an absolute date does not have an impact. This happens when the imposed absolute date is equal to the late milestone computed finish date.

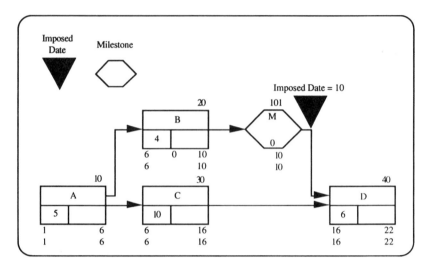

Figure 4 Absolute Required Date Impact on Project Schedule

For large-scale or complex projects, many scheduled or imposed dates could be issued. The contractor should therefore keep track of all these dates to identify all information related to the assigned scheduled date. An example of a scheduled date log is shown in Figure 5.

SCHEDULED DATE LOG				
Date.........................			Project................................	
Affected Activity or Event Code	Activity/Event Description	Imposed Date	Type (NE, NL, or Absolute) *	Imposed by
1500	Finishing	04/09/93	NL	Owner

* NE = Not Earlier Than
NL = Not Later Than
Absolute = Not Earlier & Not Later Than

Figure 5 Scheduled Date Log

148

CONTRACT WBS A kind of PBS but more contract-oriented family tree, such as design contracts, construction contracts, and operation and maintenance contracts. A contract WBS is developed for better management of various contracts during the initial stage of a project. Each contract is managed based on the contract WBS.

A contract WBS is developed for better management of various contracts during the initial stage of a project. Each contract is managed based on the contract WBS. For example, the contract WBS developed for a commercial building is shown in Figure 1.

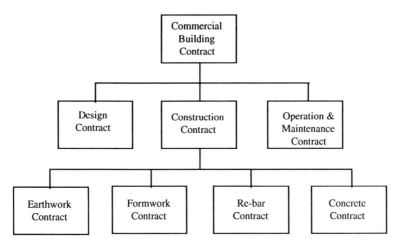

Figure 1 Contract WBS for Commercial Building

COST PERFORMANCE INDEX (CPI) Provides the cost efficiency of the work accomplished. The CPI is defined as the ratio of the cumulative, budgeted cost of work performed (BCWP) divided by the cumulative, actual cost of work performed (ACWP).

The cost performance index (CPI) is one of the C/SCSC elements and represents a performance measurement of over cost or under cost of the budget established for the work accomplished to date based on the earned value *(see Cost and schedule control system criteria).* While variances such as the CV provide significant information about the progress of the project to data date (time now) in absolute terms, an index such as the CPI provides additional insight into the status of the project in relative terms.

At a given time now, the CPI is calculated by using the formula

$$CPI = BCWP/ACWP$$

(See Budgeted cost of work performed and *Actual cost of work performed.)* Interpretation of the CPI is as follows:

VARIANCE	Less Than 1.00	Equal to 1.00	More Than 1.00
CPI	Over Cost	On Cost	Under Cost

For example, a CPI of 0.90 (90%) indicates that 90 contract dollars' worth of planned work has been completed for every 100 actual dollars spent.

To develop the CPI in a project, the following steps are generally taken:

Step 1: Use the original PBS of a project.

Step 2: Use the original cost estimation of the project based on the PBS.

Step 3: Use the latest updated CPM schedule by the data date *(see Data date).*

Step 4: Calculate the BCWP of the activity level at the point of the data date (time now).

Step 5: Calculate the ACWP of the activity level at the point of the data date (time now).

Step 6: Calculate the CPI dividing the BCWP by the ACWP at the project level. The results of steps 4 to 6 are shown in the monthly cost performance report.

The preceding six steps are explained using a residential house as an example.

1. *Step 1:* PBS

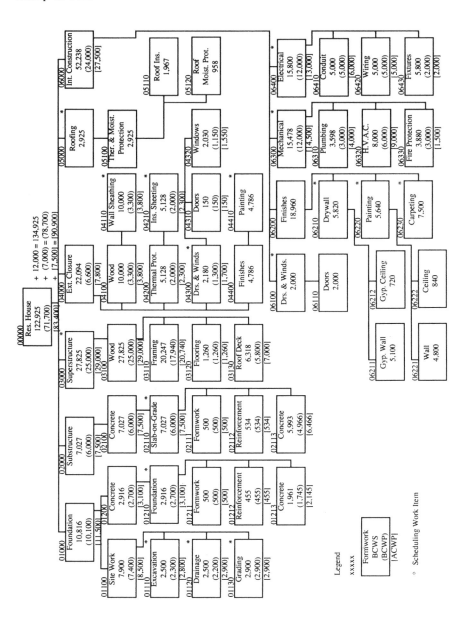

151

2. *Step 2:* Cost Estimation

DESCRIPTION	U/M	U/P	Q'TY	TOTAL
01 General Requirements				
Subtotal	L/S	12,000	1	12,000
02 Site Work				
01110 Excavation	C.Y.	1.89	1,323	2,500
01120 Drainage	L.F.	3.11	804	2,500
01130 Grading	C.Y.	1.06	2,736	2,900
Subtotal				7,900
03 Concrete				
01211 Foundation Form	SFCA	2.89	173	500
01212 Foundation Re-bar	Tons	910	0.5	455
01213 Foundation Concrete	C.Y.	103.2	19	1,961
02111 Slab-on-Grade Form	L.F.	1.38	362	500
02112 Slab-on-Grade Re-bar	Tons	890	0.6	534
02113 Slab-on-Grade Concrete	C.Y.	90.8	66	5,993
Subtotal				9,943
06 Wood				
03110 Superstructure Framing	M.B.F.	1,191	17	20,247
03120 Superstructure Flooring	S.F.	0.63	2,000	1,260
03130 Roof Decking	S.F.	5.01	1,261	6,318
04110 Wall Sheathing	S.F.	1.17	8,547	10,000
Subtotal				37,825
07 Thermal & Moisture Protection				
04210 Wall Insulating Sheeting	S.F.	0.6	8,547	5,128
05110 Roof Insulating	S.F.	1.56	1,261	1,967
05120 Roof Moisture Protection	S.F.	0.76	1,261	958
Subtotal				8,053
08 Doors & Windows				
04310 Ext. Doors	Ea.	150	1	150
04320 Ext. Windows	Ea.	203	10	2,030
06110 Int. Doors	Ea.	200	10	2,000
Subtotal				4,180
09 Finishes				
04410 Ext. Painting	S.F.	0.56	8,547	4,786
06211 Int. Gyp. Wallboard	S.F.	0.34	15,000	5,100
06212 Int. Gyp. Ceiling Board	S.F.	0.36	2,000	720
06311 Int. Wall Painting	S.F.	0.32	15,000	4,800
06312 Int. Ceiling Painting	S.F.	0.42	2,000	840
06230 Carpeting	S.Y.	5	1,500	7,500
Subtotal				23,746
15 Mechanical				
06310 Plumbing	L.F.	3.98	904	3,598
06320 H.V.A.C.	L/S	8,000	1	8,000
06330 Fire Protection	Ea.	776	5	3,880
Subtotal				15,478
16 Electrical				
06410 Electrical Conduit	L/S	5,000	1	5,000
06420 Electrical Wiring	L/S	5,000	1	5,000
06430 Electrical Fixtures	L/S	5,800	1	5,800
Subtotal				15,800
GRAND TOTAL				134,925

3. *Step 3:* Updated CPM Network Schedule

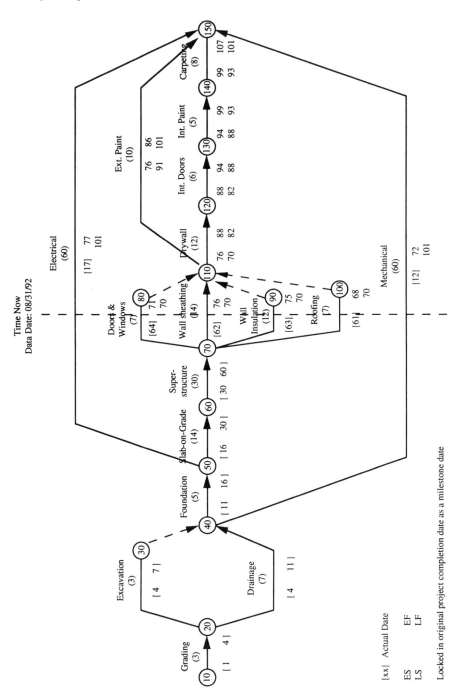

153

4. *Steps 4 to 6:* Monthly Cost Performance Report

MONTHLY COST PERFORMANCE REPORT: CPM ITEM

PROJECT: RESIDENTIAL HOUSE

ITEM		CURRENT PERIOD									CUMULATIVE TO DATE									AT COMPLETION		
		BUDGET COST		ACWP	VARIANCE			INDEX			BUDGET COST		ACWP	VARIANCE			INDEX					
PBS	CPM	BCWS	BCWP	ACWP	SV	CV	AV	SPI	CPI	PC	BCWS	BCWP	ACWP	SV	CV	AV	SPI	CPI	PC	BAC	EAC	ACV
01130	10 - 20										2,900	2,900	2,900	0	0	0	1.00	1.00	1.00	2,900	2,900	
01110	20 - 30										2,500	2,300	2,800	-200	-500	-300	0.92	0.82	0.92	2,500	2,800	-300
01120	20 - 40										2,500	2,200	2,800	-300	-600	-300	0.88	0.79	0.88	2,500	2,800	-300
01210	40 - 50										2,916	2,700	3,100	-216	-400	-184	0.93	0.87	0.93	2,916	3,100	-184
06300	40 - 150	5,160	5,000	5,200	-160	-200	-40	0.97	0.96	0.32	13,416	12,000	14,500	-1,416	-2,500	-1,084	0.89	0.83	0.78	15,478	16,562	-1,084
02110	50 - 60										7,027	6,000	7,500	-1,027	-1,500	-473	0.85	0.80	0.85	7,027	7,500	-473
03000	60 - 70	14,903	13,900	15,000	-1,003	-1,100	-97	0.93	0.93	0.50	27,825	25,000	29,000	-2,825	-4,000	-1,175	0.90	0.86	0.90	27,825	29,000	-1,175
06400	50 - 150	5,260	5,100	5,300	-160	-200	-40	0.97	0.96	0.32	12,361	12,000	13,000	-361	-1,000	-639	0.97	0.92	0.76	15,800	16,439	-639
04300	70 - 80	1,555	1,300	1,700	-255	-400	-145	0.84	0.76	0.60	1,555	1,300	1,700	-255	-400	-145	0.84	0.76	0.60	2,180	2,325	-145
04210	70 - 90	2,135	2,000	2,300	-135	-300	-165	0.94	0.87	0.39	2,135	2,000	2,300	-135	-300	-165	0.94	0.87	0.39	5,128	5,293	-165
04110	70 - 110	3,570	3,300	3,800	-270	-500	-230	0.92	0.87	0.33	3,570	3,300	3,800	-270	-500	-230	0.92	0.87	0.33	10,000	10,230	-230
05000	70 - 100																			2,925	2,925	0
06210	110 - 120																			5,820	5,820	0
04410	110 - 150																			4,786	4,786	0
06100	120 - 130																			2,000	2,000	0
06220	130 - 140																			5,640	5,640	0
06230	140 - 150																			7,500	7,500	0
SUBTOTAL		32,583	30,600	33,300	-1,983	-2,700	-717	0.94	0.92	0.25	78,705	71,700	83,400	-7,005	-11,700	-4,695	0.91	0.86	0.58	122,925	127,620	-4,695
GEN. REQ.		2,400	2,400	2,500		-100	-100	1.00	0.96	0.20	7,200	7,000	7,500	-200	-500	-300	0.97	0.93	0.58	12,000	12,300	-300
TOTAL		34,983	33,000	35,800	-1,983	-2,800	-817	0.94	0.92	0.24	85,905	78,700	90,900	-7,205	-12,200	-4,995	0.92	0.87	0.58	134,925	139,920	-4,995

Note: The project at the data date has 0.92 in the current period and 0.87 cumulatively as the cost performance index (CPI). It shows that the project is over cost because the CPI is less than 1. The same comment is valid for all work items except for item 01130, which is on schedule (CPI=1).

154

COST AND SCHEDULE CONTROL SYSTEM CRITERIA (C/SCSC) The procedures for monitoring project cost performance based on Instruction DODI 7000.2 issued by Department of Defense in December 1967.

1. *History.* After the critical path method (CPM) and project evaluation and review technique (PERT) were introduced to measure project performance and to prevent frequent cost and schedule overruns, a new system was developed. PERT/Cost, which proposed a framework for planning and control of both time and cost, was introduced by 1961. However, PERT/Cost was not implemented widely because it required a lot of detailed knowledge to use.

In December 1967, the Department of Defense (DOD) published Instruction DODI 7000.2, Performance Measurement for Selected Acquisitions. In addition, C/SCSC was introduced to monitor cost performance of defense contractors, and the Cost/Schedule Control System Criteria Joint Implementation Guide (JIG) was published to implement the criteria uniformly.

In 1980, the Department of Energy (DOE), National Aeronautics and Space Administration (NASA), and Department of Transportation (DOT) prepared similar C/SCSC implementation documents applied to all projects. Now, C/SCSC is also being applied in both Canadian and Australian industry.

2. *Elements.* Before we discuss the C/SCSC, the earned value concept should be understood, and a work breakdown structure and network-based schedule should be prepared *(see Earned value* and *Contract WBS).* Basically, the C/SCSC has three elements: data analysis terms, variance, and index. The budget cost of work schedule (BCWS), budget cost of work performed (BCWP), actual cost of work performed (ACWP), budget at completion (BAC), and estimate at completion (EAC) are included in the elements of the data analysis terms. Accounting variance (AV), cost variance (CV), and schedule variance (SV) are included in the elements of the variance. Percent complete (PC), cost performance index (CPI), and schedule performance Index (SPI) are included in the elements of the index. For detailed information, *see Budgeted cost of work schedule, Budgeted cost of work performed, Actual cost of work performed, Budget at completion, Estimate at completion, Accounting variance, Cost variance, Schedule variance, Percent complete, Cost performance index,* and *Schedule performance index.*

3. *Implementation.* To develop the C/SCSC in a project, the following steps are generally used:

Step 1: Develop the PBS (WBS) of the project.
Step 2: Schedule each work item of the PBS (WBS).
Step 3: Assign an approved budget cost to each work item.
Step 4: Develop a cumulative, time-phased BCWS for the project, adding up the cumulative budgets of all work items.
Step 5: Periodically, establish a cumulative, time-phased BCWP multiplying the budget cost by the work completed. BCWP is the earned value.

155

Step 6: Periodically, establish a cumulative, time-phased ACWP multiplying the actual cost by the work completed.

Step 7: Produce monthly progress reports comparing the ACWP with the BCWS and BCWP.

Figure 1 shows the project cost/schedule progress report.

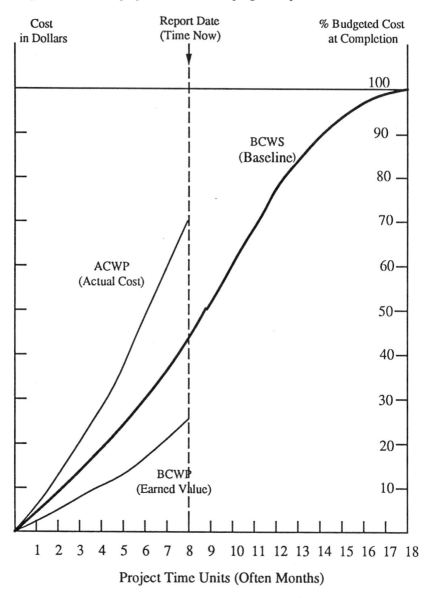

Figure 1 Project Cost/Schedule Progress Report

COST SLOPE The rate of increase of direct cost with the reduction in activity duration. This varies with duration and is a property of the activity crashing process.

The cost slope is the basic information for accomplishing the time-cost trade-off analysis *(see Time-cost trade-off, Normal activity time cost point,* and *Crash activity time cost point).* The cost of slope is used to decide which activity is most economical when accelerating the project. It is defined and applied based on several assumptions: Each activity has a time-cost relationship and a linear variation in the time and cost (linear approximation).

1. *Definition.* The cost slope is defined as follows (Figure 1):

$$\text{cost slope} \quad = \quad \frac{\text{crash cost} - \text{normal cost}}{\text{normal time} - \text{crash time}}$$

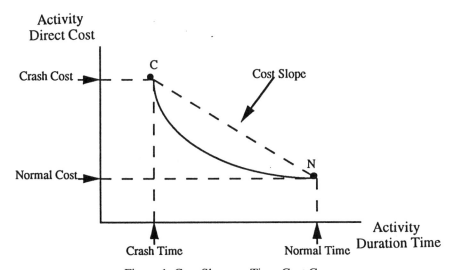

Figure 1 Cost Slope on Time-Cost Curve

a. *Normal time:* normal activity duration, which is estimated based on the usual amount of labor and equipment
b. *Normal cost:* direct expense of an activity, which is estimated based on normal time
c. *Crash time:* minimum activity duration, which is estimated based on overtime, multishifts, and additional equipment
d. *Crash cost:* normal cost plus the extra cost incurred in performing crash time

2. *Application.* Compare the cost slopes (Figure 2) and decide on an activity for acceleration. In the figure, activity A has a steeper slope than activity B. When the project is accelerated in activities A and B, the increase of cost in activity A would be

157

greater than that in activity B. So it is apparent that activity B is the better choice for acceleration because the slope of activity B is less steep.

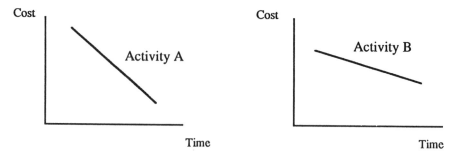

Figure 2 Cost Slopes of Activity A and Activity B

COST VARIANCE (CV) The difference between budgeted cost of work performed (BCWP) and the actual cost of work performed (ACWP) at any point over the life of the project.

The cost variance (CV) is one of the C/SCSC elements and represents a performance measure of cost overrun or underrun of the budget established for the work accomplished to date based on the earned value *(see Cost and schedule control system criteria).* At a given time now, the CV is calculated by using the formula

$$CV = BCWP - ACWP$$

(See Budgeted cost of work performed and Actual cost of work perform.) Interpretation of the CV is as follows:

VARIANCE	-	0	+
CV	Over Cost	On Cost	Under Cost

As shown in the table, if a (–) value of the CV is found, the cause should be identified. For example, the following can be possible:

1. Technical problems requiring allocation of extra resources
2. Inaccuracies in the original estimate of the work
3. Lower productivity than expected
4. Unexpected increases in material, labor, or equipment costs

To develop the CV in a project, the following steps are generally taken:

Step 1: Use the original PBS of a project.
Step 2: Use the original cost estimation of the project based on the PBS.
Step 3: Use the latest updated CPM schedule by the data date *(see Data date).*
Step 4: Calculate the BCWP of the activity level at the point of the data date (time now).
Step 5: Calculate the ACWP of the activity level at the point of the data date (time now).
Step 6: Calculate the CV subtracting the ACWP from the BCWP at the project level. The results of steps 3 to 6 are shown in the monthly cost performance report.
Step 7: Draw the CV on the graph using the time-phased and cumulative earned values by the data date.

The preceding seven steps are explained using a residential house as an example.

1. *Step 1:* PBS

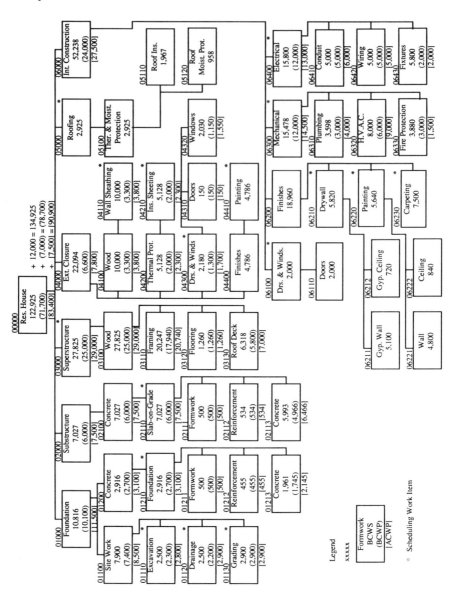

160

2. *Step 2:* Cost Estimation

DESCRIPTION	U/M	U/P	Q'TY	TOTAL
01 General Requirements				
Subtotal	L/S	12,000	1	12,000
02 Site Work				
01110 Excavation	C.Y.	1.89	1,323	2,500
01120 Drainage	L.F.	3.11	804	2,500
01130 Grading	C.Y.	1.06	2,736	2,900
Subtotal				7,900
03 Concrete				
01211 Foundation Form	SFCA	2.89	173	500
01212 Foundation Re-bar	Tons	910	0.5	455
01213 Foundation Concrete	C.Y.	103.2	19	1,961
02111 Slab-on-Grade Form	L.F.	1.38	362	500
02112 Slab-on-Grade Re-bar	Tons	890	0.6	534
02113 Slab-on-Grade Concrete	C.Y.	90.8	66	5,993
Subtotal				9,943
06 Wood				
03110 Superstructure Framing	M.B.F.	1,191	17	20,247
03120 Superstructure Flooring	S.F.	0.63	2,000	1,260
03130 Roof Decking	S.F.	5.01	1,261	6,318
04110 Wall Sheathing	S.F.	1.17	8,547	10,000
Subtotal				37,825
07 Thermal & Moisture Protection				
04210 Wall Insulating Sheeting	S.F.	0.6	8,547	5,128
05110 Roof Insulating	S.F.	1.56	1,261	1,967
05120 Roof Moisture Protection	S.F.	0.76	1,261	958
Subtotal				8,053
08 Doors & Windows				
04310 Ext. Doors	Ea.	150	1	150
04320 Ext. Windows	Ea.	203	10	2,030
06110 Int. Doors	Ea.	200	10	2,000
Subtotal				4,180
09 Finishes				
04410 Ext. Painting	S.F.	0.56	8,547	4,786
06211 Int. Gyp. Wallboard	S.F.	0.34	15,000	5,100
06212 Int. Gyp. Ceiling Board	S.F.	0.36	2,000	720
06311 Int. Wall Painting	S.F.	0.32	15,000	4,800
06312 Int. Ceiling Painting	S.F.	0.42	2,000	840
06230 Carpeting	S.Y.	5	1,500	7,500
Subtotal				23,746
15 Mechanical				
06310 Plumbing	L.F.	3.98	904	3,598
06320 H.V.A.C.	L/S	8,000	1	8,000
06330 Fire Protection	Ea.	776	5	3,880
Subtotal				15,478
16 Electrical				
06410 Electrical Conduit	L/S	5,000	1	5,000
06420 Electrical Wiring	L/S	5,000	1	5,000
06430 Electrical Fixtures	L/S	5,800	1	5,800
Subtotal				15,800
GRAND TOTAL				134,925

3. *Step 3:* Updated CPM Network Schedule

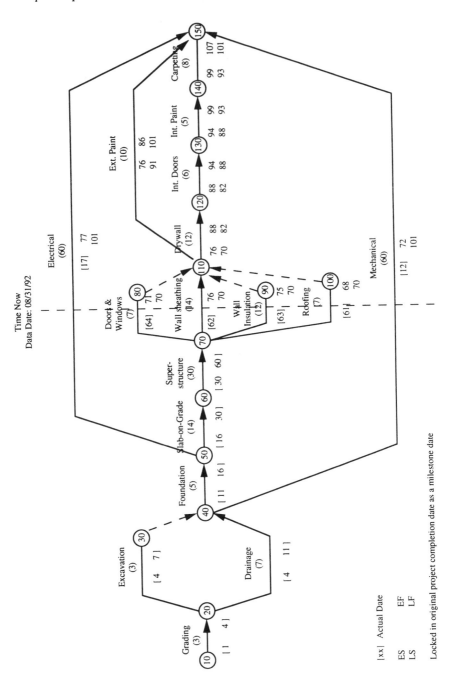

4. *Steps 4 to 6:* Monthly Cost Performance Report

MONTHLY COST PERFORMANCE REPORT: CPM ITEM

PROJECT: RESIDENTIAL HOUSE   REPORT DATE: 08/31/92

ITEM		CURRENT PERIOD									CUMULATIVE TO DATE									AT COMPLETION		
		BUDGET COST		ACWP	VARIANCE			INDEX		PC	BUDGET COST		ACWP	VARIANCE			INDEX		PC			
PBS	CPM	BCWS	BCWP		SV	CV	AV	SPI	CPI		BCWS	BCWP		SV	CV	AV	SPI	CPI		BAC	EAC	ACV
01130	10 - 20										2,900	2,900	2,900	0	0	0	1.00	1.00	1.00	2,900	2,900	0
01110	20 - 30										2,500	2,300	2,800	-200	-500	-300	0.92	0.82	0.92	2,500	2,800	-300
01120	20 - 40										2,500	2,200	2,800	-300	-600	-300	0.88	0.79	0.88	2,500	2,800	-300
01210	40 - 50										2,916	2,700	3,100	-216	-400	-184	0.93	0.87	0.93	2,916	3,100	-184
06300	40 - 150	5.160	5.000	5.200	-160	-200	-40	0.97	0.96	0.32	13.416	12,000	14.500	-1,416	-2,500	-1,084	0.89	0.83	0.78	15,478	16,562	-1,084
02110	50 - 60										7,027	6,000	7,500	-1,027	-1,500	-473	0.85	0.80	0.85	7,027	7,500	-473
03000	60 - 70	14.903	13.900	15.000	-1.003	-1.100	-97	0.93	0.93	0.50	27.825	25,000	29,000	-2,825	-4,000	-1,175	0.90	0.86	0.90	27,825	29,000	-1,175
06400	50 - 150	5.260	5.100	5.300	-160	-200	-40	0.97	0.96	0.32	12.361	12,000	13,000	-361	-1,000	-639	0.97	0.92	0.76	15,800	16,439	-639
04300	70 - 80	1.555	1.300	1.700	-255	-400	-145	0.84	0.76	0.60	1.555	1,300	1,700	-255	-400	-145	0.84	0.76	0.60	2,180	2,325	-145
04210	70 - 90	2.135	2.000	2.300	-135	-300	-165	0.94	0.87	0.39	2.135	2,000	2,300	-135	-300	-165	0.94	0.87	0.39	5,128	5,293	-165
04110	70 - 110	3.570	3.300	3.800	-270	-500	-230	0.92	0.87	0.33	3.570	3,300	3,800	-270	-500	-230	0.92	0.87	0.33	10,000	10,230	-230
05000	70 - 100																			2,925	2,925	0
06210	110-120																			5,820	5,820	0
04410	110-150																			4,786	4,786	0
06100	120-130																			2,000	2,000	0
06220	130-140																			5,640	5,640	0
06230	140-150																			7,500	7,500	0
SUBTOTAL		32.583	30.600	33.300	-1.983	-2.700	-717	0.94	0.92	0.25	78.705	71.700	83.400	-7,005	-11.700	-4.695	0.91	0.86	0.58	122.925	127.620	-4,695
GEN. REQ.		2.400	2.400	2.500		-100	-100	1.00	0.96	0.20	7.200	7,000	7,500	-200	-500	-300	0.97	0.93	0.58	12.000	12,300	-300
TOTAL		34.983	33.000	35.800	-1.983	-2.800	-817	0.94	0.92	0.24	85.905	78.700	90.900	-7.205	-12.200	-4.995	0.92	0.87	0.58	134.925	139.920	-4.995

Note: The project at the data date has -2,800 in the current period and -12,200 cumulatively as the cost variance (CV). It shows that the project is over cost because the CV is less than 0.

The same comment is valid for all work items except for item 01130, which is on schedule (CV=0).

163

5. *Step 7:* CV

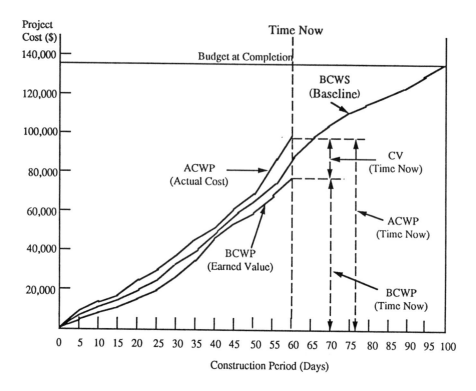

CPM *(see Critical path method)*

CPM TIME IMPACT ANALYSIS (FRAGNET ANALYSIS) *(see Fragnet)* An industry standard approach of measuring delay impacts in the current version of the project CPM schedule whereby the contractor inserts the delay activities into the latest update of the CPM schedule, performs a time analysis, and measures the discrete time impact (if any) of each of the changed conditions. These are now generally required for all publicly funded construction projects and are being included in many privately funded contract general conditions.

CRASH ACTIVITY DURATION *(see Crash activity time cost point)*

164

CRASH ACTIVITY TIME COST POINT (ULTIMATE CRASH POINT) The minimum activity duration that is technologically possible and the corresponding minimum direct cost required to achieve it.

The crash activity time cost point is the basic information for performing the time-cost trade-off analysis *(see Time-cost trade-off, Cost slope,* and *Normal activity time cost point).* Prager's mechanical analogy is used to explain the crash activity time cost point and normal activity time cost point (Figure 1). In Figure 1 a structural member represents each network activity (i,j). It is composed of a rigid sleeve of length (crash time) and a compressible rod whose natural length is Dij (normal time), like a piston. When the compressive force, fij (cost), is gradually increased to the member, the rod remains rigid until this force reaches intensity Cij (cost slope), which is referred to as the yield limit of the member ij. A force (cost) of this intensity can freely compress the rod, but the piston and a stop at the end of the sleeve make it impossible for the rod to be compressed. Any further increase of the compressive force applied to the member will be carried by the sleeve. The point at which the compressible rod reaches the rigid sleeve (crash time) by increasing the force (crash cost) is called the crash activity time cost point.

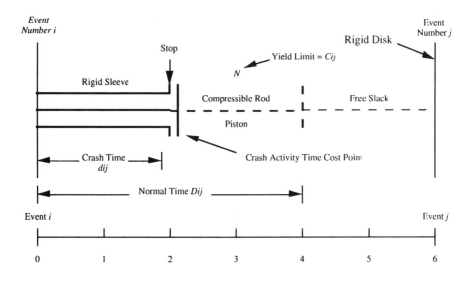

Figure 1 Crash Activity Time Cost Point Using Prager's Mechanical Analogy

In addition, the time-cost curve (Figure 2) shows the crash activity time cost point. In Figure 2, point C is indicated by coordinate d and Cd on the activity duration time axis and activity direct cost axis respectively. Point d on the time axis expresses crash time, and point Cd on the cost axis expresses crash cost. So point C is decided by the crash time and crash cost and is called the *crash activity time cost point.*

165

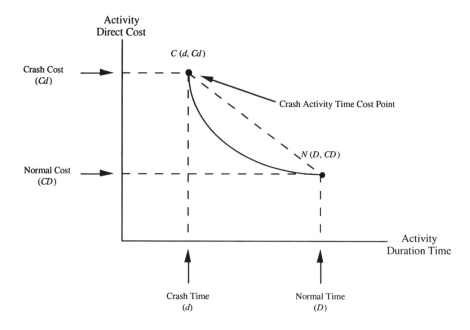

Figure 2 Crash Activity Time Cost Point on Time-Cost Curve

CRASHING *(see Time-cost trade-off)*

CRITICAL ACTIVITY Any activity on the critical path *(see Critical path).*

Critical activities are those that must be finished by the early finish date in order for the project to be started and completed by the calculated dates and with the minimum duration. Any delays to the start of critical activities will postpone the earliest completion time of the project. Identifying critical activities on the network requires CPM computation, indicating the activities' total floats (slacks) *(see Activity total float)*. Which activities become critical depends on the amount of their total slacks. In identifying the critical activities, three possible situations can occur:

1. If the latest finish date of the project is set equal to the earliest finish date, the critical activities will have zero total float (slack).
2. If the latest finish date of the project is set later than the early finish date, the activities with the least positive total float (slack) will be considered critical activities.
3. If the latest finish date of the project is set earlier than the early finish date, the activities with the most negative total float (slack) will constitute the critical path.

166

Controlling the project schedule requires management to denote close attention to critical and nearly critical activities *(see Nearly critical activity)*. Any activities found to be critical or nearly critical are significant by contributing to the overall project duration. During the course of the project execution, the critical activities may change due to various unexpected constraints *(see Constraint)* or occurrences that modify the original schedule. Therefore, great importance lies in updating the network regularly to reflect changes in the project schedule, so management can focus on the current critical and nearly critical activities. Figures 1 and 2 show arrow and precedence diagram networks with critical activities. Figure 3 shows a tabular schedule report in a precedence diagramming method sorted by total float.

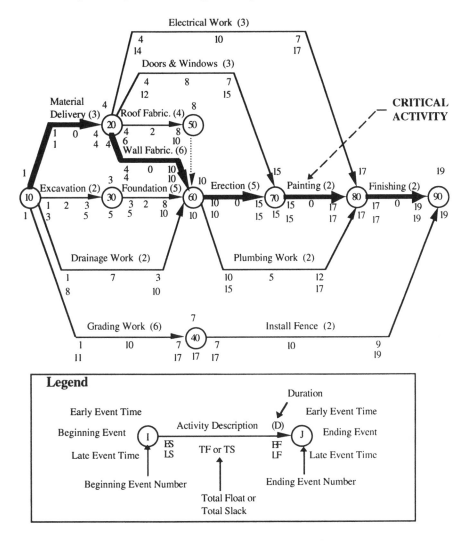

Figure 1 Arrow Diagram Network

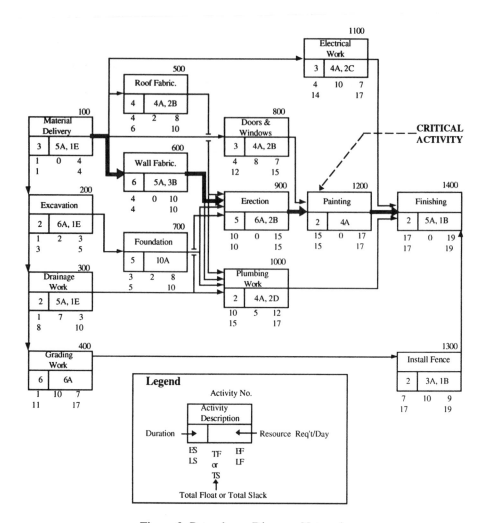

Figure 2 Precedence Diagram Network

```
--------------------------------------------------------------------------------------------------------------------
CIVIL ENGINEERING DEPARTMENT                    PRIMAVERA PROJECT PLANNER

REPORT DATE 23NOV93  RUN NO.   19            PROJECT SCHEDULE REPORT                   START DATE 13MAR93  FIN DATE  9APR93
         21:48
CLASSIC SCHEDULE REPORT - SORT BY ES, TF                                              DATA DATE  17MAR93  PAGE NO.    1
```

ACTIVITY ID	ORIG DUR	REM DUR	%	CODE	ACTIVITY DESCRIPTION	EARLY START	EARLY FINISH	LATE START	LATE FINISH	TOTAL FLOAT
100	3	3	0		MATERIAL DELIVERY	17MAR93*	19MAR93	17MAR93*	19MAR93	0
600	6	6	0		WALL FABRICATION	22MAR93	29MAR93	22MAR93	29MAR93	0
900	5	5	0		ERECTION	30MAR93	5APR93	30MAR93	5APR93	0
1200	2	2	0		PAINTING	6APR93	7APR93	6APR93	7APR93	0
1400	2	2	0		FINISHING	8APR93	9APR93	8APR93	9APR93	0
200	2	2	0		EXCAVATION	17MAR93	18MAR93	19MAR93	22MAR93	2
700	5	5	0		FOUNDATION	19MAR93	25MAR93	23MAR93	29MAR93	2
500	4	4	0		ROOF FABRICATION	22MAR93	25MAR93	24MAR93	29MAR93	2
1000	2	2	0		PLUMBING WORK	30MAR93	31MAR93	6APR93	7APR93	5
300	2	2	0		DRAINAGE WORK	17MAR93	18MAR93	26MAR93	29MAR93	7
800	3	3	0		DOORS & WINDOWS	22MAR93	24MAR93	1APR93	5APR93	8
400	6	6	0		GRADING WORK	17MAR93	24MAR93	31MAR93	7APR93	10
1100	3	3	0		ELECTRICAL WORK	22MAR93	24MAR93	5APR93	7APR93	10
1300	2	2	0		INSTALL FENCE	25MAR93	26MAR93	8APR93	9APR93	10

Figure 3 Tabular Schedule Report, Precedence Diagramming Method

169

CRITICAL FACTOR (1) At any given data date *(see Data date),* the ratio between the number of remaining critical activities and the total number of remaining activities. (2) At any given data date, the ratio between the sum of remaining critical activities' duration and the sum of all remaining activities' duration.

The critical factor is used to monitor how crucial the project status is regarding the overall project schedule before and during construction. Determining a critical factor requires CPM network calculations identifying critical activities *(see Critical activity)* as a basis for the critical factor computation. The critical factor can be computed in two ways: (1) as the ratio between the number of remaining critical activities from the data date *(see Data date)* and the total number of remaining activities, and (2) as the ratio between the sum of remaining critical activities' duration and the sum of all remaining activities' duration from the data date.

To keep track of the project status, the critical factor must be determined regularly, usually at the updating time. A typical trend in critical factor versus project percent completion is shown in Figure 1. At the start of the project, the critical factor should be low; then as the project progresses toward completion, the critical factor will increase. This means that at the end of the project most of the remaining activities are critical or nearly critical with minimal total float. Project critical factor calculations using a precedence diagramming method prior to the start of the project and after the start of the project (updating) are shown in Figures 2 and 3.

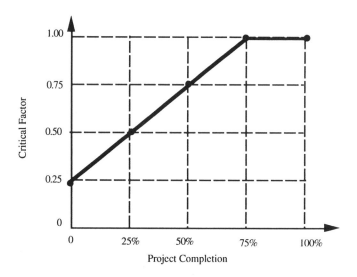

Figure 1 Typical Trend of Critical Factor vs. Project Percent Completion

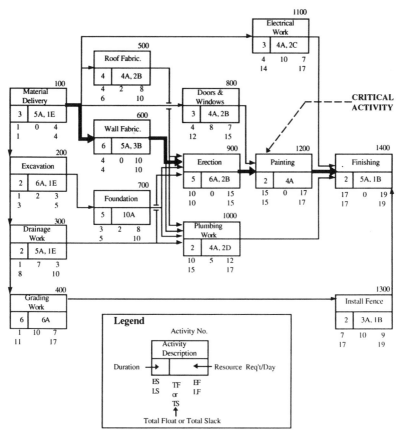

Note: Critical Factor Calculation Example at
the Project Start Date

1. Critical Factor = $\dfrac{\text{No. of Critical Activities}}{\text{Total No. of Activites}}$

    Number of Critical Activities = 5
    Total Number of Activities   = 14
    Critical Factor   = 5 /14   = 0.36

2. Critical Factor = $\dfrac{\text{Sum of Critical Activities' Duration}}{\text{Total No. of Activities' Duration}}$

    Sum of Critical Activities' Duration = 18 days
    Total No. of Activities' Duration   = 47 days
    Critical Factor   = 18 / 47   = 0.38

Figure 2 Critical Factor Prior to Project Start, Precedence Diagramming Method

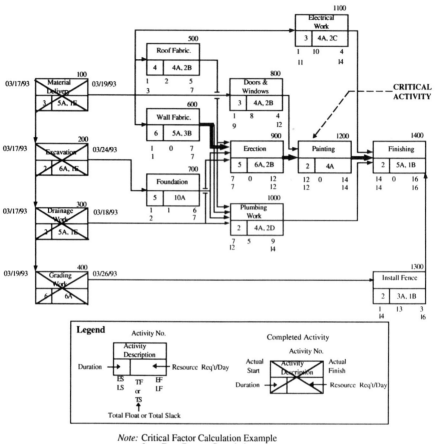

1. Critical Factor = $\dfrac{\text{No. of Critical Activities}}{\text{Total No. of Activites}}$ 2. Critical Factor = $\dfrac{\text{Sum of Critical Activities' Duration}}{\text{Total No. of Activities' Duration}}$

Number of Critical Activities = 4   Sum of Critical Activities' Duration = 15 days
Total Number of Activities = 10   Total No. of Activities' Duration = 34 days
Critical Factor   = 4 /10   = 0.40   Critical Factor   = 15 / 34   = 0.44

Figure 3 Critical Factor After Project Start, Precedence Diagramming Method

CRITICAL PATH The particular sequence of activities in a CPM network that has the least slack (total float) *(see Activity total float)* and is, therefore, the longest path through the network. (Assume that no dates are imposed on the network.)

A critical path is a chain of critical activities *(see Critical activity)* in a network from start to finish. Any delay in the duration of the activities in a critical path will cause the project earliest finish date to be delayed. The critical path is determined after the network *(see Forward pass computation* and *Backward pass computation)* has been established. Which path in the network will become critical depends on the duration of activities in the network. It is not uncommon to have more than one critical path in the network. To identify a critical path, one has to realize that three possible situations can occur:

1. The latest finish date of the project is set equal to the earliest finish date. Then the critical path will be the path in which all activities have zero total float (slack).
2. The latest finish date of the project is set later than the earliest finish date. Then the critical path will be the chain of the activities with the least positive total float (slack).
3. The latest finish date of the project is set earlier than the earliest finish date. Then the critical path will be the chain of the activities with the maximum negative total float (slack).

Unlike an arrow diagram network, the relationships between activities in a precedence diagram network can be more than just finish to start *(see Dependency)*. Therefore, the arrow representing the relationships between activities must be carefully considered in a network computation to locate a critical path. Figures 1 and 2 show the network in arrow and precedence diagrams with a critical path.

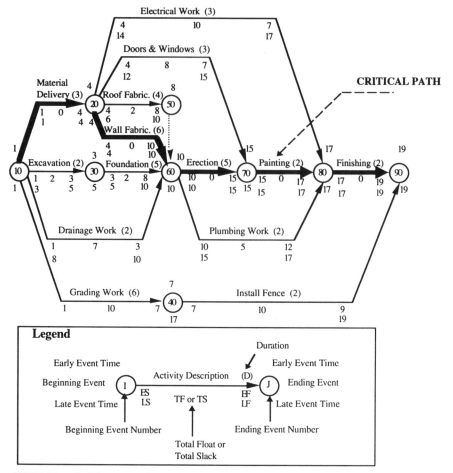

Electrical Work (3)

| 4 | 10 | 7 |
| 14 | | 17 |

Doors & Windows (3)

| 4 | 8 | 7 |
| 12 | | 15 |

CRITICAL PATH

Material
Delivery (3) — Roof Fabric. (4) — Wall Fabric. (6) — Excavation (2) — Foundation (5) — Erection (5) — Painting (2) — Finishing (2)

Drainage Work (2)

| 1 | 7 | 3 |
| 8 | | 10 |

Plumbing Work (2)

| 10 | 5 | 12 |
| 15 | | 17 |

Grading Work (6)

| 1 | 10 | 7 |
| | | 17 |

Install Fence (2)

| 7 | 10 | 9 |
| 7 | | 19 |

Legend

Duration

Early Event Time — Activity Description (D) — Early Event Time

Beginning Event — I — J — Ending Event

| | ES | TF or TS | EF | |
| Late Event Time | LS | | LF | Late Event Time |

Beginning Event Number — Ending Event Number

Total Float or
Total Slack

NOTE: The schedule computation in time units is based on the following assumption:

Activity Duration

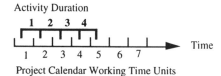

Time

Project Calendar Working Time Units

Activity with 4 time units starts at working time unit 1 and is finished at working time unit 5.

Figure 1 Arrow Diagram Network

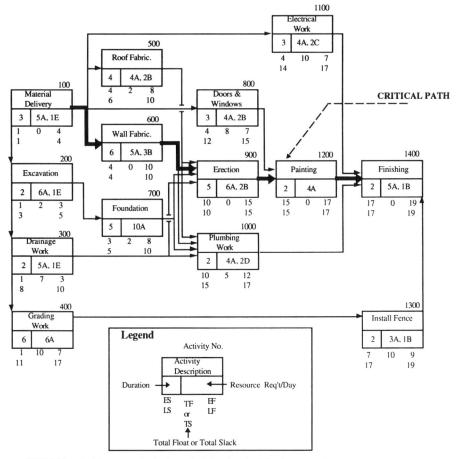

Legend

Activity No.

Activity
Description

Duration → ← Resource Req't/Day

ES TF EF
LS or LF
    TS

Total Float or Total Slack

NOTE: The schedule computation in time units is based on the following assumption:

Activity Duration

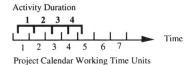

1 2 3 4

Time

1 2 3 4 5 6 7

Project Calendar Working Time Units

Activity with 4 time units starts at working time unit 1 and is finished at working time unit 5.

Figure 2 Precedence Diagram Network

175

CRITICAL PATH METHOD (1) A graphical presentation of a planned sequence of activities using an arrow or precedence technique *(see Arrow diagramming method and Precedence diagramming method),* which shows the interrelationship of the elements comprising a project to determine the length of a project and to identify activities, events, and constraints that are on the critical path. (2) A method for analyzing a dependency activity chart to determine early start and finish, late start and finish, and float, identifying activities or events that are on the critical path.

Following is a summary of the history and development of the critical path method.

1956: The critical path method (CPM) was initiated by the E.I. du Pont de Nemours Company of Newark, Delaware. The objective was to study possible application of a new management technique that could be used as a technical information service and a common language vehicle to communicate cost and technical information among projects.

1957: In January, John W. Mauchly and James E. Kelly, Jr. of Remington Rand (UNIVAC) and Morgan R. Walker of Du Pont formed a group at the UNIVAC Applications Research Center (UARC) to revise the original concept of the project, developed in 1956 by Du Pont. In April, a complete basis for CPM was available. Using this basis, the project schedule showed the shortest project duration, called the "main chain" of the network. The term "critical path" was introduced by PERT *(see PERT)* developers.

In July, the first network of 61 jobs (activities), 8 timing restraints, and 16 dummies plus all related data was developed from a small mixing and packaging plant selected as a test case. By late October, the results of the test case run by UNIVAC I were completed. It was found that the speed and capability of handling the construction schedule were the problem. A program for more rapid computers was written for UNIVAC 1103A.

In December, a live test was prepared for the construction of hydrogen and anhydrous ammonia facilities at the Repauno Works in New Jersey. An 846-arrow network of construction activities was developed for this project.

1958: In February, the construction schedule for the live test project was run using a UNIVAC 1103A computer located at Palo Alto. Cost and labor analyses were included in this test; great success was achieved. In May, the result was presented to the department managers at du Pont.

1959: In early 1959, Mauchly's group at Remington Rand was dissolved. CPM was presented to the public in March.

1965-1970: Colleges and universities offered a civil engineering program incorporating CPM into their undergraduate and graduate curricula.

1967: In December, the Department of Defense issued the widely cited Instruction DODI 7000.2, CPM cost and schedule control systems criteria (C/SCSC), for project progress measurement and control.

1968: Du Pont's engineering department adopted CPM as its standard.

1980-Present: Even though personal computers as a mass movement have existed since 1975, CPM software development for personal computer systems has blossomed since the introduction of IBM personal computers in October 1981. During this period, CPM has been used as a standard tool for construction control in both the public and private sectors. Moreover, during this period, CPM has been used in construction litigation citing delays as a cause of damages.

CPM has been proved to be a successful technique for the planning and scheduling of construction projects as well as production planning, research, and development. CPM has been widely used as a tool in the systematic management of complex projects. The technique is useful in planning and scheduling, coordinating, and controlling. The main concerns in CPM involve time, cost, and resource management to better utilize available resources in a project. Presently, the use of CPM in a complex project requires computer hardware and CPM-oriented software extensively available in the marketplace. To optimize the usefulness of CPM, updating and reanalyzing of project conditions are required due to the constant changes in a project. CPM can be performed by either an arrow diagram or a precedence diagram *(see Arrow diagramming method* and *Precedence diagramming method).* Samples of CPM networks and computations are shown in Figures 1 and 2.

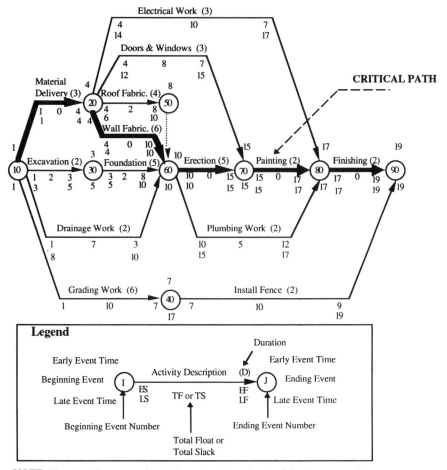

Electrical Work (3)

| 4 | 10 | 7 |
| 14 | | 17 |

Doors & Windows (3)

| 4 | 8 | 7 |
| 12 | | 15 |

CRITICAL PATH

Material Delivery (3)

Roof Fabric. (4)

Wall Fabric. (6)

Excavation (2)

Foundation (5)

Erection (5)

Painting (2)

Finishing (2)

Drainage Work (2)

| 1 | 7 | 3 |
| 8 | | 10 |

Plumbing Work (2)

| 10 | 5 | 12 |
| 15 | | 17 |

Grading Work (6)

Install Fence (2)

| 1 | 10 | 7 | 7 | 10 | 9 |
| | | 17 | | | 19 |

Legend

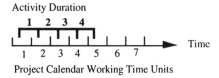

Duration

Early Event Time

Beginning Event

Activity Description

(D)

Early Event Time

Ending Event

Late Event Time

ES
LS

TF or TS

EF
LF

Late Event Time

Beginning Event Number

Ending Event Number

Total Float or
Total Slack

NOTE: The schedule computation in time units is based on the following assumption:

Activity Duration

```
  1  2  3  4
```

Time

```
  1  2  3  4  5  6  7
```

Project Calendar Working Time Units

Activity with 4 time units starts at working time unit 1 and is finished at working time unit 5.

Figure 1 Arrow Diagramming Method

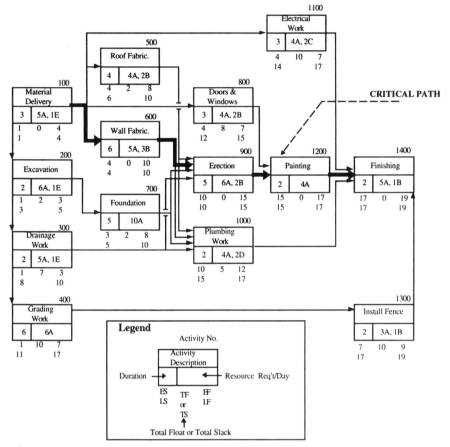

CRITICAL PATH

Legend

Activity No.

Activity Description

Duration →

← Resource Req't/Day

ES TF EF
LS or LF
TS

Total Float or Total Slack

NOTE: The schedule computation in time units is based on the following assumption:

Activity Duration

1 2 3 4

1 2 3 4 5 6 7 → Time

Project Calendar Working Time Units

Activity with 4 time units starts at working time unit 1 and is finished at working time unit 5.

Figure 2 Precedence Diagramming Method

179

CRITICAL SCHEDULE A schedule in which all or nearly all of the activities have no float, such that delay to any activity will result in a delay to the project *(see Nearly critical activity* and *Nearly critical path)*.

C/SCSC *(see Cost schedule control system criteria)*

CYCLE *(see Loop)*

D

DAILY CALENDAR(S) The start and stop working times for each day, Monday through Sunday.

The daily calendar allows a scheduler to create calendars with specifically numbered start and stop times for each day, Monday through Sunday. Once created, a daily calendar can be associated with projects, increasing the flexibility of scheduling, especially when hours and minutes are used to specify duration. Normally, any CPM-oriented software allows more than one daily calendar to be created, but only one daily calendar can be assigned to one calendar.

DANGLE Any activity with no predecessors and/or successors in the network that is not identified for data processing as a start or end activity.

In a CPM computerized time computation, some CPM-oriented software packages require that each activity have no predecessor or successor to be identified as a start or finish activity, respectively. Otherwise, they will be detected when performing the time computation as a start or finish dangle. These activities must be corrected before the computation can be completed. Typically, various types of dangle may be found:

1. *Start dangle.* An activity is not designated as a start and has no predecessor. Figure 1 shows a precedence diagram network with a start dangle.

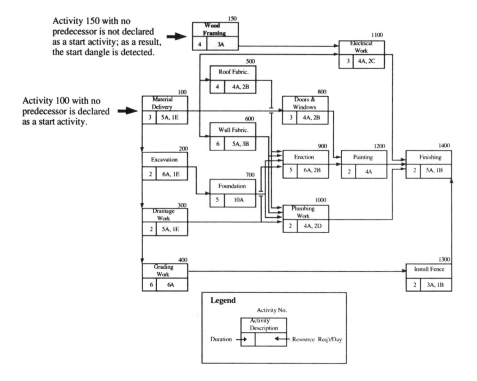

Figure 1 Precedence Diagram Network with Start Dangle

182

2. *End dangle.* An activity is not designated as a finish and has no successor. Figure 2 shows a precedence diagram network with an end dangle.

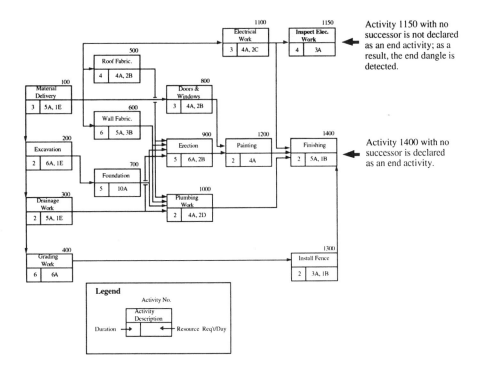

Figure 2 Precedence Diagram Network with End Dangle

183

3. *Start end dangle.* An activity has been designated as a start activity but has no successor. Figure 3 shows a precedence diagram network with a start end dangle.

Activity 50 is declared as a start activity but has no successor; as a result, this activity is detected as a start end dangle.

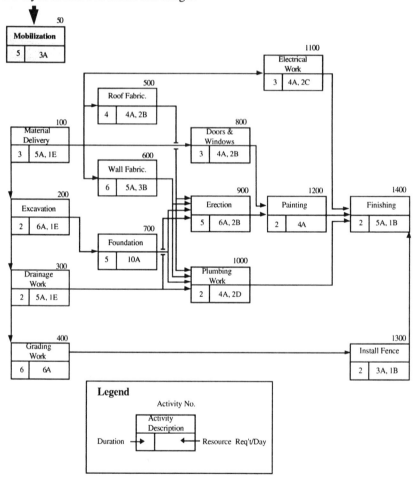

Figure 3 Precedence Diagram with Start End Dangle

184

4. *End start dangle.* An activity has been designated as an end activity but has no predecessor. Figure 4 shows a precedence diagram network with an end start dangle.

Activity 1500 is declared as an end activity but has no predecessor; as a result, this activity is detected as an end start dangle.

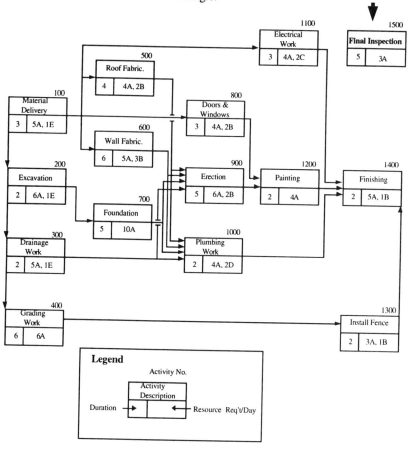

Figure 4 Precedence Diagram Network with End Start Dangle

DATA DATE (TIME NOW) The calendar date through which the project has been updated and the network revised, or progressed.

The data date is the date that the project information is revised. The computer needs the data date for updating the project status, project progress, and project sched-

ule computation. For an in-progress activity on the data date, a revised early finish date for the activity is the sum of the data date and its remaining duration. The remaining duration can be obtained in one of two ways: using the number of days from the activity's remaining duration, or by converting from a percentage to complete the activity multiplied by the original duration or latest revised activity duration. Figures 1 (for arrow diagram technique activity codes) and 2 (for precedence diagram technique activity codes) illustrate a revised schedule computation with the data date or time now, March 19, 1993, represented by an asterisk (*).

```
-------------------------------------------------------------------------------------------
CIVIL ENGINEERING DEPARTMENT              PRIMAVERA PROJECT PLANNER

REPORT DATE 23NOV93  RUN NO.   13         PROJECT SCHEDULE REPORT              START DATE 17MAR93   FIN DATE 12APR93
                19:33
ACTUAL VS. PLANNED BAR CHART                                                  DATA DATE 19MAR93    PAGE NO.    1


                                                                              DAILY-TIME PER.    1

-------------------------------------------------------------------------------------------
........ACTIVITY DESCRIPTION............          22   29   05   12   19   26   03   10   17   24
PRED   SUCC  OD  PCT   CODES   FLOAT  SCHEDULE     MAR  MAR  APR  APR  APR  APR  MAY  MAY  MAY  MAY
----   ----  ---- ---  ----------- -----  --------   93   93   93   93   93   93   93   93   93   93
-------------------------------------------------------------------------------------------

DRAINAGE WORK                           CURRENT   AA*  .     .    .    .    .    .    .    .    .
    10    60   2    0 100               PLANNED   EE*  .     .    .    .    .    .    .    .    .

MATERIAL DELIVERY                       CURRENT   AA*  .     .    .    .    .    .    .    .    .
    10    20   3    0 100               PLANNED   EEE  .     .    .    .    .    .    .    .    .

EXCAVATION                              CURRENT   AAE..EE    .    .    .    .    .    .    .    .
    10    30   2    3   0        0      PLANNED   EE*  .     .    .    .    .    .    .    .    .

GRADING WORK                            CURRENT   AAE..EEE   .    .    .    .    .    .    .    .
    10    40   6    4  33       11      PLANNED   EEE..EEE   .    .    .    .    .    .    .    .

DOORS AND WINDOWS                       CURRENT   E..EE  .    .    .    .    .    .    .    .    .
    20    70   3    3   0       10      PLANNED   *  EEE  .    .    .    .    .    .    .    .

ELECTRICAL WORK                         CURRENT   E..EE  .    .    .    .    .    .    .    .    .
    20    80   3    3   0       12      PLANNED   *  EEE  .    .    .    .    .    .    .    .

ROOF FABRICATION                        CURRENT   E..EEE  .    .    .    .    .    .    .    .   .
    20    50   4    4   0        4      PLANNED   *  EEEE  .    .    .    .    .    .    .    .

WALL FABRICATION                        CURRENT   E..EEEEE  .    .    .    .    .    .    .    .
    20    60   6    6   0        2      PLANNED   *  EEEEE..E  .    .    .    .    .    .    .

FOUNDATION                              CURRENT   *  . EEE..EE    .    .    .    .    .    .    .
    30    60   5    5   0        0      PLANNED   E..EEEE  .    .    .    .    .    .    .    .

DUMMY 1                                 CURRENT   *  . E    .    .    .    .    .    .    .    .
    50    60   0    0   0        4      PLANNED   *  . E    .    .    .    .    .    .    .    .

INSTALL FENCE                           CURRENT   *  . EE   .    .    .    .    .    .    .    .
    40    90   2    2   0       11      PLANNED   *  . EE   .    .    .    .    .    .    .    .

PLUMBING WORK                           CURRENT   *  .    . EE   .    .    .    .    .    .    .
    60    80   2    2   0        5      PLANNED   *  .    .EE   .    .    .    .    .    .    .

ERECTION                                CURRENT   *  .    . EEE..EE    .    .    .    .    .    .
    60    70   5    5   0        0      PLANNED   *  .    .EEEE..E    .    .    .    .    .    .

PAINTING                                CURRENT   *  .    .    . EE   .    .    .    .    .    .
    70    80   2    2   0        0      PLANNED   *  .    .    .EE   .    .    .    .    .    .

FINISHING                               CURRENT   *  .    .    . E..E   .    .    .    .    .    .
    80    90   2    2   0        0      PLANNED   *  .    .    . EE   .    .    .    .    .    .
```

Figure 1 Time Now Shown on a Bar Chart Based on Arrow Diagram Technique

```
----------------------------------------------------------------------------------------------------------------
CIVIL ENGINEERING DEPARTMENT                    PRIMAVERA PROJECT PLANNER

REPORT DATE 23NOV93  RUN NO.  12                PROJECT SCHEDULE REPORT                 START DATE 13MAR93  FIN DATE 12APR93
            19:39
ACTUAL VS. PLANNED BAR CHART                                                           DATA DATE 19MAR93   PAGE NO.   1

                                                                                              DAILY-TIME PER.   1
----------------------------------------------------------------------------------------------------------------
.............ACTIVITY DESCRIPTION.............          22    29    05    12    19    26    03    10    17    24
ACTIVITY ID  OD   RD  PCT   CODES    FLOAT   SCHEDULE   MAR   MAR   APR   APR   APR   APR   MAY   MAY   MAY   MAY
-----------  ----  ----  ---  -------------  -----   --------   93    93    93    93    93    93    93    93    93    93
                                                           ------------------------------------------------------
DRAINAGE WORK                                 CURRENT  AA*    .     .     .     .     .     .     .     .     .     .
        300   2   0 100                       PLANNED  EE*    .     .     .     .     .     .     .     .     .     .

MATERIAL DELIVERY                             CURRENT  AA*    .     .     .     .     .     .     .     .     .     .
        100   3   0 100                       PLANNED  EEE    .     .     .     .     .     .     .     .     .     .

EXCAVATION                                    CURRENT  AAE..EE      .     .     .     .     .     .     .     .     .
        200   2   3   0                  0    PLANNED  EE*    .     .     .     .     .     .     .     .     .     .

GRADING WORK                                  CURRENT  AAE..EEE     .     .     .     .     .     .     .     .     .
        400   6   4  33                 11    PLANNED  EEE..EEE     .     .     .     .     .     .     .     .     .

DOORS & WINDOWS                               CURRENT  E..EE        .     .     .     .     .     .     .     .     .
        800   3   3   0                 10    PLANNED  *   EEE      .     .     .     .     .     .     .     .     .

ELECTRICAL WORK                               CURRENT  E..EE        .     .     .     .     .     .     .     .     .
       1100   3   3   0                 12    PLANNED  *   EEE      .     .     .     .     .     .     .     .     .

ROOF FABRICATION                              CURRENT  E..EEE       .     .     .     .     .     .     .     .     .
        500   4   4   0                  4    PLANNED  *   EEEE     .     .     .     .     .     .     .     .     .

WALL FABRICATION                              CURRENT  E..EEEEE     .     .     .     .     .     .     .     .     .
        600   6   6   0                  2    PLANNED  *   EEEEE..E      .     .     .     .     .     .     .     .

FOUNDATION                                    CURRENT  *   .  EEE..EE     .     .     .     .     .     .     .     .
        700   5   5   0                  0    PLANNED  E..EEEE      .     .     .     .     .     .     .     .     .

INSTALL FENCE                                 CURRENT  *   .  EE    .     .     .     .     .     .     .     .     .
       1300   2   2   0                 11    PLANNED  *   .  EE    .     .     .     .     .     .     .     .     .

PLUMBING WORK                                 CURRENT  *   .     .  EE    .     .     .     .     .     .     .     .
       1000   2   2   0                  5    PLANNED  *   .     .EE      .     .     .     .     .     .     .     .

ERECTION                                      CURRENT  *   .     .  EEE..EE      .     .     .     .     .     .     .
        900   5   5   0                  0    PLANNED  *   .     .EEEE..E        .     .     .     .     .     .     .

PAINTING                                      CURRENT  *   .     .     .  EE     .     .     .     .     .     .     .
       1200   2   2   0                  0    PLANNED  *   .     .     .EE        .     .     .     .     .     .     .

FINISHING                                     CURRENT  *   .     .     .  E..E        .     .     .     .     .     .
       1400   2   2   0                  0    PLANNED  *   .     .     .  EE     .     .     .     .     .     .     .
```

Figure 2 Time Now Shown on Bar Chart
Based on Precedence Diagram Technique

187

DELAY, CONCURRENT (CONCURRENT DELAY) The occurrence of two or more delays arising from independent causes and affecting a project during the same or overlapping time periods. Concurrent delays may act jointly to affect a single activity or path, or may act independently to affect multiple activities or paths.

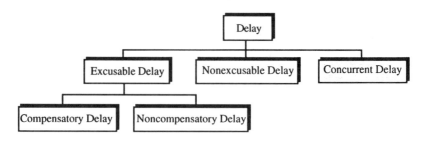

Figure 1 Delay Classifications

Concurrent delays, shown in Figure 1, are two or more delays that occur at least to some degree simultaneously. As used in construction law, the term refers to the situation when there is more than one delay occurring at the same time, each of which, if it had occurred alone, would have affected the project completion date.

Courts determine the legal impact of concurrent delays by examining the responsibility for the concurrent delays and determining whether the parties are seeking compensation or an extension of time. The concurrent delays can be more than one type of delay. With respect to contractor recovery for concurrent delays, the delays must be solely the owner's responsibility. Similarly, if the owner can clearly distinguish the contractor's responsibility for concurrent delays, the owner can collect liquidated damages. In general, when excusable and nonexcusable delays *(see Delay, excusable* and *Delay, nonexcusable)* are concurrent, the contractor ought to be entitled to an extension of construction time. In case of concurrent compensatory and noncompensatory delays *(see Delay, excusable compensatory* and *Delay, excusable noncompensatory),* the contractor should be entitled to a time extension but not to damages. For the contractor to collect damages, the owner would have to cause both (or all) compensatory delay.

If the concurrent delays consist of delays attributed to both the owner and contractor, some cases hold that neither can recover damages for the other's act. Some endeavor should be made to apportion the concurrent delays between parties. Inadequate documentation may, however, make apportionment impossible. If concurrent delays cannot be apportioned, neither the owner nor the contractor can recover delay damages.

DELAY, EXCUSABLE COMPENSATORY (EXCUSABLE COMPENSATORY DELAY) A delay that entitles the contractor to extended field office costs and perhaps extended home office costs, as well as additional project time.

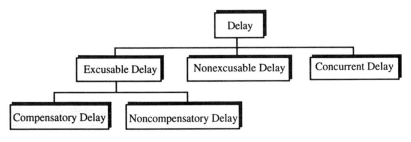

Figure 1 Delay Classifications

Excusable compensatory delays, shown in Figure 1, are due to acts or omissions of the owner or owner's representatives. This type of delay entitles a contractor to additional compensation for costs of delays and time of project completion. A delay can be compensable solely by causing damages for the contractor. Typically, excusable compensatory delays are attributable to change order *(see Change order)* or to owner's actions that change the contracted requests.

In the contract, the compensation provision may allow extension of time or compensation for additional costs, but frequently the extension of time is the sole remedy for delays. In this case, if the contractor seeks compensation, they have to file a lawsuit for delay damage cost. Examples of excusable compensatory delay are shown below.

Delay caused by owner
    1. Failure to provide a project site
    2. Late notice to proceed
    3. Failure to provide proper financing
    4. Failure to provide furnished materials or components
    5. Interfering with or obstruction of work on the project

Delay caused by architect/engineer
    1. Defective plans and specifications
    2. Failure to provide drawings on schedule
    3. Delay in review or approval of shop drawings
    4. Delay in change orders
    5. Stop-work order

DELAY, EXCUSABLE (EXCUSABLE DELAY) A delay that entitles the contractor to additional time for completion of the contract work, generally arising from causes beyond the contractor's control. Excusable delays may be classified further as excusable compensatory delays and excusable noncompensatory delays. Whether a delay is classified as compensatory or noncompensatory depends primarily on the terms of the contract.

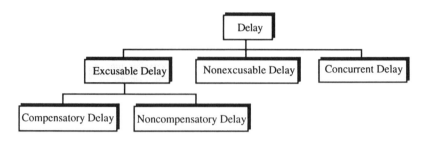

Figure 1 Delay Classifications

An excusable delay as shown in Figure 1 can occur due to various factors, which can be classified into two categories: (1) beyond the control or without the fault of either party *(see Delay, excusable noncompensatory)* and (2) within the owner's or owner's representative's control *(see Delay, excusable compensatory)*. The first case is one that will serve to justify an extension of contract performance time. The latter will allow the contractor both time extension and additional cost.

When delays are excusable, the contractor will not be subject to liquidated damages, nor can the contractor be terminated for default due to such delays. Liquidated damages constitute the specified amount that a contractor will pay an owner for nonexcused late completion. Whether the contractor can recover the delay cost for an excusable delay depends on whether the delay is compensatory or noncompensatory or whether it is concurrent with other delays *(see Delay, concurrent)*. Examples of excusable delays caused by different factors are shown.

Delay caused by owner
    1. Failure to provide a project site
    2. Late notice to proceed
    3. Failure to provide proper financing
    4. Failure to provide owner's furnished materials or components
    5. Interfering with or obstructing work on the project

Delay caused by architect/engineer
    1. Defective plans and specifications
    2. Failure to provide drawings on schedule
    3. Delay in review or approval of shop drawings
    4. Delay in change orders
    5. Stop-work order

<u>Delay not caused by any party or participant</u>
  1. Acts of God
  2. Act of a public enemy
  3. Unusual delays in transportation, such as a freight embargo
  4. Epidemics
  5. Unusual weather conditions (force majure)
  6. Strikes

The above-mentioned delays are normally excusable for the contractor and may lead to a time extension, but to recover delay damages, the delays should be caused by the owner or owner's representatives.

DELAY, EXCUSABLE NONCOMPENSATORY (EXCUSABLE NONCOMPENSATORY DELAY) A delay that entitles the contractor to additional time for completion of the contract work but no additional compensation.

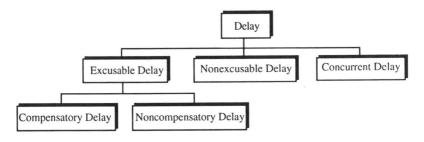

Figure 1 Delay Classifications

Excusable noncompensatory delays, shown in Figure 1, are not caused by the owner, designer, contractor, subcontractors, suppliers, or other parties in the design and construction process. Because this delay is beyond the control of any of the parties, contract and case laws generally minimize the risk to all parties by a compromise: The contractor's late completion will be allowed equal to the amount of delay, but no additional compensation will be awarded. Most contracts contain written statements that deal specifically with this type of delay. Examples of excusable noncompensatory delay are shown below.

<u>Delay not caused by any party or participant</u>
  1. Acts of God
  2. Act of a public enemy
  3. Unusual delays in transportation, such as a freight embargo
  4. Epidemics
  5. Unusual weather conditions
  6. Strikes

DELAY, NONEXCUSABLE (NONEXCUSABLE DELAY) A delay that does not entitle the contractor to either additional time for completion of the contract work or additional compensation. Such a delay may be nonexcusable due to the contractor's failure to meet its contractual obligations or due to the terms of the contract.

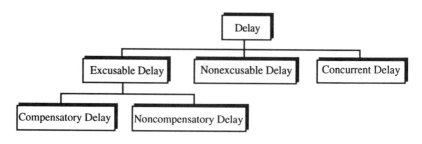

Figure 1 Delay Classifications

A nonexcusable delay is within the contractor's control and could have been avoided. This type of delay does not allow the contractor to recover any additional time or cost. Conversely, such delay could be compensable to the owner in the form of liquidated or actual damages paid by the contractor for late completion or increased cost to accelerate the work. Furthermore, the nonexcusable delay may constitute a breach of the construction contract by the contractor and may justify the termination of the construction contract.

The owner normally is in a difficult position to identify the nonexcusable delays at the early stages because he seldom maintains the construction schedule with sufficient detail to pinpoint the contractor's delay. This type of delay, therefore, is identified when the dispute arises. A contractor, on the other hand, is more likely to maintain the detailed schedule, so he is in a better position to monitor job progress and identify delays which are attributable to the owner. Examples of nonexcusable delay are shown below.

Delay caused by contractor
   1. Slow mobilization
   2. Inadequate labor force
   3. Strike caused by unfair labor practice
   4. Poor workmanship
   5. Late delivery of materials and components
   6. Failure to coordinate multiple subcontractors

DELAY (LAG) Any enforced or imposed time gap between the completion of an activity and the start of its succeeding activity(s), or any enforced or imposed increase in the duration of an activity (*see Lead* and *Lag*).

Delay is applied to the relationship between two activities *(see Dependency).* An activity with a delay relationship must wait until the period of delay has expired before beginning or completing. In other words, lag or delay is the condition of waiting for a prescribed period before an activity can be started or completed. Normally, delay will be specified by a positive value, and it is neither practical nor recommended to use negative delay.

Delay can be applied to the activity relationship for both an arrow diagram method and a precedence diagram method (ADM and PDM) *(see Arrow diagramming method* and *Precedence diagramming method).* Since project activities must wait a certain period of time before action can occur, delay utilization creates more realistic schedules. Examples of delay include waiting for concrete to be cured, for inspection, and for equipment or crews to be moved from another activity.

Delays can be applied to four types of dependencies *(see Dependency):* (1) finish to start, (2) start to start, (3) finish to finish, and (4) start to finish *(see Dependency, finish to start, Dependency, start to start, Dependency, finish to finish,* and *Dependency, start to finish).* The most commonly encountered relationship with delay is finish to start.

By assigning delay to the relationship between two activities, activities that do not require resources and are classified as waiting can be eliminated. This results in a network diagram with fewer activities that is easier to read and quicker in calculations and printouts. However, the assignment of delay should be performed by an experienced scheduler who possesses a good understanding and knowledge of work sequences. In the precedence diagramming method, delays will be imposed on the relationship between activities as shown in Figure 1. The delay interpretation of the network in Figure 1 can be:

A = start activity—no activity precedes it.
B = can be started only after 1 time unit has elapsed after the start of activity A or later.
C = can be started only after 2 time units have elapsed after the completion of activity A and can be finished only 1 time unit after the start of activity B or later.
D = finish activity—no activity succeeds it. Activity D can be started only after the completion of activity B and can be finished only after 3 time units have elapsed after the completion of activity C.

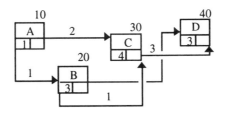

Figure 1 Delay in Precedence Diagram Network

In the arrow diagram method, delays will be introduced by using dummy activities *(see Dummy activity)*, as shown in Figure 2. The delay interpretation of a network in Figure 2 can be:

A = start activity—no activity precedes it.
B = can be started only after 1 time unit has elapsed after the start of activity A or later.
C = can be started only after 2 time units have elapsed after the completion of activity A and can be finished only after 1 time unit after the start of activity B or later.
D = finish activity—no activity succeeds it. Activity D can be started only after the completion of activity B and can be finished only after 3 time units have elapsed after the completion of activity C.

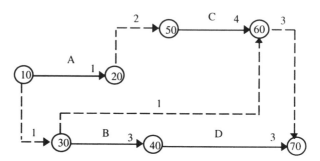

Figure 2 Delay in Arrow Diagram Network

DEPENDENCY, FINISH TO FINISH (FF) A relationship that restricts the finish of an activity until the finish of the immediately preceding activity has taken place or until a specified elapsed time (lag) *(see Delay)*.

The finish-to-finish relationship is a common relationship in network construction and is used to show the relationship between the completion of two activities. It sets a constraint that the succeeding activity cannot be finished until the preceding activity is completed with or without elapsed time. Due to the limitation of some CPM-oriented software packages, the number of finish-to-finish relationships for one activity may be limited. Therefore, before establishing this relationship between activities, the software should be reviewed to determine its limitations. Figure 1 shows a portion of a precedence diagram network indicating finish-to-finish relationships between two activities. The finish-to-finish relationship of the sample network in Figure 1 can be interpreted as follows:

1. Installing the fourth-floor slab formwork can be finished only 2 days after completion of stripping the first-floor slab formwork.

2. Installing the fifth-floor slab formwork can be finished only 2 days after completion of stripping the second-floor slab formwork.

Sample Network: Elevated Concrete Slabs

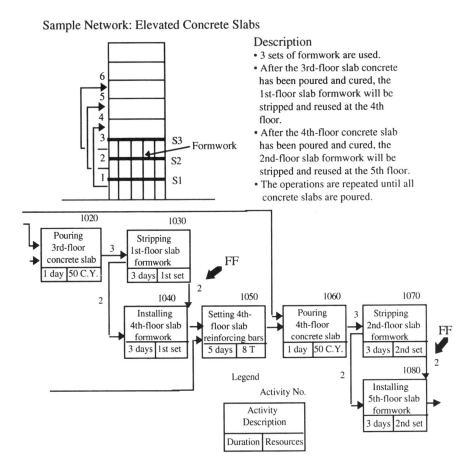

Description
• 3 sets of formwork are used.
• After the 3rd-floor slab concrete has been poured and cured, the 1st-floor slab formwork will be stripped and reused at the 4th floor.
• After the 4th-floor concrete slab has been poured and cured, the 2nd-floor slab formwork will be stripped and reused at the 5th floor.
• The operations are repeated until all concrete slabs are poured.

Figure 1 Precedence Diagram Network for Elevated Concrete Slabs

DEPENDENCY, FINISH TO START (FS) A relationship that restricts the start of an activity until the completion of the immediately preceding activity has taken place or after a specified elapsed time (lag) *(see Delay).*

The finish-to-start relationship is the most common relationship in a CPM network, in both arrow and precedence diagramming methods, especially in construction-related tasks. It sets a constraint that the succeeding activity cannot be started until the preceding activity(s) is completed with or without the specified elapsed

time. This type of constraint is typical for construction activities. Figure 1 shows a portion of a precedence diagram network indicating finish-to-start relationships between two activities. The finish-to-start relationship of the sample network in Figure 1 can be interpreted as follows:

1. Stripping the first-floor slab formwork cannot be started until 3 days after completion of pouring the third-floor concrete slab.
2. Setting the fourth-floor slab reinforcing bars cannot be started until completion of the fourth-floor slab formwork installation.
3. Pouring the fourth-floor concrete slab cannot be started until completion of setting the fourth-floor slab reinforcing bars.

Sample Network: Elevated Concrete Slabs

Description
• 3 sets of formwork are used.
• After the 3rd-floor slab concrete has been poured and cured, the 1st-floor slab formwork will be stripped and reused at the 4th floor.
• After the 4th-floor concrete slab has been poured and cured, the 2nd-floor slab formwork will be stripped and reused at the 5th floor.
• The operations are repeated until all concrete slabs are poured.

Figure 1 Precedence Diagram Network for Elevated Concrete Slabs

196

DEPENDENCY, START TO FINISH (SF) A relationship that restricts the finish of an activity until the start of the immediately preceding activity has taken place or after a specified elapsed time (lag) *(see Delay)*.

The start-to-finish relationship is not a common relationship in network construction. It sets a constraint that one activity cannot be finished until the preceding activity is started with or without elapsed time. It is recommended that this type of relationship not be used in network construction. Due to the limitation of some CPM-oriented software packages, the number of start-to-finish relationships for one activity may be limited or not be allowed. Therefore, before establishing this relationship between activities, the software should be reviewed to determine its limitations. Figure 1 shows a portion of a precedence diagram network indicating start-to-finish relationships between two activities. The start-to-start relationship of the sample network in Figure 1 can be interpreted as: The completion of activity B can take place only after d time units have elapsed from the start of activity A.

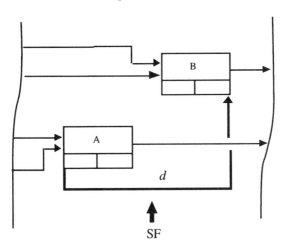

Figure 1 Portion of Precedence Diagram Network
with Start-to-Finish Relationship

DEPENDENCY, START TO START (SS) A relationship that restricts the start of an activity until the start of the immediately preceding activity has taken place or after a specified elapsed time (lag) *(see Delay)*.

The start-to-start relationship is a common relationship in a precedence diagramming method. It sets a constraint that one activity cannot start until the preceding activity has started with or without elapsed time. In construction-related tasks, some related activities may not require completion of the preceding activity in order to start; instead, they may be started concurrently (no delay), or an activity can be

197

started some elapsed time after the start of its preceding activity. Figure 1 shows a portion of a precedence diagram network indicating start-to-start relationships between two activities. The start-to-start relationship of the sample network in Figure 1 can be interpreted as follows:

1. Installation of the fourth-floor slab formwork cannot be started until 2 days after the start of stripping the first-floor slab formwork.
2. Installation of the fifth-floor slab formwork cannot be started until 2 days after the start of stripping the second-floor slab formwork.

Sample Network: Elevated Concrete Slabs

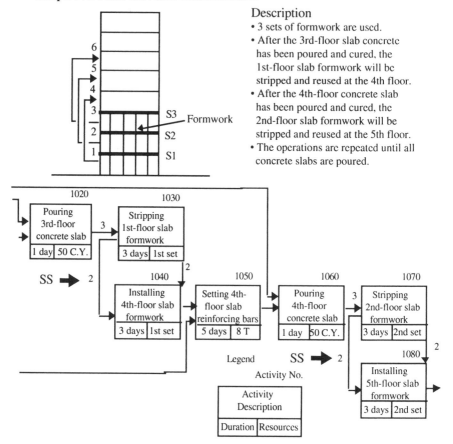

Description
• 3 sets of formwork are used.
• After the 3rd-floor slab concrete has been poured and cured, the 1st-floor slab formwork will be stripped and reused at the 4th floor.
• After the 4th-floor concrete slab has been poured and cured, the 2nd-floor slab formwork will be stripped and reused at the 5th floor.
• The operations are repeated until all concrete slabs are poured.

Figure 1 Precedence Diagram Network for Elevated Concrete Slabs

DEPENDENCY CHART *(see Precedence diagramming method)*

DEPENDENCY DIAGRAM *(see Precedence diagramming method)*

DEPENDENCY (RELATIONSHIPS) The logical links between activities in a precedence network (precedence logic diagram) *(see Precedence diagramming method).*

The dependency between two activities is shown by a connecting line from one to another with an arrow showing the direction. The precedence diagramming method allows four ways of displaying dependency (logic) relationships between two activities: (1) finish to start (FS), (2) finish to finish (FF), (3) start to start (SS), and (4) start to finish (SF). However, some CPM-oriented software does not allow all four types of relationships to be incorporated in the network; therefore, the software should be reviewed to determine its capability before the activity relationships are established. To simplify the network, activity relationships in a precedence diagramming method allow the delay (lag) in time units to be incorporated with the dependency *(see Delay)* between two activities. There are four types of activity relationships or dependency in a precedence network, as shown in Figure 1.

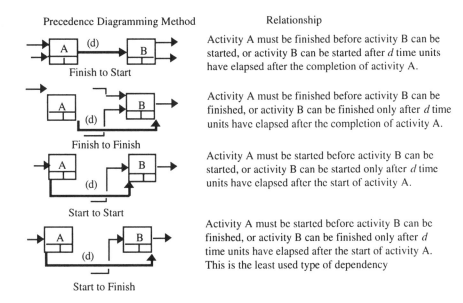

Precedence Diagramming Method

Finish to Start

Activity A must be finished before activity B can be started, or activity B can be started after *d* time units have elapsed after the completion of activity A.

Finish to Finish

Activity A must be finished before activity B can be finished, or activity B can be finished only after *d* time units have elapsed after the completion of activity A.

Start to Start

Activity A must be started before activity B can be started, or activity B can be started only after *d* time units have elapsed after the start of activity A.

Start to Finish

Activity A must be started before activity B can be finished, or activity B can be finished only after *d* time units have elapsed after the start of activity A. This is the least used type of dependency

Figure 1 Four Types of Dependency in Precedence Diagram

Figure 2 shows all four types of dependency in a sample of precedence diagram network. The relationships between activities in Figure 2 can be interpreted as follows:

A = start activity—no activity precedes it.
B = can be started only after 1 time unit has elapsed after the start of activity A.
C = can be started only after 2 time units elapsed after the completion of activity A and can be finished only after 1 time unit after the start of activity B.
D = finish activity—no activity succeeds it. Activity D can be started only after the completion of activity B, and can be completed only after 3 time units have elapsed after the finish of activity C.

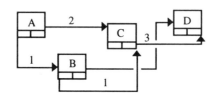

Figure 2 Precedence Diagram Network

DETAILED SCHEDULE This type of schedule displays the lowest level of detail necessary to manage and control the project from conception through job completion.

A detailed schedule contains the amount of detail about a project required by management at the level at which the project can be closely controlled. Adequate detailed information for each activity in this type of schedule includes accurate logic or relationships, and allocated resources, cost, and duration. The detailed schedule in a construction project should represent every major part of the project, such as mechanical, electrical, and general contractor activities. The level of detail should be based on scheduling specifications *(see Scheduling specifications)* provided at the invitation to bid or negotiation, such as logic diagram content, number of activities, maximum activity duration, and maximum activity cost.

The detailed schedule development is based on the project breakdown structure (PBS) *(see Project breakdown structure)* using the lowest level of PBS, called work packages *(see Work package),* to schedule activities. To control the project in detail, one work package will represent an activity in the network. A CPM network developed using either the arrow diagram method (ADM) or the precedence diagramming method (PDM) *(see Arrow diagramming method* or *Precedence diagramming method)* is normally used for detailed schedule construction.

The detailed schedule and its related information will be used at a field level as a basis for as-planned cost, schedule, and resource control of the project. The detailed schedule should be developed and submitted to the owner or the owner's representative within the number of days specified in the contract after notice of the contract award. The schedule typically includes a complete network diagram and computerized analysis. Since there is an overlap between a detailed and a preliminary schedule *(see Preliminary schedule)* at the time when the detailed schedule is submitted, some part of the preliminary schedule will be shown in the detailed schedule as the as-built portion. Figure 1 shows the time span of a detailed schedule related to other schedules. Figure 2 shows the detailed schedule development process.

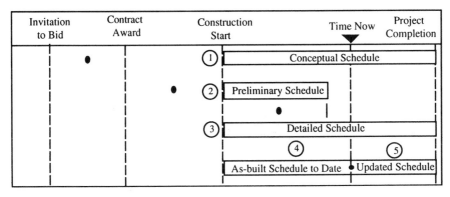

● Submitted Date

Figure 1 Time Span of Project Schedules

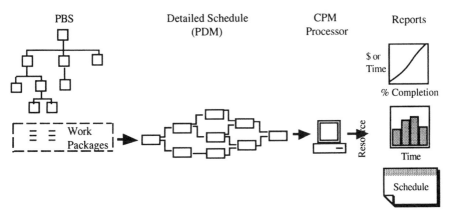

Figure 2 Detailed Schedule Development Process

DIRECT COST The costs of material, equipment, and labor resources that may be classified as resulting from the actual performance of a specific work activity or project. The direct costs for a project are the sum of the direct costs of all the activities or work items.

The direct costs for a project are the sum of the direct costs of all the activities or work items, not including the indirect costs *(see Activity indirect costs).* The direct costs of each activity or work item include the costs for material, labor, equipment, and subcontracts related to the activities or work items. For example, calculation of the direct costs for a project from work items or activities is shown by a project breakdown structure (Figure 1), work packages (Figure 2), cost estimates (Figure 3), and CPM network (Figure 4).

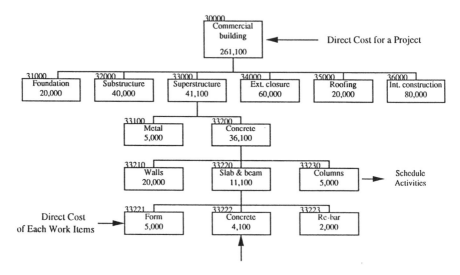

Figure 1 Project Breakdown Structure with Direct Cost for Commercial Building

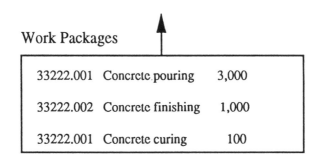

Figure 2 Work Packages Related with Work Item 332222, "Concrete"

Cost Estimate

Item Code	Description	Quantity	Material	Labor	Equipment	Subcontr.	Total
33222.001	Concrete pouring	30 C.Y.	1,500	500	1,000	0	3,000
33222.002	Concrete finishing	60 S.F.	0	500	500	0	1,000
33222.003	Concrete curing	60 S.F.	0	100	100	0	100
	Subtotal	30 C.Y.	1,500	1,100	1,500	0	4,100

Direct Costs

Figure 3 Direct Cost Estimate at Work Package Level

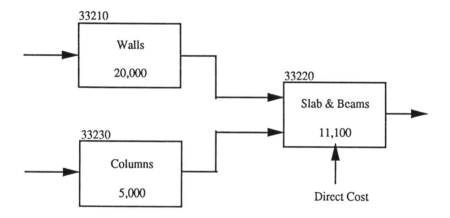

Figure 4 Fragnet of Concrete Operations

203

DISTRIBUTED TOTAL FLOAT (DTF) *(see Allocated total float)*

DUE DATE (PROJECT) The allowable completion date.

The due dates of a project include both contract interim due dates and contract final due date. The contract interim due dates are determined for a contractor to complete each phase of the work required in each period of time until completion of the project. The contract interim due dates are used to indicate whether the project is in progress as planned toward timely project completion. The last interim due date will be the contractor's time to complete the project. The contract final due date is assigned for the owner's management to finish up after project completion.

Normally, the "not later than" contract date *(see Contract date)* represents the due dates and affects the backward pass calculation. The contract must clearly specify all the interim due dates and the contract final due date. Basically, the contractor's intended completion date is before the contract final due date as shown in Figure 1.

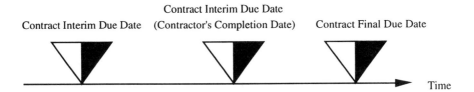

Figure 1 Time Line of Project Showing Stages of Due Dates

Examples of using the contract due dates for the computation in different situations of networks are shown below for both arrow and precedence diagram techniques.

1. The early finish date of the project is the same as the contract final due date (Figures 2 and 3).
2. The early finish date of the project is earlier than the contract final due date (Figures 4 and 5).
3. The early finish date of the project is later than the project contract due date (Figures 6 and 7). Negative floats are also shown on the network.

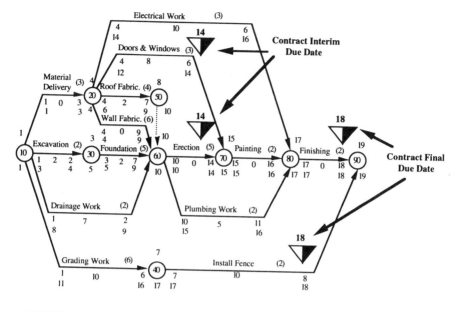

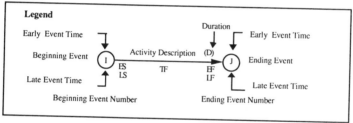

Legend

Early Event Time
Beginning Event
Late Event Time
Beginning Event Number

Duration
Activity Description (D)
ES   TF   EF
LS   LF

Early Event Time
Ending Event
Late Event Time
Ending Event Number

NOTE: The schedule computation in time units is based on the following assumption:

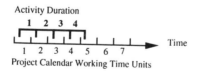

Activity Duration
1 2 3 4

Time

1 2 3 4 5 6 7
Project Calendar Working Time Units

Activity with 4 time units starts at working time unit 1 and is finished at working time unit 5.

Figure 2 Schedule Computation of Project Whose Early Finish Date Is
Same as Contract Final Due Date in Arrow Diagram

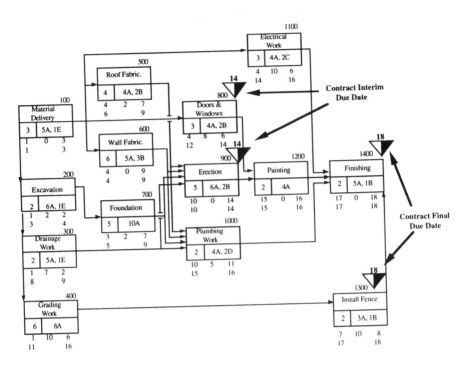

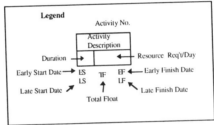

Legend

Activity No.

Activity Description

Duration →

Early Start Date → ES TF EF ← Early Finish Date
              LS   LF
Late Start Date ↗   ↑   ↖ Late Finish Date
              Total Float

NOTE: The schedule computation in time units is based on the following assumption:

Activity Duration

Time

Project Calendar Working Time Units

Activity with 4 time units starts at working time unit 1 and is finished at working time unit 5.

Figure 3 Schedule Computation of Project Whose Early Finish Date Is Same as Contract Final Due Date in Precedence Diagram

206

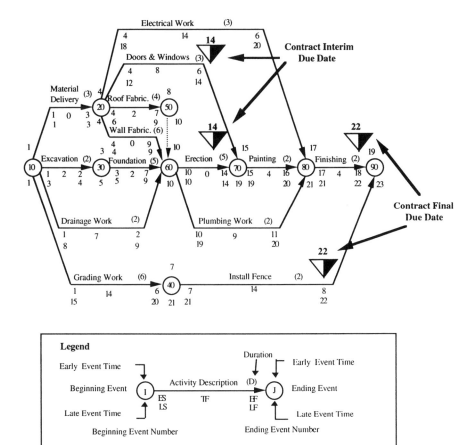

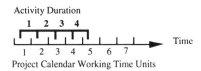

NOTE: The schedule computation in time units is based on the following assumption:

Activity Duration

1 2 3 4

Time

1 2 3 4 5 6 7

Project Calendar Working Time Units

Activity with 4 time units starts at working time unit 1 and is finished at working time unit 5.

Figure 4 Schedule Computation of Project Whose Early Finish Date Is Earlier
Than Contract Final Due Date in Arrow Diagram

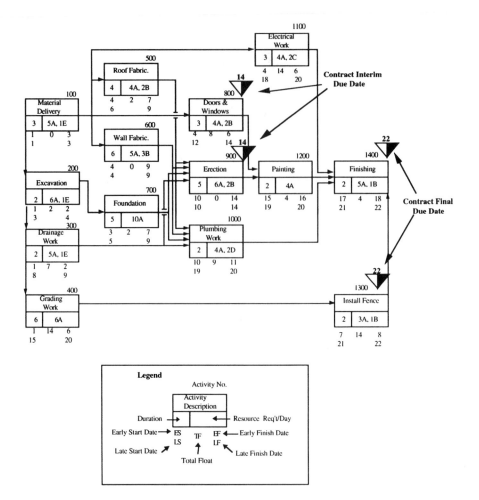

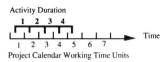

NOTE: The schedule computation in time units is based on the following assumption:

Activity Duration

Time

Project Calendar Working Time Units

Activity with 4 time units starts at working time unit 1 and is finished at working time unit 5.

Figure 5 Schedule Computation of Project Whose Early Finish Date Is Earlier Than Contract Final Due Date in Precedence Diagram

208

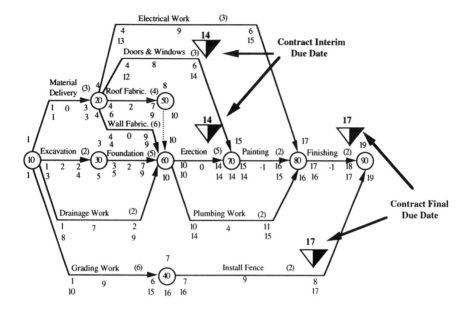

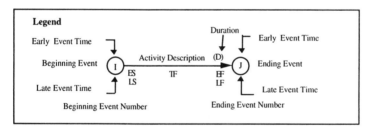

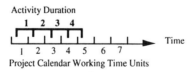

NOTE: The schedule computation in time units is based on the following assumption:

Activity with 4 time units starts at working time unit 1 and is finished at working time unit 5.

Figure 6 Schedule Computation of Project Whose Early Finish Date Is
Later Than Contract Final Due Date in Arrow Diagram

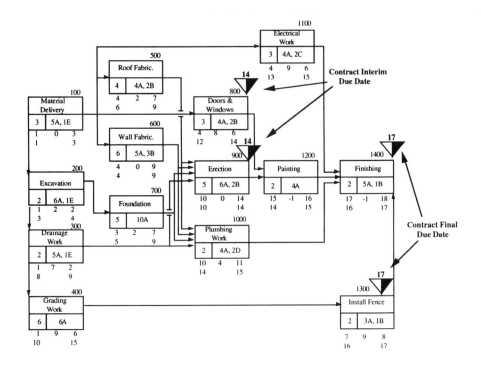

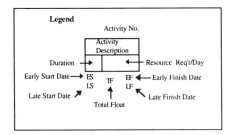

Figure 7 Schedule Computation of Project Whose Early Finish Date Is Later Than Contract Final Due Date in Precedence Diagram

DUMMY ACTIVITY (RESTRAINTS) An ADM relationship that provides a logical or imposed link between other activities in the arrow diagramming method. The duration of a dummy activity is usually zero with no cost and resource requirements.

A dummy activity is used in an arrow diagramming method to show a logical sequence of events (activities). It is represented by a dashed line or broken arrow. Dummy activities do not require cost or resource. When a network is being constructed, dummy activities are added. A dummy activity is used because an arrow representing an activity should have unique start and end events. Figure 1 shows the situation when the dummy activity should be introduced.

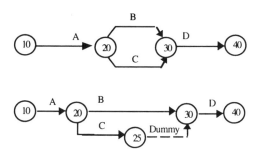

Figure 1 Dummy Activity

In Figure 1, activities B and C are started after activity A is completed. Activity D is started after activities B and C are completed. If no dummy is used, both activities B and C will have the same start and end event numbers, which is not allowed and will lead to an error in the computerized CPM calculations. To correct this situation, at least one dummy should be introduced for each arrow activity to have a unique pair of events.

There are three types of constraints (dummies):

1. *Technological constraints (hard constraints).* For example, the concrete slab cannot be poured until the slab formwork, reinforcing bars, and embedded conduits are installed. Figure 2 shows a dummy activity imposing a technological constraint.

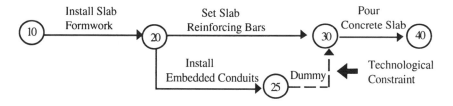

Figure 2 Dummy Activity Imposing Technological Constraint

2. *Managerial constraints.* For example, due to the limitation of concrete crews, the concrete slab can be poured after the slab formwork, reinforcing bars, and embedded conduits are installed and the concrete elevator shafts are poured. Figure 3 shows a dummy activity imposing a managerial constraint.

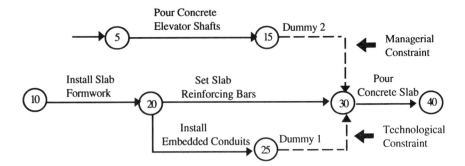

Figure 3 Dummy Activity Imposing Managerial Constraints

3. *Constraints directed by outside of a project.* For example, after the OSHA inspection, the concrete slab cannot be poured because the proper protection (guardrails) has not been provided. Therefore, the concrete slab can be poured only after the guardrails and slab formwork are installed and the concrete elevator shafts are poured. Figure 4 shows a dummy activity imposing an outside-directed constraint.

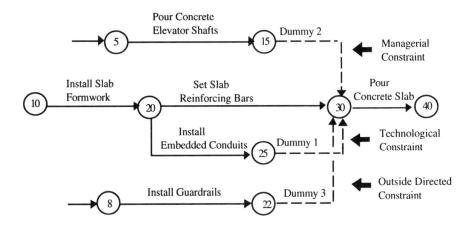

Figure 4 Dummy Activity Imposing Outside-Directed Constraint

212

DUMMY FINISH ACTIVITY An activity included in a network for the sole purpose of creating a common finish for all activities of the network.

A CPM network may have one or multiple finish activities because there may be a need for different late finish dates. The network can be arranged to have a finish activity such as "contract complete," called a dummy finish activity, connecting all other finish activities to this dummy finish activity. The concept can be used in both arrow and precedence diagramming. The finish activity usually has zero (0) duration. Figure 1 shows an arrow diagram network with multiple finish activities, Figure 2 an arrow diagram network with a dummy finish activity, Figure 3 a precedence diagram network with multiple finish activities, and Figure 4 a precedence diagram network with a dummy finish activity.

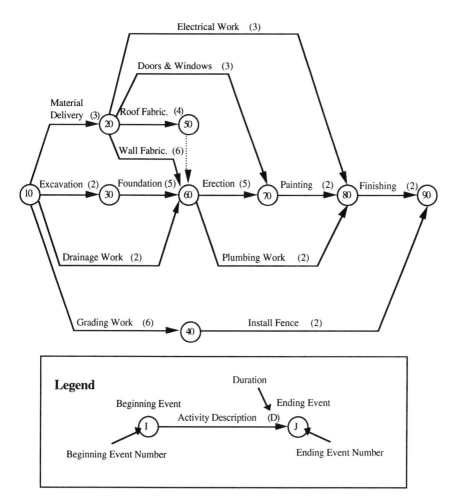

Figure 1 Arrow Diagram Network with Multiple Finish Activities

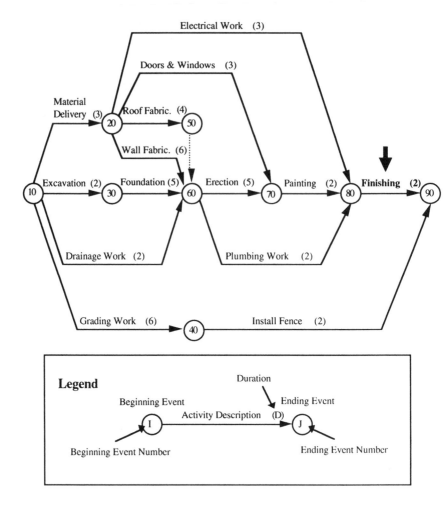

Figure 2 Arrow Diagram Network with Dummy Finish Activity

214

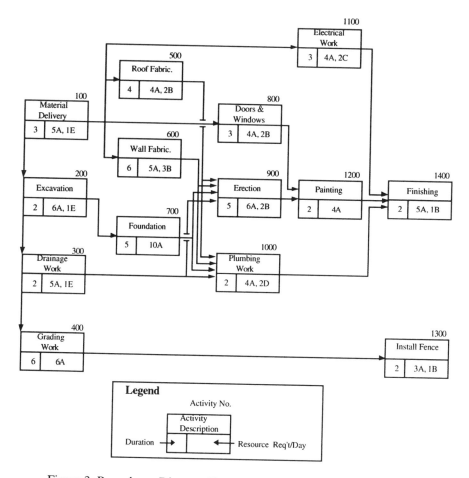

Figure 3 Precedence Diagram Network with Multiple Finish Activities

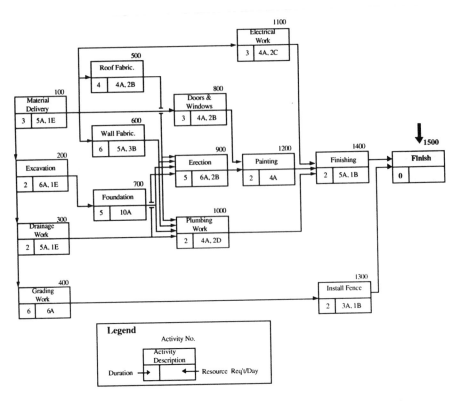

Figure 4 Precedence Diagram Network with Dummy Finish Activity

DUMMY START ACTIVITY An activity included in a network for the sole purpose of creating a common start for all the following activities.

A CPM network may have one or multiple start activities because there may be a need for different starting dates. The network can be arranged to have only one start activity, called a *dummy start activity,* connecting all start activities to this dummy start activity. The dummy start activity is different from the dummy activity (restraint) and is not represented by a dashed line or broken arrow; instead, a solid line is used. The concept can be used in both the arrow and precedence diagram methods. Most of the time it represents the notice to proceed or contractor mobilization. The dummy start activity usually has zero (0) duration. Figure 1 shows an arrow diagram network with multiple start activities, Figure 2 an arrow diagram network with a dummy start activity, Figure 3 a precedence diagram network with multiple start activities, and Figure 4 a precedence diagram network with a dummy start activity.

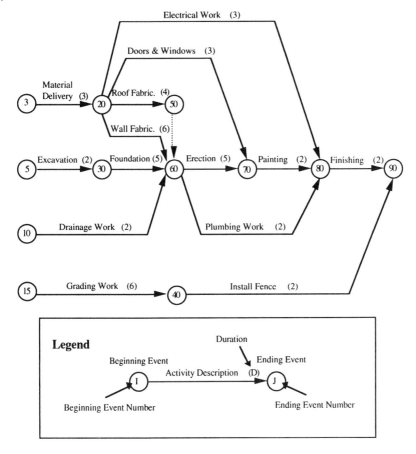

Figure 1 Arrow Diagram Network with Multiple Start Activities

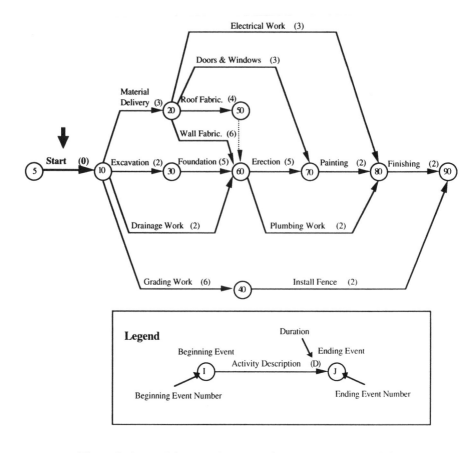

Figure 2 Arrow Diagram Network with Dummy Start Activity

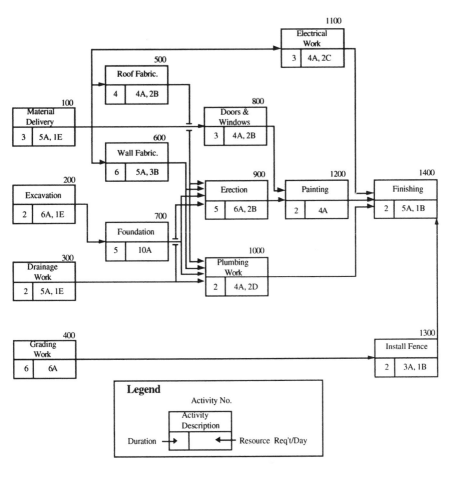

Figure 3 Precedence Diagram Network with Multiple Start Activities

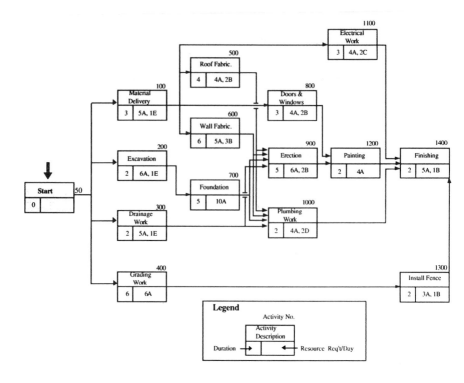

Figure 4 Precedence Diagram Network with Dummy Start Activity

E

EARLY COMPLETION SCHEDULE A project schedule that shows a late finish date (project) prior to the project finish date specified by the contract.

EARLY DATE *(see Early start date* and *Early finish date)*

EARLY EVENT TIME (EET) The earliest time at which an event in an ADM CPM network may occur.

In an arrow diagram network, an early event time can basically be used to identify an activity's early start time *(see Early start date)* except activities with "not earlier than" contract dates *(see Contract date)*. The early event time is calculated from forward pass calculation *(see Forward pass computation)*. The early event time can be recorded when the early event times have been computed for every node that precedes it and the preceding activity's durations have been added. If more than one activity precedes it, several trial values must be calculated. The largest trial value is selected as the early event time to be recorded.

Figure 1 illustrates the early event time calculation in the arrow diagram technique. For this example, event 60 is immediately preceded by the following activities:

$$20–60 \text{ with early finish time} = 10$$
$$30–60 \text{ with early finish time} = 8$$
$$10–60 \text{ with early finish time} = 3$$
$$20–50 \text{ with early finish time} = 8$$

The greatest early finish time (10), which is from activity 20–60, will indicate the early event time for event 60.

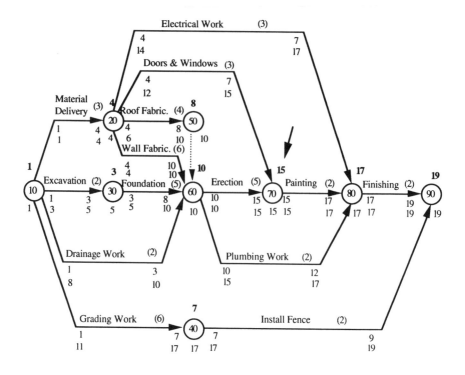

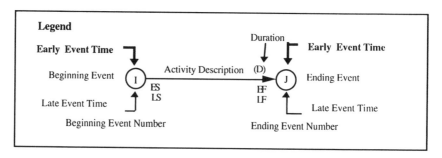

NOTE: The schedule computation in time units is based on the following assumption:

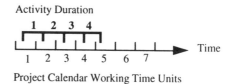

Project Calendar Working Time Units

Activity with 4 time units starts at working time unit 1 and is finished at working time unit 5.

Figure 1 Early Event Times Shown in Arrow Diagram Method

222

EARLY FINISH DATE (ACTIVITY) (EF) The earliest time at which an activity can be completed within project constraints.

Early finish = early start + activity duration

The early finish date (EF) is the earliest possible time that an activity can be completed without interfering with completion of any of the preceding activities. The early finish date is calculated by a forward pass calculation *(see Forward pass computation)* by adding the activity's early start date and its duration. Basically, an activity's early finish date will also be the early start date of its immediate successor. However, if the successor has more than one preceding activity, the greatest early finish date will be considered as the successor's early start date. The "not earlier than" contract date *(see Contract date)* and delay or lag of any activity must be taken into account in the early finish date computation.

The following list shows the various conditions for obtaining early finish dates.

1. The early finish date of activity A is equal to its early start date plus duration as shown in Figures 1 and 2.

$$EF(A) = ES(A) + duration$$
$$= 2 + 5 = 7$$

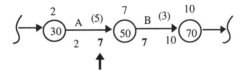

Figure 1 EF in Arrow Diagram

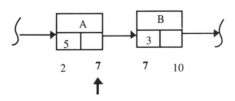

Figure 2 EF in Precedence Diagram

223

2. Activities A, B, and C have their own early finish dates by adding each activity's own early start date and duration, but only the greatest early finish date will be the early start date of activity D, their direct successor (Figures 3 and 4).

$$EF(A) = ES(A) + \text{duration} = 2 + 5 = 7$$
$$EF(B) = ES(B) + \text{duration} = 3 + 3 = 6$$
$$EF(C) = ES(C) + \text{duration} = 1 + 7 = 8$$

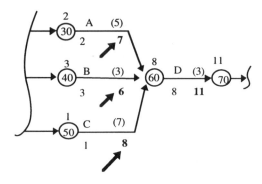

Figure 3 EF in Arrow Diagram

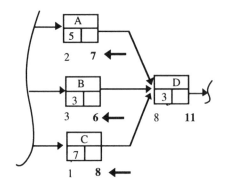

Figure 4 EF in Precedence Diagram

3. The early finish date of activity A cannot be earlier than the "not earlier than" imposed date. The EF(A) will be the larger number between the imposed date and the summation of its early start date and its duration is 7 (Figures 5 and 6).

Summation: $EF(A) = ES(A) + \text{duration} = 2 + 5 = 7$
"Not earlier than" imposed date = 9

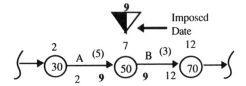

Figure 5 EF in Arrow Diagram

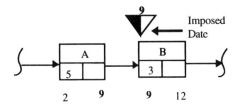

Figure 6 EF in Precedence Diagram

4. The early finish date of activity C is calculated based on the lag and dependency finish-to-finish constraint *(see Dependency, finish to finish)* (Figure 7). The early finish date of activity C is the larger number of the following:

$$EF(C) = EF(A) + lag = 7 + 2 = 9$$

or

$$EF(C) = ES(C) + duration = 8 + 3 = 11$$

Hence the early finish date of activity C is 11.

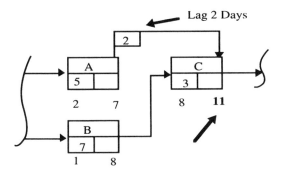

Figure 7 EF in Precedence Diagram

Figures 8 to 11 show the early finish dates of sample construction networks. Figure 8 is an arrow diagram network, and Figure 9 is a computerized report related to the arrow diagram. Figure 10 is a precedence diagram network, and Figure 11 is a computerized report related to the precedence diagram.

225

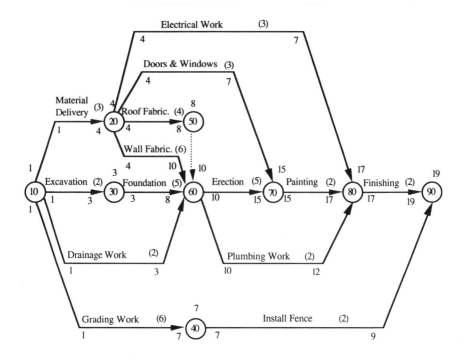

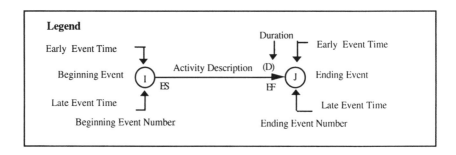

NOTE: The schedule computation in time units is based on the following assumption:

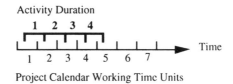

Project Calendar Working Time Units

Activity with 4 time units starts at working time unit 1 and is finished at working time unit 5.

Figure 8 Early Finish Dates in Arrow Diagram Network

CIVIL ENGINEERING DEPARTMENT   PRIMAVERA PROJECT PLANNER

REPORT DATE 23NOV93 RUN NO.   10   PROJECT SCHEDULE REPORT   START DATE 17MAR93 FIN DATE 9APR93
                  17:51
CLASSIC SCHEDULE REPORT - SORT BY ES, TF   DATA DATE 17MAR93 PAGE NO.   1

PRED	SUCC	ORIG DUR	REM DUR	%	CODE	ACTIVITY DESCRIPTION	EARLY START	EARLY FINISH	LATE START	LATE FINISH	TOTAL FLOAT
10	20	3	3	0		MATERIAL DELIVERY	17MAR93*	19MAR93	17MAR93*	19MAR93	0
10	30	2	2	0		EXCAVATION	17MAR93	18MAR93	19MAR93	22MAR93	2
10	60	2	2	0		DRAINAGE WORK	17MAR93	18MAR93	28MAR93	29MAR93	7
10	40	6	6	0		GRADING WORK	17MAR93	24MAR93	31MAR93	7APR93	10
30	60	5	5	0		FOUNDATION	19MAR93	25MAR93	23MAR93	29MAR93	2
20	60	6	6	0		WALL FABRICATION	22MAR93	29MAR93	22MAR93	29MAR93	0
20	50	4	4	0		ROOF FABRICATION	22MAR93	25MAR93	24MAR93	29MAR93	2
20	70	3	3	0		DOORS AND WINDOWS	22MAR93	24MAR93	1APR93	5APR93	8
20	80	3	3	0		ELECTRICAL WORK	22MAR93	24MAR93	5APR93	7APR93	10
40	90	2	2	0		INSTALL FENCE	25MAR93	26MAR93	8APR93	9APR93	10
50	60	0	0	0		DUMMY 1	26MAR93	25MAR93	30MAR93	29MAR93	2
60	70	5	5	0		ERECTION	30MAR93	5APR93	30MAR93	5APR93	0
60	80	2	2	0		PLUMBING WORK	30MAR93	31MAR93	6APR93	7APR93	5
70	80	2	2	0		PAINTING	6APR93	7APR93	6APR93	7APR93	0
80	90	2	2	0		FINISHING	8APR93	9APR93	8APR93	9APR93	0

Figure 9 Early Finish Dates in Tabular Format of Arrow Diagram Method

Note: Most existing software packages calculate the early and late activity dates as follows:

- Early finish = early start + activity duration – 1 day.
- Early start of a successor = early finish of immediate predecessor + 1 day.
- Late start = late finish – activity duration + 1 day.
- Late finish of a predecessor = late start of direct successor – 1 day.
- First project day is day 1.

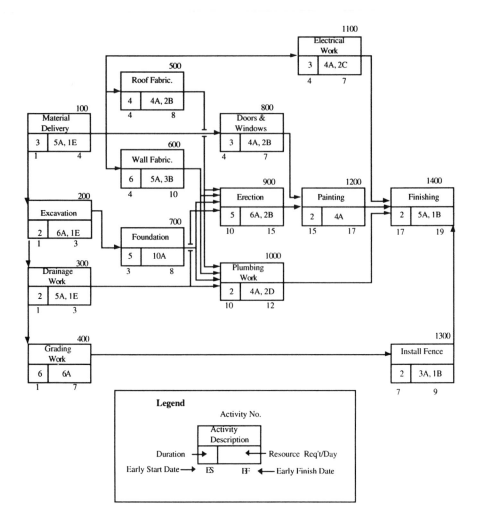

NOTE: The schedule computation in time units is based on the following assumption:

Activity Duration

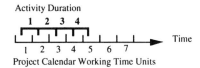

Activity with 4 time units starts at working time unit 1 and is finished at working time unit 5.

Figure 10 Early Finish Dates in Precedence Diagram Network

228

CIVIL ENGINEERING DEPARTMENT   PRIMAVERA PROJECT PLANNER

REPORT DATE 23NOV93   RUN NO.   14   PROJECT SCHEDULE REPORT   START DATE 13MAR93   FIN DATE   9APR93

          18:04

CLASSIC SCHEDULE REPORT - SORT BY ES, TF   DATA DATE   17MAR93   PAGE NO.   1

ACTIVITY ID	ORIG DUR	REM DUR	%	CODE	ACTIVITY DESCRIPTION	EARLY START	EARLY FINISH	LATE START	LATE FINISH	TOTAL FLOAT
100	3	3	0		MATERIAL DELIVERY	17MAR93*	19MAR93	17MAR93*	19MAR93	0
200	2	2	0		EXCAVATION	17MAR93	18MAR93	19MAR93	22MAR93	2
300	2	2	0		DRAINAGE WORK	17MAR93	18MAR93	26MAR93	29MAR93	7
400	6	6	0		GRADING WORK	17MAR93	24MAR93	31MAR93	7APR93	10
700	5	5	0		FOUNDATION	19MAR93	25MAR93	23MAR93	29MAR93	2
600	6	6	0		WALL FABRICATION	22MAR93	29MAR93	22MAR93	29MAR93	0
500	4	4	0		ROOF FABRICATION	22MAR93	25MAR93	24MAR93	29MAR93	2
800	3	3	0		DOORS & WINDOWS	22MAR93	24MAR93	1APR93	5APR93	8
1100	3	3	0		ELECTRICAL WORK	22MAR93	24MAR93	5APR93	7APR93	10
1300	2	2	0		INSTALL FENCE	25MAR93	26MAR93	8APR93	9APR93	10
900	5	5	0		ERECTION	30MAR93	5APR93	30MAR93	5APR93	0
1000	2	2	0		PLUMBING WORK	30MAR93	31MAR93	6APR93	7APR93	5
1200	2	2	0		PAINTING	6APR93	7APR93	6APR93	7APR93	0
1400	2	2	0		FINISHING	8APR93	9APR93	8APR93	9APR93	0

Figure 11 Early Finish Dates in Tabular Format of Precedence Diagram Method

Note: Most existing CPM software packages calculate the early and late activity dates as follows:

- Early finish = early start + activity duration – 1 day.
- Early start of a successor = early finish of immediate predecessor + 1 day.
- Late start = late finish – activity duration + 1 day.
- Late finish of a predecessor = late start of direct successor – 1 day.
- First project day is day 1.

EARLY FINISH DATE (PROJECT) (EF$_p$) The earliest time at which a project can be completed, based on a forward pass computation of the current schedule *(see Early completion schedule)*.

An early finish date of a project (EF$_p$) is the same date as the early finish date *(see Early finish date)* of the end activity(s) *(see End activity)* of the project. A project can have either one or multiple end activities, depending on its logic. For a project with multiple end activities, the early finish dates may or may not be the same, and the latest one can be considered as the early finish date of the project.

Figures 1 and 2 illustrate a single end activity project network using the arrow diagram and precedence diagram techniques, respectively. Figures 3 and 4 show computerized schedule reports of a single end activity project as activity codes of the arrow diagram and precedence diagram techniques, respectively. The early finish date of the sample project is April 5, 1993.

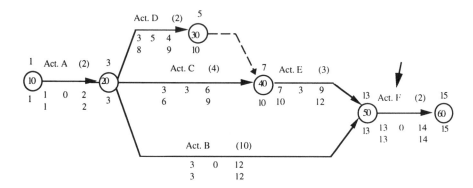

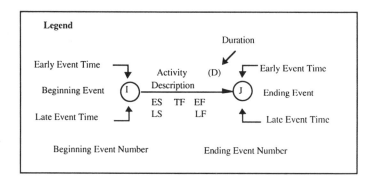

Figure 1 Arrow Diagram Network with Single End Activity

230

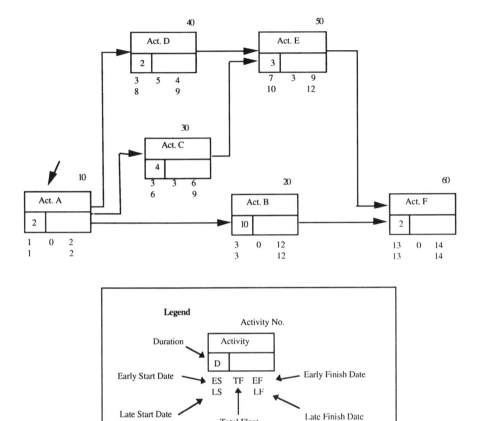

Figure 2 Precedence Diagram Network with Single End Activity

231

CIVIL ENGINEERING DEPARTMENT   PRIMAVERA PROJECT PLANNER

REPORT DATE 23NOV93 RUN NO.   4   PROJECT SCHEDULE REPORT   START DATE 17MAR93   FIN DATE 5APR93
           18:33
CLASSIC SCHEDULE REPORT - SORT BY ES, TF   DATA DATE 17MAR93   PAGE NO.   1

PRED	SUCC	ORIG DUR	REM DUR	%	CODE	ACTIVITY DESCRIPTION	EARLY START	EARLY FINISH	LATE START	LATE FINISH	TOTAL FLOAT
10	20	2	2	0		ACTIVITY A	17MAR93	18MAR93	17MAR93	18MAR93	0
20	50	10	10	0		ACTIVITY B	19MAR93	1APR93	19MAR93	1APR93	0
20	40	4	4	0		ACTIVITY C	19MAR93	24MAR93	24MAR93	29MAR93	3
20	30	2	2	0		ACTIVITY D	19MAR93	22MAR93	26MAR93	29MAR93	5
30	40	0	0	0		DUMMY	23MAR93	22MAR93	30MAR93	29MAR93	5
40	50	3	3	0		ACTIVITY E	25MAR93	29MAR93	30MAR93	1APR93	3
50	60	2	2	0		ACTIVITY F	2APR93	5APR93	2APR93	5APR93	0

Figure 3 Computerized Schedule Report (Single-End-Activity Project)
Sorted by Earliest Start, Arrow Diagram Technique

Note: Most existing CPM software packages calculate the early and late activity
dates as follows:

- Early finish = early start + activity duration – 1 day.
- Early start of a successor = early finish of immediate predecessor + 1 day.
- Late start = late finish – activity duration + 1 day.
- Late finish of a predecessor = late start of direct successor – 1 day.
- First project day is day 1.

CIVIL ENGINEERING DEPARTMENT   PRIMAVERA PROJECT PLANNER

REPORT DATE 23NOV93 RUN NO.   2   PROJECT SCHEDULE REPORT   START DATE 17MAR93 FIN DATE 5APR93
            18:46
CLASSIC SCHEDULE REPORT - SORT BY ES, TF   DATA DATE 17MAR93 PAGE NO.   1

| ACTIVITY | ORIG | REM | | | ACTIVITY DESCRIPTION | EARLY | EARLY | LATE | LATE | TOTAL |
ID	DUR	DUR	%	CODE		START	FINISH	START	FINISH	FLOAT
10	2	2	0		ACTIVITY A	17MAR93	18MAR93	17MAR93	18MAR93	0
20	10	10	0		ACTIVITY B	19MAR93	1APR93	19MAR93	1APR93	0
30	4	4	0		ACTIVITY C	19MAR93	24MAR93	24MAR93	29MAR93	3
40	2	2	0		ACTIVITY D	19MAR93	22MAR93	26MAR93	29MAR93	5
50	3	3	0		ACTIVITY E	25MAR93	29MAR93	30MAR93	1APR93	3
60	2	2	0		ACTIVITY F	2APR93	5APR93	2APR93	5APR93	0

Figure 4 Computerized Schedule Report (Single-End-Activity Project)
Sorted by Earliest Start, Precedence Diagram Technique

*Note: M*ost *e*xisting CPM software packages calculate the early and late activity dates as follows:

- Early finish = early start + activity duration − 1 day.
- Early start of a successor = early finish of immediate predecessor + 1 day.
- Late start = late finish − activity duration + 1 day.
- Late finish of a predecessor = late start of direct successor − 1 day.
- First project day is day 1.

Figures 5 and 6 illustrate a multiple-end-activity project network using the arrow diagram and precedence diagram techniques, respectively. In this example, the project contains two end activities: Finishing (80–90) and Install Fence (40–90). Figures 7 and 8 show computerized schedule reports of a multiple-start-activity project as activity codes of the arrow diagram and precedence diagram techniques, respectively. Note that the latest finish date of the end activities is April 9, 1993, which is the early finish date of the activity Finishing.

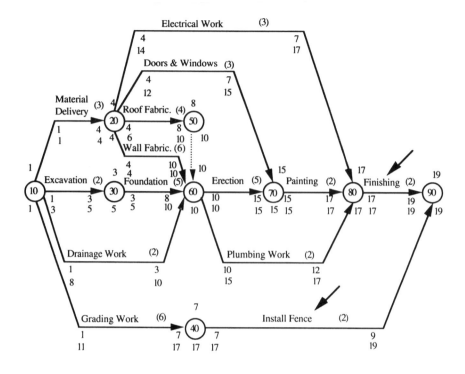

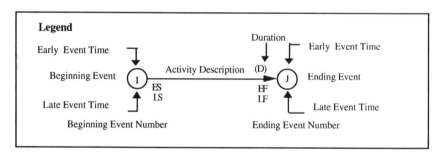

Legend

Early Event Time

Beginning Event

Late Event Time

Beginning Event Number

Duration

Activity Description (D)

ES
LS

Early Event Time

Ending Event

EF
LF

Late Event Time

Ending Event Number

NOTE: The schedule computation in time units is based on the following assumption:

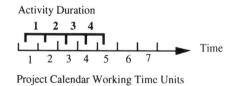

Activity Duration

1 2 3 4

Time

1 2 3 4 5 6 7

Project Calendar Working Time Units

Activity with 4 time units starts at working time unit 1 and is finished at working time unit 5.

Figure 5 Arrow Diagram Network with Multiple End Activities

234

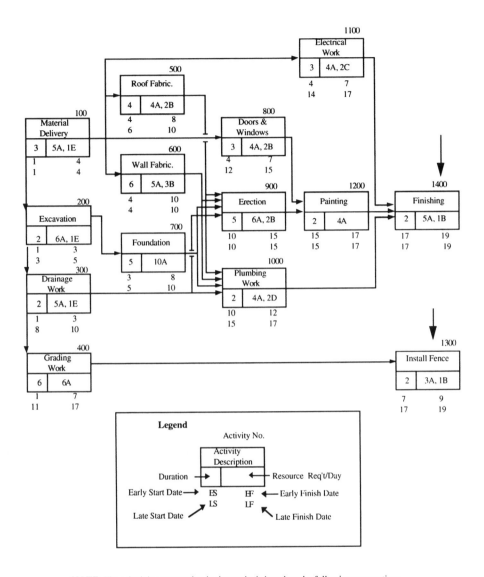

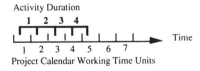

NOTE: The schedule computation in time units is based on the following assumption:

Activity Duration

Activity with 4 time units starts at working time unit 1 and is finished at working time unit 5.

Figure 6 Precedence Diagram Network with Multiple End Activities

235

CIVIL ENGINEERING DEPARTMENT   PRIMAVERA PROJECT PLANNER

REPORT DATE 23NOV93 RUN NO.   10   PROJECT SCHEDULE REPORT   START DATE 17MAR93 FIN DATE 9APR93
            17:51
CLASSIC SCHEDULE REPORT - SORT BY ES, TF   DATA DATE 17MAR93 PAGE NO.   1

PRED	SUCC	ORIG DUR	REM DUR	%	CODE	ACTIVITY DESCRIPTION	EARLY START	EARLY FINISH	LATE START	LATE FINISH	TOTAL FLOAT
10	20	3	3	0		MATERIAL DELIVERY	17MAR93*	19MAR93	17MAR93*	19MAR93	0
10	30	2	2	0		EXCAVATION	17MAR93	18MAR93	19MAR93	22MAR93	2
10	60	2	2	0		DRAINAGE WORK	17MAR93	18MAR93	26MAR93	29MAR93	7
10	40	6	6	0		GRADING WORK	17MAR93	24MAR93	31MAR93	7APR93	10
30	60	5	5	0		FOUNDATION	19MAR93	25MAR93	23MAR93	29MAR93	2
20	60	6	6	0		WALL FABRICATION	22MAR93	29MAR93	22MAR93	29MAR93	0
20	50	4	4	0		ROOF FABRICATION	22MAR93	25MAR93	24MAR93	29MAR93	2
20	70	3	3	0		DOORS AND WINDOWS	22MAR93	24MAR93	1APR93	5APR93	8
20	80	3	3	0		ELECTRICAL WORK	22MAR93	24MAR93	5APR93	7APR93	10
40	90	2	2	0		INSTALL FENCE	25MAR93	26MAR93	8APR93	9APR93	10
50	60	0	0	0		DUMMY 1	26MAR93	25MAR93	30MAR93	29MAR93	2
60	70	5	5	0		ERECTION	30MAR93	5APR93	30MAR93	5APR93	0
60	80	2	2	0		PLUMBING WORK	30MAR93	31MAR93	6APR93	7APR93	5
70	80	2	2	0		PAINTING	6APR93	7APR93	6APR93	7APR93	0
80	90	2	2	0		FINISHING	8APR93	9APR93	8APR93	9APR93	0

Figure 7 Computerized Schedule Report (Multiple-End-Activity Project)
Sorted by Earliest Start, Arrow Diagram Technique

Note: Most existing CPM software packages calculate the early and late activity
dates as follows:

- Early finish = early start + activity duration – 1 day.
- Early start of a successor = early finish of immediate predecessor + 1 day.
- Late start = late finish – activity duration + 1 day.
- Late finish of a predecessor = late start of direct successor – 1 day.
- First project day is day 1.

236

CIVIL ENGINEERING DEPARTMENT   PRIMAVERA PROJECT PLANNER

REPORT DATE 23NOV93 RUN NO.   14   PROJECT SCHEDULE REPORT   START DATE 13MAR93 FIN DATE 9APR93
              18:04
CLASSIC SCHEDULE REPORT - SORT BY ES, TF   DATA DATE 17MAR93 PAGE NO.   1

ACTIVITY ID	ORIG DUR	REM DUR	% CODE	ACTIVITY DESCRIPTION	EARLY START	EARLY FINISH	LATE START	LATE FINISH	TOTAL FLOAT
100	3	3	0	MATERIAL DELIVERY	17MAR93*	19MAR93	17MAR93*	19MAR93	0
200	2	2	0	EXCAVATION	17MAR93	18MAR93	19MAR93	22MAR93	2
300	2	2	0	DRAINAGE WORK	17MAR93	18MAR93	26MAR93	29MAR93	7
400	6	6	0	GRADING WORK	17MAR93	24MAR93	31MAR93	7APR93	10
700	5	5	0	FOUNDATION	19MAR93	25MAR93	23MAR93	29MAR93	2
600	6	6	0	WALL FABRICATION	22MAR93	29MAR93	22MAR93	29MAR93	0
500	4	4	0	ROOF FABRICATION	22MAR93	25MAR93	24MAR93	29MAR93	2
800	3	3	0	DOORS & WINDOWS	22MAR93	24MAR93	1APR93	5APR93	8
1100	3	3	0	ELECTRICAL WORK	22MAR93	24MAR93	5APR93	7APR93	10
1300	2	2	0	INSTALL FENCE	25MAR93	26MAR93	8APR93	9APR93	10
900	5	5	0	ERECTION	30MAR93	5APR93	30MAR93	5APR93	0
1000	2	2	0	PLUMBING WORK	30MAR93	31MAR93	6APR93	7APR93	5
1200	2	2	0	PAINTING	6APR93	7APR93	6APR93	7APR93	0
1400	2	2	0	FINISHING	8APR93	9APR93	8APR93	9APR93	0

Figure 8 Computerized Schedule Report (Multiple-End-Activity Project) Sorted by Earliest Start, Precedence Diagram Technique

Note: Most existing CPM software packages calculate the early and late activity dates as follows:

- Early finish = early start + activity duration – 1 day.
- Early start of a successor = early finish of immediate predecessor + 1 day.
- Late start = late finish – activity duration + 1 day.
- Late finish of a predecessor = late start of direct successor – 1 day.
- First project day is day 1.

237

EARLY START DATE (ACTIVITY) (ES) The earliest time at which an activity can be started within the project constraints.

The early start date (ES) is the earliest possible time that an activity can begin without interfering with completion of any of the preceding activities. The early start date is computed using a forward pass calculation *(see Forward pass computation)* and normally is equal to the immediate predecessor's early finish date *(see Early finish date)* except for an activity with multiple immediate predecessors. In this case, the greatest early finish date of the immediate predecessors will be the activity's early start date. Also, the "not earlier than" contract date *(see Contract date)* and delay or lag constraints must be considered for early start date computation.
The following shows various situations for calculating early start dates.

1. For an arrow diagram network, the early start date of activity B is equal to the early event time *(see Early event time)* of event 50 (Figure 1).

$$ES(B) = EET \text{ of event } 50 = 7$$

For a precedence diagram network, the early start date of activity B is equal to the early finish date of activity A, the immediate predecessor (Figure 2).

$$ES(B) = EF(A) = 7$$

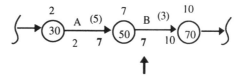

Figure 1 ES in Arrow Diagram

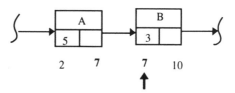

Figure 2 ES in Precedence Diagram

2. For an arrow diagram network, the early start date of an activity D is equal to the early event time of event 60 (Figure 3).

$$ES(D) = EET \text{ of event } 60 = 8$$

For a precedence diagram network, the early start date of activity D is equal to the greatest early finish date of its multiple immediate predecessors (Figure 4).

ES(D) = maximum number among EF(A), EF(B), and EF(C)
EF(A) = 7; EF(B) = 6; EF(C) = 8

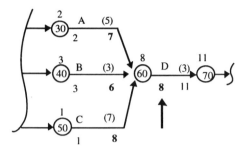

Figure 3 ES in Arrow Diagram

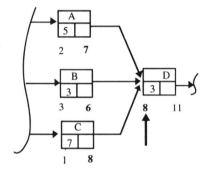

Figure 4 ES in Precedence Diagram

3. The early start date of activity B is equal to 9, the "not earlier than" contract date, even though the early finish date of its immediate predecessor is 7. ES(B) is the larger number between the "not earlier than" contract date for an activity to start and its direct predecessor's early finish date. Examples are shown in Figures 5 and 6 using the arrow and precedence diagram methods, respectively.

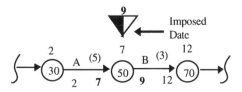

Figure 5 ES in Arrow Diagram

239

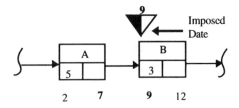

Figure 6 ES in Precedence Diagram

4. The early start date of activity C is equal to 4, as shown in Figure 7, which is calculated from the dependency start-to-start constraint *(see Dependency , start to start).*

ES(C) = larger number between ES(A) + lag and EF(B)
ES(A) + lag = 2 + 2 = 4
EF(B) = 3

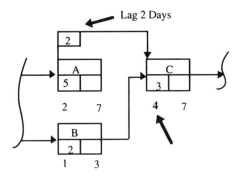

Figure 7 ES in Precedence Diagram

Figures 8 to 11 show the early start dates of a sample construction network. Figure 8 is an arrow diagram network, and Figure 9 is a computerized report related to the arrow diagram network. Figure 10 illustrates a precedence diagram network, and Figure 11 illustrates the computerized report related to the precedence diagram network.

240

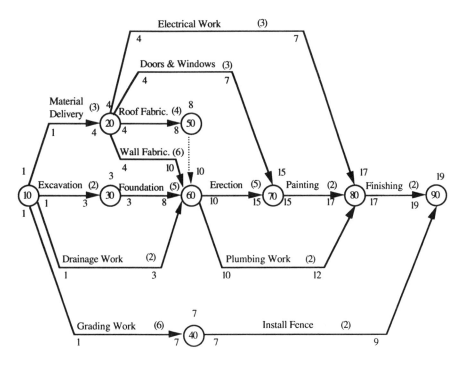

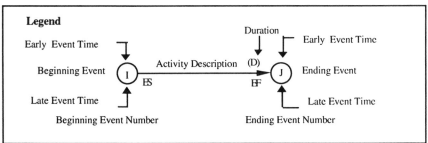

Legend

Early Event Time

Beginning Event

Late Event Time

Beginning Event Number

Duration

Activity Description (D)

ES   EF

Early Event Time

Ending Event

Late Event Time

Ending Event Number

NOTE: The schedule computation in time units is based on the following assumption:

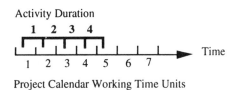

Activity Duration

Project Calendar Working Time Units

Activity with 4 time units starts at working time unit 1 and is finished at working time unit 5.

Figure 8 Early Start Dates in Arrow Diagram Network

CIVIL ENGINEERING DEPARTMENT   PRIMAVERA PROJECT PLANNER

REPORT DATE 23NOV93   RUN NO.   10   PROJECT SCHEDULE REPORT   START DATE 17MAR93   FIN DATE   9APR93
        17:51
CLASSIC SCHEDULE REPORT - SORT BY ES, TF   DATA DATE   17MAR93   PAGE NO.   1

		ORIG	REM			ACTIVITY DESCRIPTION	EARLY	EARLY	LATE	LATE	TOTAL
PRED	SUCC	DUR	DUR	%	CODE		START	FINISH	START	FINISH	FLOAT
10	20	3	3	0		MATERIAL DELIVERY	17MAR93*	19MAR93	17MAR93*	19MAR93	0
10	30	2	2	0		EXCAVATION	17MAR93	18MAR93	19MAR93	22MAR93	2
10	60	2	2	0		DRAINAGE WORK	17MAR93	18MAR93	26MAR93	29MAR93	7
10	40	6	6	0		GRADING WORK	17MAR93	24MAR93	31MAR93	7APR93	10
30	60	5	5	0		FOUNDATION	19MAR93	25MAR93	23MAR93	29MAR93	2
20	60	6	6	0		WALL FABRICATION	22MAR93	29MAR93	22MAR93	29MAR93	0
20	50	4	4	0		ROOF FABRICATION	22MAR93	25MAR93	24MAR93	29MAR93	2
20	70	3	3	0		DOORS AND WINDOWS	22MAR93	24MAR93	1APR93	5APR93	8
20	80	3	3	0		ELECTRICAL WORK	22MAR93	24MAR93	5APR93	7APR93	10
40	90	2	2	0		INSTALL FENCE	25MAR93	26MAR93	8APR93	9APR93	10
50	60	0	0	0		DUMMY 1	26MAR93	25MAR93	30MAR93	29MAR93	2
60	70	5	5	0		ERECTION	30MAR93	5APR93	30MAR93	5APR93	0
60	80	2	2	0		PLUMBING WORK	30MAR93	31MAR93	6APR93	7APR93	5
70	80	2	2	0		PAINTING	6APR93	7APR93	6APR93	7APR93	0
80	90	2	2	0		FINISHING	8APR93	9APR93	8APR93	9APR93	0

Figure 9   Early Start Dates in Tabular Format of Arrow Diagram Method

Note: Most existing CPM software packages calculate the early and late activity dates as follows:

- Early finish = early start + activity duration – 1 day.
- Early start of a successor = early finish of immediate predecessor + 1 day.
- Late start = late finish – activity duration + 1 day.
- Late finish of a predecessor = late start of direct successor – 1 day.
- First project day is day 1.

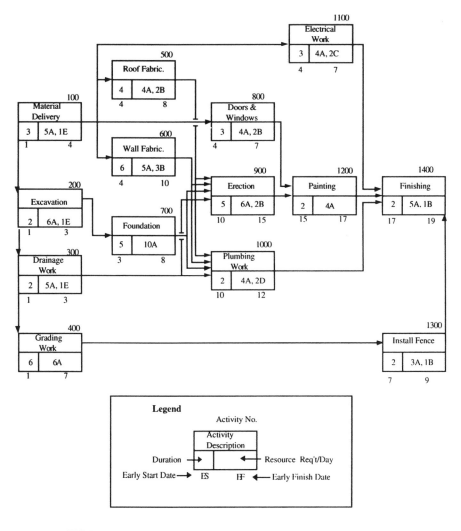

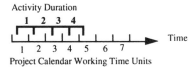

NOTE: The schedule computation in time units is based on the following assumption:

Activity Duration

Project Calendar Working Time Units

Activity with 4 time units starts at working time unit 1 and is finished at working time unit 5.

Figure 10 Early Start Dates in Precedence Diagram Network

243

CIVIL ENGINEERING DEPARTMENT PRIMAVERA PROJECT PLANNER

REPORT DATE 23NOV93 RUN NO. 14 PROJECT SCHEDULE REPORT START DATE 13MAR93 FIN DATE 9APR93
18:04

CLASSIC SCHEDULE REPORT - SORT BY ES, TF DATA DATE 17MAR93 PAGE NO. 1

ACTIVITY ID	ORIG DUR	REM DUR	%	CODE	ACTIVITY DESCRIPTION	EARLY START	EARLY FINISH	LATE START	LATE FINISH	TOTAL FLOAT
100	3	3	0		MATERIAL DELIVERY	17MAR93*	19MAR93	17MAR93*	19MAR93	0
200	2	2	0		EXCAVATION	17MAR93	18MAR93	19MAR93	22MAR93	2
300	2	2	0		DRAINAGE WORK	17MAR93	18MAR93	26MAR93	29MAR93	7
400	6	6	0		GRADING WORK	17MAR93	24MAR93	31MAR93	7APR93	10
700	5	5	0		FOUNDATION	19MAR93	25MAR93	23MAR93	29MAR93	2
600	6	6	0		WALL FABRICATION	22MAR93	29MAR93	22MAR93	29MAR93	0
500	4	4	0		ROOF FABRICATION	22MAR93	25MAR93	24MAR93	29MAR93	2
800	3	3	0		DOORS & WINDOWS	22MAR93	24MAR93	1APR93	5APR93	8
1100	3	3	0		ELECTRICAL WORK	22MAR93	24MAR93	5APR93	7APR93	10
1300	2	2	0		INSTALL FENCE	25MAR93	26MAR93	8APR93	9APR93	10
900	5	5	0		ERECTION	30MAR93	5APR93	30MAR93	5APR93	0
1000	2	2	0		PLUMBING WORK	30MAR93	31MAR93	6APR93	7APR93	5
1200	2	2	0		PAINTING	6APR93	7APR93	6APR93	7APR93	0
1400	2	2	0		FINISHING	8APR93	9APR93	8APR93	9APR93	0

Figure 11 Early Start Dates in Tabular Format of Precedence Diagram Method

Note: Most existing CPM software packages calculate the early and late activity dates as follows:

- Early finish = early start + activity duration – 1 day.
- Early start of a successor = early finish of immediate predecessor + 1 day.
- Late start = late finish – activity duration + 1 day.
- Late finish of a predecessor = late start of direct successor – 1 day.
- First project day is day 1.

EARLY START DATE (PROJECT) (ES$_p$) The earliest time at which a project can be started.

An early start date of a project (ES$_p$) is the same date as the early start date of start activity(s) *[see Early start date (activity) and Start activity(s)]* of the project. A project can have either one start activity or multiple start activities, depending on its logic. For the project with multiple start activities without imposed dates *(see Contract date)* all start activities will possess the same early start date, which is also considered as the early start date of the project.

Figures 1 and 2 illustrate a single-start-activity project network by using the arrow diagram and precedence diagram techniques, respectively. Figures 3 and 4 show computerized schedule reports of a single-start-activity project as activity codes of the arrow diagram and precedence diagram techniques, respectively. The early start date of this project is March 17, 1993.

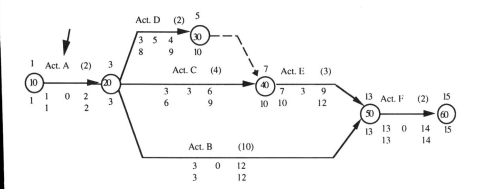

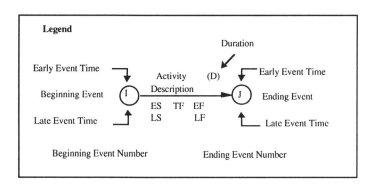

Figure 1 Arrow Diagram Network with Single Start Activity

245

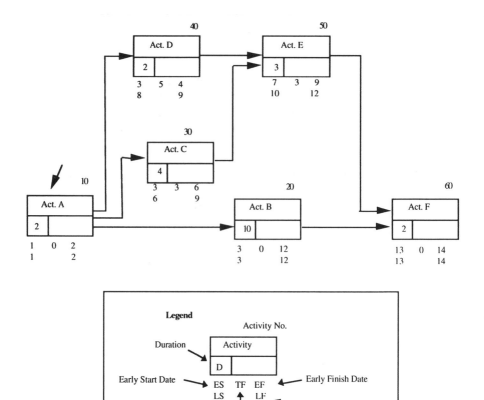

Figure 2 Precedence Diagram Network with Single Start Activity

----
CIVIL ENGINEERING DEPARTMENT   PRIMAVERA PROJECT PLANNER

REPORT DATE 23NOV93 RUN NO.   4   PROJECT SCHEDULE REPORT   START DATE 17MAR93 FIN DATE 5APR93
          18:33
CLASSIC SCHEDULE REPORT - SORT BY ES, TF   DATA DATE 17MAR93 PAGE NO.   1

PRED	SUCC	ORIG DUR	REM DUR	%	CODE	ACTIVITY DESCRIPTION	EARLY START	EARLY FINISH	LATE START	LATE FINISH	TOTAL FLOAT
10	20	2	2	0		ACTIVITY A	17MAR93	18MAR93	17MAR93	18MAR93	0
20	50	10	10	0		ACTIVITY B	19MAR93	1APR93	19MAR93	1APR93	0
20	40	4	4	0		ACTIVITY C	19MAR93	24MAR93	24MAR93	29MAR93	3
20	30	2	2	0		ACTIVITY D	19MAR93	22MAR93	26MAR93	29MAR93	5
30	40	0	0	0		DUMMY	23MAR93	22MAR93	30MAR93	29MAR93	5
40	50	3	3	0		ACTIVITY E	25MAR93	29MAR93	30MAR93	1APR93	3
50	60	2	2	0		ACTIVITY F	2APR93	5APR93	2APR93	5APR93	0

Figure 3 Computerized Schedule Report (Single-Start-Activity Project)
Sorted by Earliest Start, Arrow Diagram Technique

246

Note: Most existing CPM software packages calculate the early and late activity dates as follows:

- Early finish = early start + activity duration – 1 day.
- Early start of a successor = early finish of immediate predecessor + 1 day.
- Late start = late finish – activity duration + 1 day.
- Late finish of a predecessor = late start of direct successor – 1 day.
- First project day is day 1.

```
------------------------------------------------------------------------------------------------------------
CIVIL ENGINEERING DEPARTMENT                    PRIMAVERA PROJECT PLANNER

REPORT DATE 23NOV93  RUN NO.    2        PROJECT SCHEDULE REPORT          START DATE 17MAR93  FIN DATE  5APR93
                     18:46
CLASSIC SCHEDULE REPORT - SORT BY ES, TF                                 DATA DATE  17MAR93  PAGE NO.    1
```

ACTIVITY ID	ORIG DUR	REM DUR	%	CODE	ACTIVITY DESCRIPTION	EARLY START	EARLY FINISH	LATE START	LATE FINISH	TOTAL FLOAT
10	2	2	0		ACTIVITY A	17MAR93	18MAR93	17MAR93	18MAR93	0
20	10	10	0		ACTIVITY B	19MAR93	1APR93	19MAR93	1APR93	0
30	4	4	0		ACTIVITY C	19MAR93	24MAR93	24MAR93	29MAR93	3
40	2	2	0		ACTIVITY D	19MAR93	22MAR93	26MAR93	29MAR93	5
50	3	3	0		ACTIVITY E	25MAR93	29MAR93	30MAR93	1APR93	3
60	2	2	0		ACTIVITY F	2APR93	5APR93	2APR93	5APR93	0

Figure 4 Computerized Schedule Report (Single-Start-Activity Project)
Sorted by Earliest Start, Precedence Diagram Technique

Note: Most existing CPM software packages calculate the early and late activity dates as follows:

- Early finish = early start + activity duration – 1 day.
- Early start of a successor = early finish of immediate predecessor + 1 day.
- Late start = late finish – activity duration + 1 day.
- Late finish of a predecessor = late start of direct successor – 1 day.
- First project day is day 1.

Figures 5 and 6 illustrate a network with multiple start activities using the arrow diagram and precedence diagram techniques, respectively. This sample project has four start activities: Material Delivery, Excavation, Drainage, and Grading Work. Figures 7 and 8 show computerized schedule reports of a multiple-start-activity project as activity codes of the arrow diagram and precedence diagram techniques, respectively. The early start date of the project is March 17, 1993.

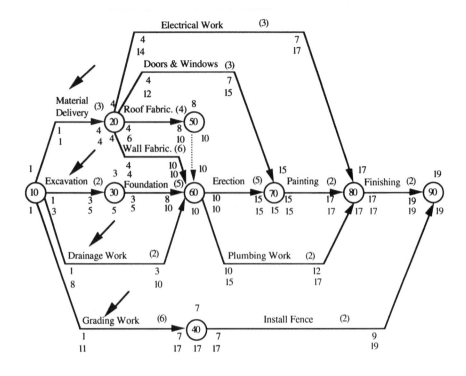

Electrical Work (3)

Doors & Windows (3)

Material Delivery (3)

Roof Fabric. (4)

Wall Fabric. (6)

Excavation (2)

Foundation (5)

Erection (5)

Painting (2)

Finishing (2)

Drainage Work (2)

Plumbing Work (2)

Grading Work (6)

Install Fence (2)

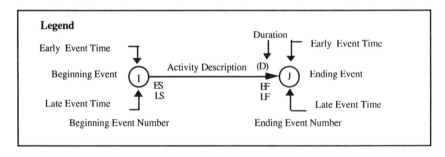

Legend

Early Event Time

Beginning Event

Late Event Time

Beginning Event Number

Duration

Early Event Time

Activity Description (D)

Ending Event

Ending Event Number

ES
LS

EF
LF

NOTE: The schedule computation in time units is based on the following assumption:

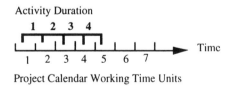

Activity Duration

Time

Project Calendar Working Time Units

Activity with 4 time units starts at working time unit 1 and is finished at working time unit 5.

Figure 5 Arrow Diagram Network with Multiple Start Activities

248

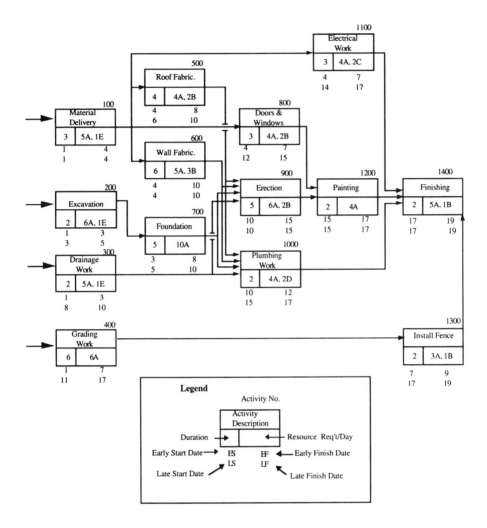

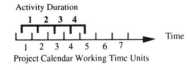

NOTE: The schedule computation in time units is based on the following assumption:

Activity Duration

Activity with 4 time units starts at working time unit 1 and is finished at working time unit 5.

Figure 6 Precedence Diagram Network with Multiple Start Activities

249

CIVIL ENGINEERING DEPARTMENT   PRIMAVERA PROJECT PLANNER

REPORT DATE 23NOV93   RUN NO.   10   PROJECT SCHEDULE REPORT   START DATE 17MAR93   FIN DATE   9APR93
     17:51

CLASSIC SCHEDULE REPORT - SORT BY ES, TF   DATA DATE   17MAR93   PAGE NO.   1

PRED	SUCC	ORIG DUR	REM DUR	%	CODE	ACTIVITY DESCRIPTION	EARLY START	EARLY FINISH	LATE START	LATE FINISH	TOTAL FLOAT
10	20	3	3	0		MATERIAL DELIVERY	17MAR93*	19MAR93	17MAR93*	19MAR93	0
10	30	2	2	0		EXCAVATION	17MAR93	18MAR93	19MAR93	22MAR93	2
10	60	2	2	0		DRAINAGE WORK	17MAR93	18MAR93	26MAR93	29MAR93	7
10	40	6	6	0		GRADING WORK	17MAR93	24MAR93	31MAR93	7APR93	10
30	60	5	5	0		FOUNDATION	19MAR93	25MAR93	23MAR93	29MAR93	2
20	60	6	6	0		WALL FABRICATION	22MAR93	29MAR93	22MAR93	29MAR93	0
20	50	4	4	0		ROOF FABRICATION	22MAR93	25MAR93	24MAR93	29MAR93	2
20	70	3	3	0		DOORS AND WINDOWS	22MAR93	24MAR93	1APR93	5APR93	8
20	80	3	3	0		ELECTRICAL WORK	22MAR93	24MAR93	5APR93	7APR93	10
40	90	2	2	0		INSTALL FENCE	25MAR93	26MAR93	8APR93	9APR93	10
50	60	0	0	0		DUMMY 1	26MAR93	25MAR93	30MAR93	29MAR93	2
60	70	5	5	0		ERECTION	30MAR93	5APR93	30MAR93	5APR93	0
60	80	2	2	0		PLUMBING WORK	30MAR93	31MAR93	6APR93	7APR93	5
70	80	2	2	0		PAINTING	6APR93	7APR93	6APR93	7APR93	0
80	90	2	2	0		FINISHING	8APR93	9APR93	8APR93	9APR93	0

Figure 7   Computerized Schedule Report (Multiple-Start-Activity Project) Sorted by Earliest Start, Arrow Diagram Technique

Note: Most existing CPM software packages calculate the early and late activity dates as follows:

- Early finish = early start + activity duration – 1 day.
- Early start of a successor = early finish of immediate predecessor + 1 day.
- Late start = late finish – activity duration + 1 day.
- Late finish of a predecessor = late start of direct successor – 1 day.
- First project day is day 1.

CIVIL ENGINEERING DEPARTMENT   PRIMAVERA PROJECT PLANNER

REPORT DATE 23NOV93 RUN NO.   14   PROJECT SCHEDULE REPORT   START DATE 13MAR93 FIN DATE 9APR93
           18:04
CLASSIC SCHEDULE REPORT - SORT BY ES, TF   DATA DATE 17MAR93 PAGE NO.   1

| ACTIVITY | ORIG | REM | | | ACTIVITY DESCRIPTION | EARLY | EARLY | LATE | LATE | TOTAL |
ID	DUR	DUR	%	CODE		START	FINISH	START	FINISH	FLOAT
100	3	3	0		MATERIAL DELIVERY	17MAR93*	19MAR93	17MAR93*	19MAR93	0
200	2	2	0		EXCAVATION	17MAR93	18MAR93	19MAR93	22MAR93	2
300	2	2	0		DRAINAGE WORK	17MAR93	18MAR93	26MAR93	29MAR93	7
400	6	6	0		GRADING WORK	17MAR93	24MAR93	31MAR93	7APR93	10
700	5	5	0		FOUNDATION	19MAR93	25MAR93	23MAR93	29MAR93	2
600	6	6	0		WALL FABRICATION	22MAR93	29MAR93	22MAR93	29MAR93	0
500	4	4	0		ROOF FABRICATION	22MAR93	25MAR93	24MAR93	29MAR93	2
800	3	3	0		DOORS & WINDOWS	22MAR93	24MAR93	1APR93	5APR93	8
1100	3	3	0		ELECTRICAL WORK	22MAR93	24MAR93	5APR93	7APR93	10
1300	2	2	0		INSTALL FENCE	25MAR93	26MAR93	8APR93	9APR93	10
900	5	5	0		ERECTION	30MAR93	5APR93	30MAR93	5APR93	0
1000	2	2	0		PLUMBING WORK	30MAR93	31MAR93	6APR93	7APR93	5
1200	2	2	0		PAINTING	6APR93	7APR93	6APR93	7APR93	0
1400	2	2	0		FINISHING	8APR93	9APR93	8APR93	9APR93	0

Figure 8 Computerized Schedule Report (Multiple-Start-Activity Project) Sorted by Earliest Start, Precedence Diagram Technique

Note: Most existing CPM software packages calculate the early and late activity dates as follows:

- Early finish = early start + activity duration – 1 day.
- Early start of a successor = early finish of immediate predecessor + 1 day.
- Late start = late finish – activity duration + 1 day.
- Late finish of a predecessor = late start of direct successor – 1 day.
- First project day is day 1.

251

EARNED VALUE The performance measurement to report the status of a project in terms of both cost and time at a given data date. It is called the budgeted cost of work performed (BCWP) *(see Budgeted cost of work performed).*

At a given time, the earned value is obtained by using the formula regardless of actual cost:

earned value = budget unit price x actual quantity in place

The example in Figure 1 shows how to calculate earned value from daily reports.

WORK ITEMS	Unit of Meas.	BUDGET			ACTUAL			EARNED VALUE		
		Unit Price	Total Quantity	Cost	Unit Price	Quantity to Date	Cost	Budget Unit Price	Quantity to Date	Earned Value
Concrete, Grade Beam	C.Y.	50.00	100	5,000	55.00	80	4,400	50.00	80	4,000

Figure 1 Daily Report

252

END ACTIVITY(S) An activity specifically nominated to have no successors *(see Dummy finish activity).*

In CPM network scheduling, there can be more than one end activity. In case of multiple end activities, the end activity with the latest early finish date will determine the length of the project, while other end activities will possess some total floats if no late finish dates are imposed *(see Contract date).* Figures 1 and 2 show arrow and precedence diagram networks with two end activities.

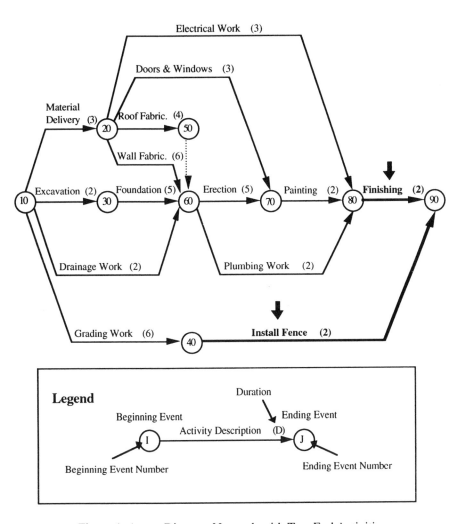

Figure 1 Arrow Diagram Network with Two End Activities

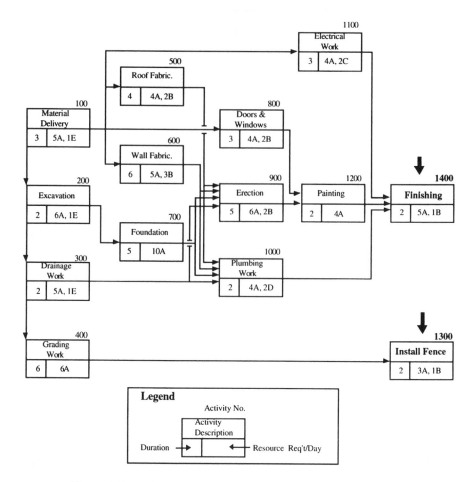

Figure 2 Precedence Diagram Network with Two End Activities

ENDING EVENT (SUCCEEDING EVENT) An event that signifies the completion of an activity or activities.

An ending event can represent the joint completion of more than one activity and can also be the beginning event *(see Beginning event)* for one or more activities. When activities terminating at an end event are completed, all activities commencing at that event can be started. For network computation, the maximum early finish time of activities leading into an end event will generate the early start time of activities leading out of that end event. Figure 1 is an arrow diagram network indicating ending events.

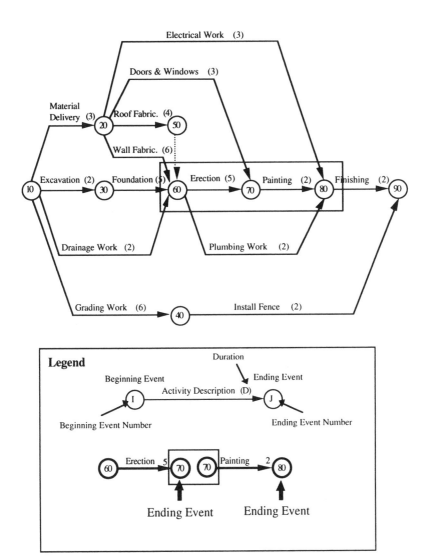

Figure 1 Arrow Diagram Network Indicating Ending Events

255

ENDING NETWORK EVENT(S) The event that signifies the end of the network. This event has no succeeding events.

An ending network event represents the ending point of the project in an arrow diagram network. The critical path or total float in the network depends on how the ending network events are arranged. There are two types of ending network events: (1) one ending event for the network, and (2) multiple ending events for the network.

1. *One ending network event.* This type of ending network event is the most commonly used in network construction. There can be one or more activities leading to this ending network event. For the network computation, the late finish time *(see Late finish date)* is set equally for all ending activities. The following computations can occur in a network with one ending event:

    a. *Late project end event time = early project end event time.* In this situation, the critical path will be the one along which activities have zero total float.

    b. *Late project end event time > early project end event time.* In this situation, the critical path will be the one along which activities have positive total float equal to the difference between the late project end event time and the early project end event time.

    c. *Late project end event time < early project end event time.* In this situation, the critical path will be the one along which activities have negative total floats equal to the difference between the late project end event time and the early project end event time.

The example of an arrow diagram network with one ending event is shown in Figure 1.

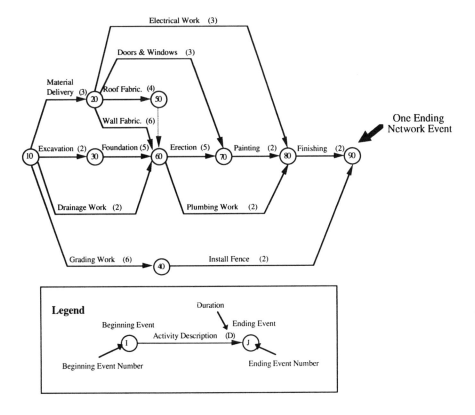

Figure 1 Arrow Diagram Network with One Ending Network Event

2. *Multiple ending network events.* A network with multiple ending events is used when parts of the project may be required to be completed at different times. This type of end network event can be used on a long-range project consisting of several subprojects requiring multiple ending events. For example, the development of multicomplex office buildings may require different facilities to be occupied at different times; therefore, the completion of each building will be set according to the essential date. The following computation can occur or be considered for a network with multiple end events:

a. *All late project end event times = greatest early project end event time.* In this situation, the critical path will be the one with the greatest computed early project end event time and zero total float.

b. *All late project end event times = each end activity early project end event time. In t*his situation, there will be multiple critical paths along which activities have zero total float.

c. *All late project end event times ≠ each end activity early project end event time. In* this situation, the late project end event times will be greater or smaller than each end activity early project end event time. There will be different paths with positive and/or negative total float. The true critical path will be the one with the least positive total float or the most negative total float.

An example of an arrow diagram network with multiple ending events is shown in Figure 2.

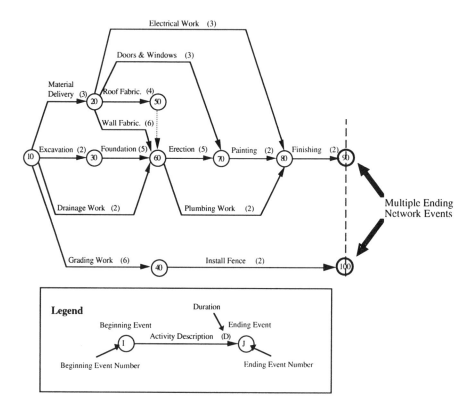

Figure 2 Arrow Diagram Network with Multiple Ending Network Events

ESTIMATE AT COMPLETION (EAC) The estimate at completion consists of the project's actual cost to date plus the forecasted costs for all remaining work with consideration of performance to date.

The estimate at completion (EAC) is one of the C/SCSC elements, and represents the forecasted total actual cost of the performance measurement at the project completion from the data date (time now). The EAC is calculated by multiplying the actual quantities in place by the actual recorded unit price (ACWP) until the data date (time now) plus the remaining quantities to go by the forecasted unit price after the data date (time now) *(see Cost and schedule control system criteria).*

Based upon the definition of EAC above, the following formula is recommended for calculating the EAC:

$$
\begin{aligned}
\text{EAC} \quad &= \quad \text{ACWP/PC} \\
&= \quad (\text{ACWP/BCWP}) \times \text{BAC (assuming no future change in} \\
&\qquad \text{unit price structure)} \\
&= \quad \text{BAC/CPI}
\end{aligned}
$$

(See Percent complete, Actual cost of work performed, Budgeted cost of work performed, Budget at completion, and *Cost performance index.)* To develop the EAC in a project, the following steps are generally recommended:

Step 1: Use the original PBS of a project.

Step 2: Use the original cost estimation of the project based on the PBS.

Step 3: Use the latest updated CPM by the data date *(see Data date).*

Step 4: Measure the work completed, and calculate the ACWP by multiplying the actual quantities in place periodically by the actual unit price of each work item.

Step 5: Derive the EAC of each work item by using the formula EAC = (ACWP/BCWP) x BAC and sum up the cost to calculate the EAC of a project from the viewpoint of the data date (time now). The results of steps 4 and 5 are shown in the monthly cost performance report.

Step 6: Develop the ACWP curve on the graph using the time-phased and cumulative actual costs by the data date based on the results of the step 4 until time now.

Step 7: Extend the ACWP curve to the point of project completion on the graph using the time-phased and cumulative estimated costs based on the results of the steps after time now.

The preceding seven steps are explained using a residential house as an example.

1. *Step 1:* PBS

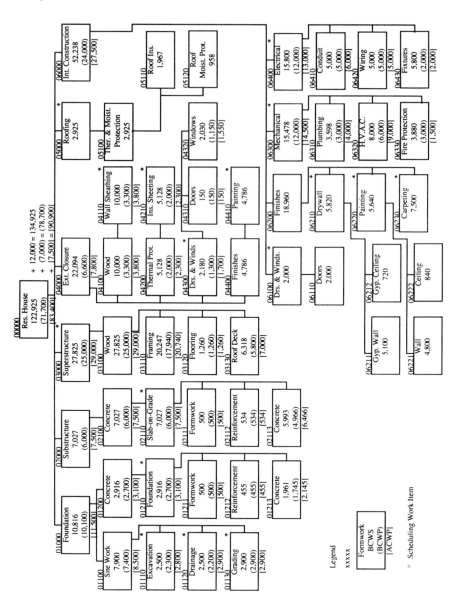

Legend

xxxxx

Formwork
BCWS
(BCWP)
[ACWP]

° Scheduling Work Item

260

2. *Step 2:* Cost Estimate

DESCRIPTION	U/M	U/P	Q'TY	TOTAL
01 General Requirements				
Subtotal	L/S	12,000	1	12,000
02 Site Work				
01110 Excavation	C.Y.	1.89	1,323	2,500
01120 Drainage	L.F.	3.11	804	2,500
01130 Grading	C.Y.	1.06	2,736	2,900
Subtotal				7,900
03 Concrete				
01211 Foundation Form	SFCA	2.89	173	500
01212 Foundation Re-bar	Tons	910	0.5	455
01213 Foundation Concrete	C.Y.	103.2	19	1,961
02111 Slab-on-Grade Form	L.F.	1.38	362	500
02112 Slab-on-Grade Re-bar	Tons	890	0.6	534
02113 Slab-on-Grade Concrete	C.Y.	90.8	66	5,993
Subtotal				9,943
06 Wood				
03110 Superstructure Framing	M.B.F.	1,191	17	20,247
03120 Superstructure Flooring	S.F.	0.63	2,000	1,260
03130 Roof Decking	S.F.	5.01	1,261	6,318
04110 Wall Sheathing	S.F.	1.17	8,547	10,000
Subtotal				37,825
07 Thermal & Moisture Protection				
04210 Wall Insulating Sheeting	S.F.	0.6	8,547	5,128
05110 Roof Insulating	S.F.	1.56	1,261	1,967
05120 Roof Moisture Protection	S.F.	0.76	1,261	958
Subtotal				8,053
08 Doors & Windows				
04310 Ext. Doors	Ea.	150	1	150
04320 Ext. Windows	Ea.	203	10	2,030
06110 Int. Doors	Ea.	200	10	2,000
Subtotal				4,180
09 Finishes				
04410 Ext. Painting	S.F.	0.56	8,547	4,786
06211 Int. Gyp. Wallboard	S.F.	0.34	15,000	5,100
06212 Int. Gyp. Ceiling Board	S.F.	0.36	2,000	720
06311 Int. Wall Painting	S.F.	0.32	15,000	4,800
06312 Int. Ceiling Painting	S.F.	0.42	2,000	840
06230 Carpeting	S.Y.	5	1,500	7,500
Subtotal				23,746
15 Mechanical				
06310 Plumbing	L.F.	3.98	904	3,598
06320 H.V.A.C.	L/S	8,000	1	8,000
06330 Fire Protection	Ea.	776	5	3,880
Subtotal				15,478
16 Electrical				
06410 Electrical Conduit	L/S	5,000	1	5,000
06420 Electrical Wiring	L/S	5,000	1	5,000
06430 Electrical Fixtures	L/S	5,800	1	5,800
Subtotal				15,800
GRAND TOTAL				134,925

3. *Step 3:* Updated CPM Network Schedule

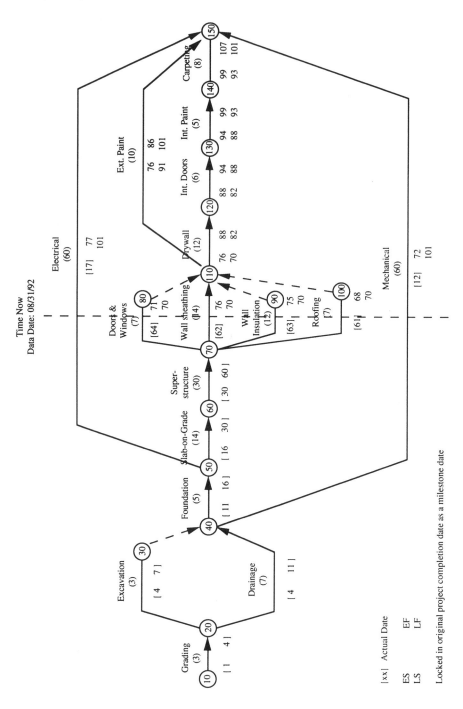

262

4. *Steps 4 and 5:* Monthly Cost Performance Report

MONTHLY COST PERFORMANCE REPORT: CPM ITEM

PROJECT : RESIDENTIAL HOUSE   REPORT DATE: 08/31/92

ITEM		CURRENT PERIOD									CUMULATIVE TO DATE									AT COMPLETION		
		BUDGET COST			VARIANCE			INDEX			BUDGET COST			VARIANCE			INDEX					
PBS	CPM	BCWS	BCWP	ACWP	SV	CV	AV	SPI	CPI	PC	BCWS	BCWP	ACWP	SV	CV	AV	SPI	CPI	PC	BAC	EAC	ACV
01130	10 - 20										2,900	2,900	2,900	0	0	0	1.00	1.00	1.00	2,900	2,900	
01110	20 - 30										2,500	2,300	2,800	-200	-500	-300	0.92	0.82	0.92	2,500	2,800	-300
01120	20 - 40										2,500	2,200	2,800	-300	-600	-300	0.88	0.79	0.88	2,500	2,800	-300
01210	40 - 50										2,916	2,700	3,100	-216	-400	-184	0.93	0.87	0.93	2,916	3,100	-184
06300	40 - 150	5,160	5,000	5,200	-160	-200	-40	0.97	0.96	0.32	13,416	12,000	14,500	-1,416	-2,500	-1,084	0.89	0.83	0.78	15,478	16,562	-1,084
02110	50 - 60										7,027	6,000	7,500	-1,027	-1,500	-473	0.85	0.80	0.85	7,027	7,500	-473
03000	60 - 70	14,903	13,900	15,000	-1,003	-1,100	-97	0.93	0.93	0.50	27,825	25,000	29,000	-2,825	-4,000	-1,175	0.90	0.86	0.90	27,825	29,000	-1,175
06400	50 - 150	5,260	5,100	5,300	-160	-200	-40	0.97	0.96	0.32	12,361	12,000	13,000	-361	-1,000	-639	0.97	0.92	0.76	15,800	16,439	-639
04300	70 - 80	1,555	1,300	1,700	-255	-400	-145	0.84	0.76	0.60	1,555	1,300	1,700	-255	-400	-145	0.84	0.76	0.60	2,180	2,325	-145
04210	70 - 90	2,135	2,000	2,300	-135	-300	-165	0.94	0.87	0.39	2,135	2,000	2,300	-135	-300	-165	0.94	0.87	0.39	5,128	5,293	-165
04110	70 - 110	3,570	3,300	3,800	-270	-500	-230	0.92	0.87	0.33	3,570	3,300	3,800	-270	-500	-230	0.92	0.87	0.33	10,000	10,230	-230
05000	70 - 100																			2,925	2,925	0
06210	110-120																			5,820	5,820	0
04410	110-150																			4,786	4,786	0
06100	120-130																			2,000	2,000	0
06220	130-140																			5,640	5,640	0
06230	140-150																			7,500	7,500	0
SUBTOTAL		32,583	30,600	33,300	-1,983	-2,700	-717	0.94	0.92	0.25	78,705	71,700	83,400	-7,005	-11,700	-4,695	0.91	0.86	0.58	122,925	127,620	-4,695
GEN. REQ.		2,400	2,400	2,500		-100	-100	1.00	0.96	0.20	7,200	7,000	7,500	-200	-500	-300	0.97	0.93	0.58	12,000	12,300	-300
TOTAL		34,983	33,000	35,800	-1,983	-2,800	-817	0.94	0.92	0.24	85,905	78,700	90,900	-7,205	-12,200	-4,995	0.92	0.87	0.58	134,925	139,920	-4,995

5. *Steps 6 and 7:* ACWP Curve and EAC

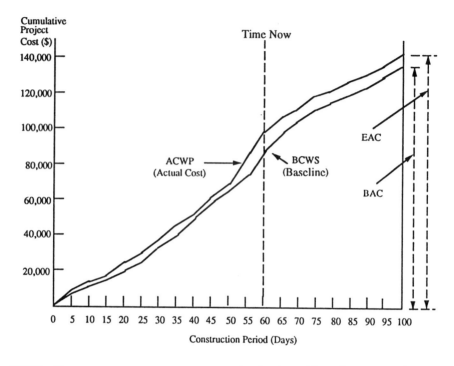

NOTE: The completion dates originally planned and estimated from time now are the same

EVENT A point in time representing the start or completion of one or more arrow activities in a CPM or PERT schedule. An event does not consume time or resources.

An event is used in a CPM arrow diagramming method to separate one activity from another and to establish the beginning and ending points of an activity *(see Beginning event* and *Ending event).* It is a specific definable accomplishment in a program plan recognizable at a particular instant in time. All preceding arrow activities leading to an event must be finished before any succeeding activities leading out of the event can be started. Two events are connected together by one arrow, representing activities to form a CPM network showing the construction plan. The graphical representation of an event is a circle with an assigned number or code that identifies the event specifically. Figure 1 shows an arrow diagram network indicating an event.

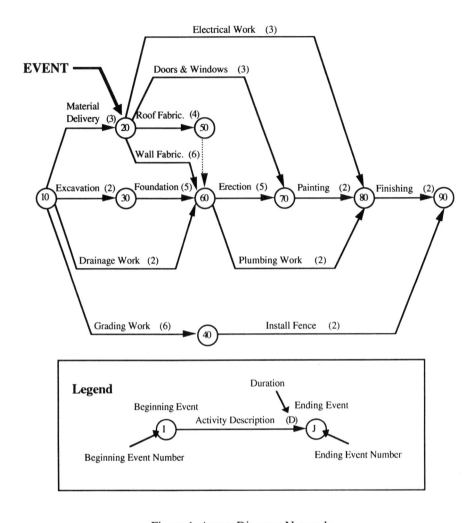

Figure 1 Arrow Diagram Network

EVENT CODE *(see Event name* and *Event number)*

EVENT NAME (DESCRIPTION) An alphanumeric description attached to an event, in a CPM arrow diagram, describing the project status at that point in time.

An event name describes a major or important event, called a milestone *(see Milestones),* that represents the goal or objective to be accomplished within a project plan. The practicality of having an event name or description in the network during

the design or implementation phase is that the specific event should be monitored closely for reliable project control. The event name will appear only in milestone reports. In the example shown in Figure 1, the selected milestone (70) is Building Dry-in.

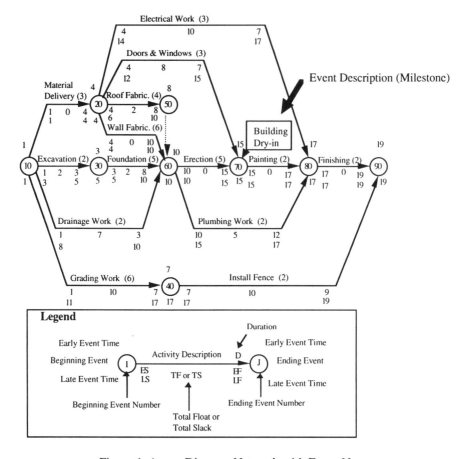

Figure 1 Arrow Diagram Network with Event Name

EVENT NUMBER (ID) An alphanumeric code assigned to an event for identification purposes.

Generally, in an arrow diagramming network, beginning and ending events *(see Beginning event and Ending event)* are identified by numbers. The main purpose of event numbers is to identify and sort activities. It is recommended that events be numbered according to the following rules to minimize the chance of loops and confusion:

266

1. The starting event number should be smaller than the ending event number.
2. The numbers should be assigned in ascending order so that the number at the arrow's head is greater than the number at the arrow's tail.
3. The event number should not be changed after the project has been started unless it is absolutely necessary.
4. The event numbering process should be performed after the network logic diagram has been completed and is ready for computation.
5. Each activity must have a unique code or pair of coded events.
6. The numbering process starts from the left (the beginning of the project) and progresses to the right toward the end of the project, from the top to the bottom of the network diagram.
7. The number between two events should not be consecutive because other activities may be inserted.

Figure 1 shows an arrow diagram network with the event numbers identified.

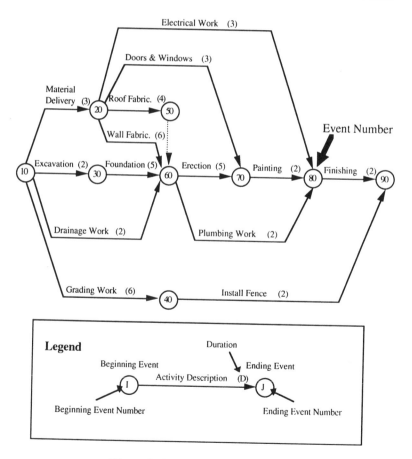

Figure 1 Arrow Diagram Network

EVENT SLACK (EVENT FLOAT)   The difference between the latest allowable event time *(TL)* and the earliest expected event time *(TE)* of PERT:

event slack = latest allowable event time *(TL)* - earliest expected event time *(TE)*

Event slack is the amount of working time units that allow an event to be delayed without extending the completion time of the project. Each case of event slack is equal to subtraction of the earliest expected event time *(TE)* from the latest allowable event time *(TL)*. Before computing the event slack, both *TL* and *TE* must be computed from forward and backward pass calculations.

The earliest expected event time *(TE)* in PERT refers to the earliest time that all activities immediately preceding the event will be completed according to the original time estimates. The latest allowable event time *(TL)* is the latest starting time for the most critical activity that succeeds the event. Basically, calculation of the *TE* in PERT is the same as that of early event time *(see Early event time)* in CPM, while the *TL* in PERT can be obtained in the same way as the late event time *(see Late event time)* is in CPM. However, the meanings are slightly different since both *TE* and *TL* are mean values resulting from nondeterministic time estimates.

Note that normally the term *event slack* is used only in PERT network because even though the arrow diagram technique in CPM does have nodes representing events, it is activity oriented and focuses on activities, not events. CPM considers just the slacks or floats of activities. On the other hand, PERT is event oriented; thus the term *event slack* refers primarily to the slack of an event in PERT. Event slack is an expected value, or mean, of a distribution of slack because both *TE* and *TL* are means of distributions. Figure 1 is an example of a PERT network, including event slacks, and Table 1 shows the computation of event slack of the sample network in Figure 1.

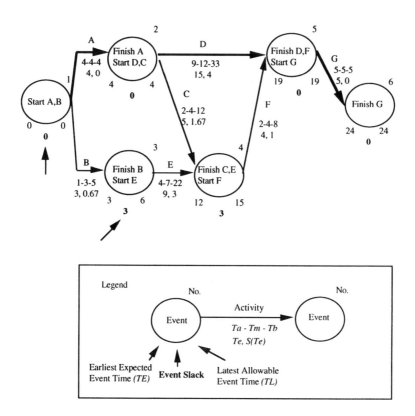

Figure 1 PERT Diagram Network Showing Event Slack

Table 1 Calculation of Event Slack from PERT Diagram

Event	Latest Allowable Event Time (TL)	Earliest Expected Event Time (TE)	Event Slack (TL − TE)
1	0	0	0
2	4	4	0
3	6	3	3
4	15	12	3
5	19	19	0
6	24	24	0

269

EVENT TIMES The time information generated through CPM calculation that identifies the start and finish times for each event in an arrow network.

An event time, consisting of an early event time *(see Early event time)* and a late event time *(see Late event time),* is used to identify the time at the starting or ending point of an activity. It may also be used to indicate the time for a milestone of the project.

In the forward pass calculation *(see Forward pass computation)* an event time specifies the early start time of an activity or its preceding activities' early finish time. For an event consisting of multiple immediately preceding activities, the greatest early finish time of the predecessors is considered as the early finish time of the event and will also become the early start time of all successors of that event.

In the backward pass calculation *(see Backward pass computation)* an event time is the late start time of an activity or its preceding latest finish time. Unlike the forward pass, the smallest early start time of the direct successor activities is considered as the late event time.

Figure 1 shows a sample of an arrow diagram network. Note that the early and late event times of node 60 are the same, which is 10, but those of node 40 are different. The early event time of node 40 is 7 and its late event time is 17.

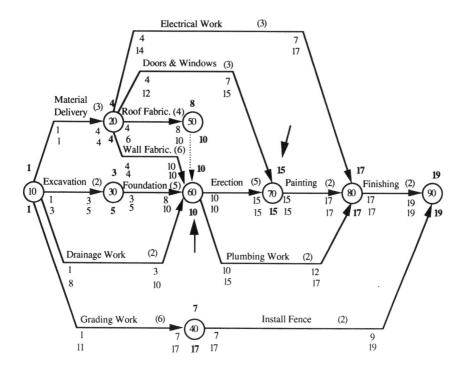

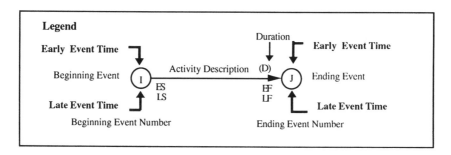

Legend

Early Event Time

Beginning Event

Activity Description (D) Duration

Early Event Time

Ending Event

Late Event Time

Late Event Time

Beginning Event Number

Ending Event Number

NOTE: The schedule computation in time units is based on the following assumption:

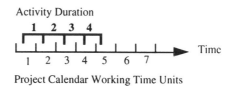

Activity Duration

Time

Project Calendar Working Time Units

Activity with 4 time units starts at working time unit 1 and is finished at working time unit 5.

Figure 1 Event Times Shown in Arrow Diagram Method

EXPECTED BEGIN DATE The begin date for an activity of an event in PERT.

The expected begin date in PERT is the date for an activity to start. It consists of an early expected begin date and a late expected begin date. The early expected begin date, which is represented by the earliest expected event time *(TE)* (Figure 1), is the earliest date for an activity to start and is the earliest date of finishing all the activity's immediately preceding activities, while the late expected begin date, which is represented by the latest allowable event time *(TL)* (Figure 1), is the latest date for an activity to start and is the latest date of finishing all the activity's immediately preceding activities of the event. Since activity durations in PERT are expected durations *[see Expected duration (activity)]*, not deterministic durations, the calculated dates on schedule will represent expected dates.

Figure 1 illustrates the PERT network with expected begin dates on activities including the schedule computation. From Figure 1, the early expected begin date of event 4 is day 12, which is the earliest expected date for activity F to start and is also the earliest expected date of all immediately preceding activities (activities C and E) to be finished. Similarly, the late expected begin date of event 4 is day 15, which is the latest expected date for activity F to start and the latest expected date of all its immediately preceding activities (activities C and E) to be finished.

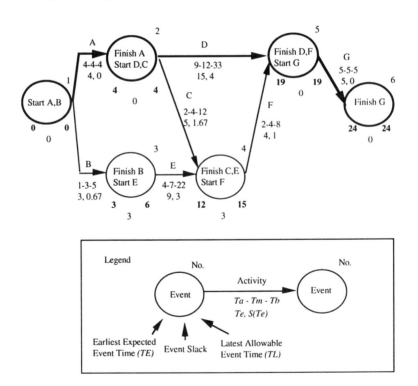

Figure 1 PERT Network with Expected Begin Dates and Schedule Calculation

272

EXPECTED COMPLETION DATE The completion date for an activity of an event in PERT.

The expected completion date in PERT is the date for an activity to be finished. It consists of the early expected completion date and the late expected completion date. The early expected completion date, which is represented by the earliest expected event time *(TE)* (Figure 1), is the earliest date for completing all the activity's immediately preceding activities and is also the earliest date for an activity to start, while the late expected begin date, which is represented by the latest allowable event time *(TL)* (Figure 1), is the latest date for completing all the activity's immediately preceding activities and also is the latest date for an activity to start. Since activity durations in PERT are expected durations *[see Expected duration (activity)]*, the mean of three time estimates, not a deterministic duration, the calculated dates on the schedule will represent expected dates.

Figure 1 illustrates the PERT network with expected completion dates on activities including the schedule computation. From Figure 1, the early expected completion date of event 4 is day 12, which is the earliest expected date for all immediately preceding activities (activities C and E) to be completed and the earliest expected date for activity F to start. Similarly, the late expected completion date of event 4 is day 15, which is the latest expected date for all immediately preceding activities (activities C and E) to be completed and is also the latest expected date for activity F to start.

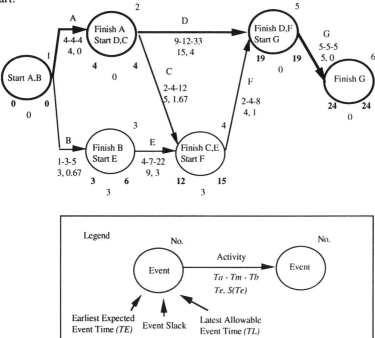

Figure 1 PERT Network with Expected Completion Dates and Schedule Calculation

273

EXPECTED DURATION (ACTIVITY) A statistically weighted estimate of the anticipated duration for an activity in PERT method.

expected duration

$$= \frac{\text{optimistic duration} + (4 \times \text{most likely duration}) + \text{pessimistic duration}}{6}$$

When a single time estimate is used, as in a summary report, the expected duration is used.

PERT is a statistical approach to schedule a project with uncertainty activity performance time. PERT applies the concept of the central limit theorem, providing a combination of activity duration distributions. The expected duration *(Te)* of each activity is calculated from three time estimates: optimistic duration *(Ta) (see Optimistic time estimate)*, most likely duration *(Tm) (see Most likely time estimate)*, and pessimistic duration *(Tb) (see Pessimistic time estimate)* . The formula to obtain the expected duration or the mean is derived from an approximation of the statistical distributions and simplified as follows:

$$\text{expected duration } (Te) = \frac{Ta + 4 \times Tm + Tb}{6}$$

The three time estimates for each activity can be displayed on distribution curves as shown in Figure 1, which also shows the location of the expected duration on the four distribution curves. Note that *Te* will be equal to *Tm* only if *Ta* and *Tb* are equally different from *Tm;* that is, *Tm - Ta = Tb - Tm* .

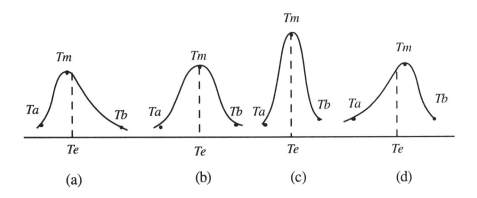

Figure 1 Three Time Estimates *(Ta, Tm,* and *Tb)* in Distribution Curves

In the PERT system, variance describes the uncertainty of how much time will be required to finish an activity. If the variance is large, which means that Ta and Tb are far apart, there is great uncertainty as to the time required to finish the activity. On the other hand, a small variance indicates very little uncertainty. The expected duration of Figure 1(b) will have larger variance than the expected duration of Figure 1(c).

$$\text{variance} = (\text{standard deviation})^2 = S^2(Te)$$

where the standard deviation, $S(Te)$, is the measure of the spread of a distribution. It is approximately one-sixth of the range of time distribution:

$$\text{standard deviation } [S(Te)] = \frac{Tb - Ta}{6}$$

Figure 2 is an example of a PERT diagram with three time estimates for each activity, Table 1 shows the calculation of Te for each activity, and Figure 3 shows a PERT diagram with the expected durations and standard deviation for each activity.

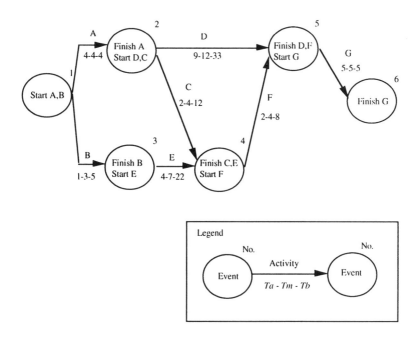

Figure 2 PERT Diagram Showing Three Time Estimates for Each Activity

Table 1 Expected Duration *(Te)*, Standard Deviation, and Variance of PERT Diagram of Figure 2

Node	Activity	Ta	Tm	Tb	Te = (Ta + 4 Tm + Tb)/6	S(Te) = [Tb-Ta]/6	S(Te) x S(Te)
			Time Estimates				
1-2	A	4	4	4	4	0	0
1-3	B	1	3	5	3	0.67	0.44
2-4	C	2	4	12	5	1.67	2.79
2-5	D	9	12	33	15	4	16
3-4	E	4	7	22	9	3	9
4-5	F	2	4	8	4	1	1
5-6	G	5	5	5	5	0	0

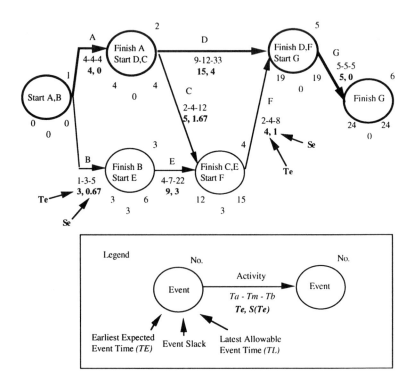

Figure 3 PERT Diagram Showing Expected Duration and Standard Deviation

F

FLOAT FACTOR The total number of total float time units for all activities in progress or not started divided by the total activity duration remaining or not started.

The float factor is an indicator of the flexibility of a project. The higher the float factor is, the more flexible the project will be. Normally, at the beginning stage of the project, the float factor is highest and it is lowered as the project gets under way. It becomes lowest at the last stage of the project, and when all activities become critical, the float factor will equal zero.

The float factor must be greater than or equal to zero and is calculated using the formula

$$\text{float factor} = \frac{\Sigma \text{ total float of all activities in progress or not started}}{\Sigma \text{ activity duration remaining or not started}}$$

At the project beginning stage, a float factor of 3 is recommended. Based on Figures 1 and 2 showing arrow diagram and precedence diagram networks, the project float factor at the beginning stage is obtained from the equation

$$\text{float factor} = \frac{56}{47}$$
$$= 1.191$$

The float factor also represents the number of float time units for each time unit of an activity of the project. For the sample project, there are 1.191 days of float for each day of a project activity.

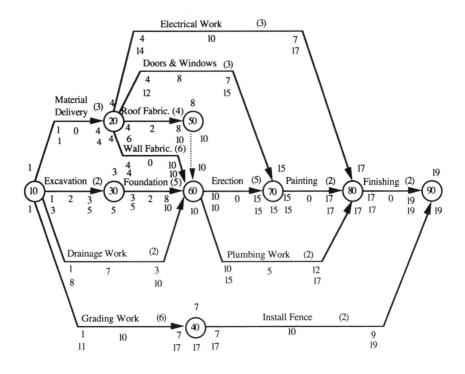

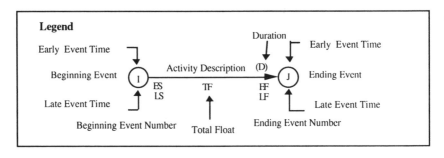

NOTE: The schedule computation in time units is based on the following assumption:

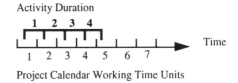

Activity Duration

Project Calendar Working Time Units

Activity with 4 time units starts at working time unit 1 and is finished at working time unit 5.

Figure 1 Sample Arrow Diagram Network

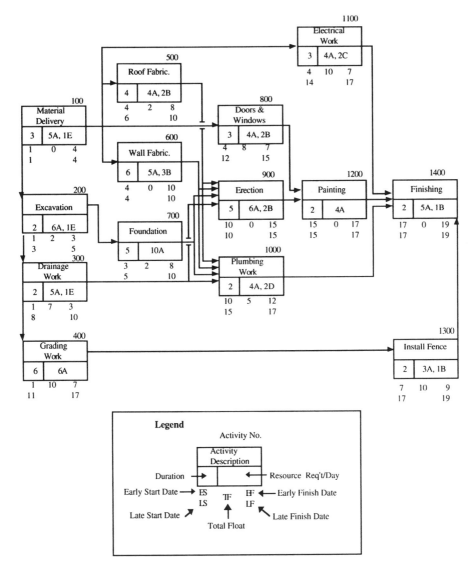

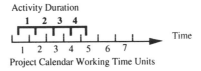

NOTE: The schedule computation in time units is based on the following assumption:

Activity Duration

Time

Project Calendar Working Time Units

Activity with 4 time units starts at working time unit 1 and is finished at working time unit 5.

Figure 2 Sample Precedence Diagram Network

FORWARD PASS COMPUTATION The network calculation that determines the earliest start and finish times, *(see Early start date* and *Early finish date)* for each activity. These calculations begin with the starting activities or events flowing sequentially in increasing time units until the ending activities or events are reached.

In a critical path method (CPM), the forward pass computation is performed after the logic diagram has been drawn and the activity duration has been estimated. The purpose of the forward pass computation is to find the earliest start (ES) and earliest finish (EF) times for all activities and events in the network. The computation is started by computing each activity's early start and finish times under the assumption that all activities start as soon as possible after all preceding activities have been completed.

The early start of the beginning activity is assumed to be equal to the first working day. The calculation proceeds from left to right or from the beginning to the end of the project. The earliest finish time of the last activity will be the earliest finish date or duration of the project. The following earliest start and finish times are calculated through the forward pass computation.

1. Earliest finish time (EF) = early start time (ES) + activity duration (D)
2. Earliest start time (ES) = maximum earliest finish of immediately preceding activity(s)

Figures 1 and 2 show arrow and precedence diagram networks using forward pass computation.

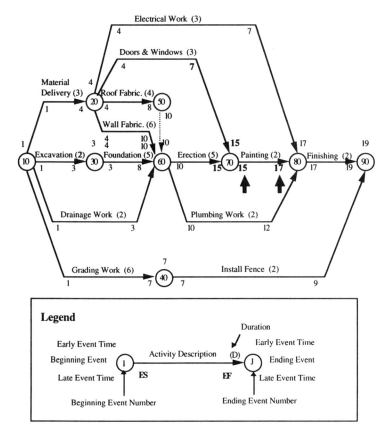

Note:

Activity: Painting 70-80

Earliest Start Time = 15 (Maximum Earliest Finish of the Immediately Preceding Activities)

Earliest Finish Time = ES + D = 15 + 2 = 17

Figure 1 Arrow Diagram Network with Forward Pass Computation

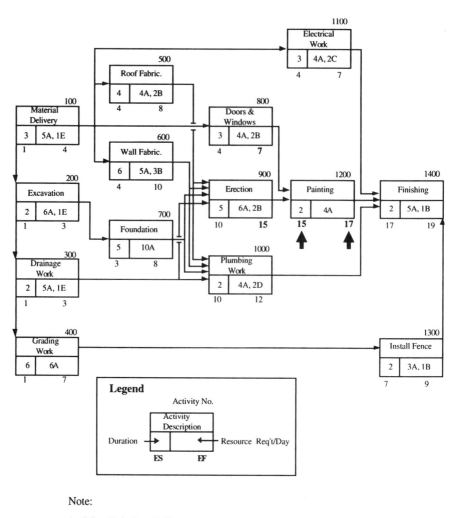

Note:

Activity: Painting 1200

Earliest Start Time = 15
(Maximum Earliest Finish of the Immediately Preceding Activities)

Earliest Finish Time = ES + D = 15 + 2 = 17

Figure 2 Precedence Diagram Network with Forward Pass Computation

FRAGNET (1) A portion of a large network, generally limited to a unique portion or area of responsibility of the project. (2) A portion of a large network used to demonstrate the impact of delays to a path or smaller portion of the entire network.

Fragnets are developed by subcontractors and integrated into a master schedule by those in charge of overall planning and scheduling. An arrow or precedence diagram may be used in developing the fragnets, but for the proper integration of fragnets and a master network, the same technique is recommended for all fragnets and for the master network.

In the construction industry, for complex projects, major subcontractors develop their subnetworks, which are later integrated into a master network. The typical contents of a complex building are site and substructure work, structural work, mechanical work, electrical work, finishing work, system startup and turnover, and so on.

To be computed on the same basis, fragnets or individual networks, as part of a larger network, need to be interconnected by common nodes, common activities, or logical constraints between them. A smaller portion of a network can also be detailed, which is called a fragment of a network, or a fragnet. Another use of fragnets is when a series of activities is scheduled to occur in a specific order, either continuously or from time to time. (It is not necessary that all activities be performed by the same organization.)

When a planning engineer identifies repetitive sequences of the same activities, time can be saved using fragnets instead of recreating them over and over. For example, for interior finishing operations in a multistory building with a similar floor plan for each story, it is wise to develop a fragnet (50 to 60 activities) for finishing a single floor and repeat it for each floor. Certainly, the number and movement of specialized trades (crews) from floor to floor must be considered. It is also recommended that fragnets be calculated and refined before being interconnected.

Most CPM software packages have the capability of storing various fragnets in a library of fragnets *(see Library network)* identified by an individual code, to be used for future projects. From the library, one can import a fragnet into a master project network and can modify activity-related data, such as description, duration, code, and so on.

A sample of a "construction fragnet" is a typical concrete pour in a complex reinforced concrete structure which requires hundreds of pourings. There is a precise sequence of pouring to meet the project logical sequence of concrete pouring operations. Figure 1 shows a typical fragnet for concrete pouring, and Figure 2 indicates the pouring sequence as dictated by the logical constraints.

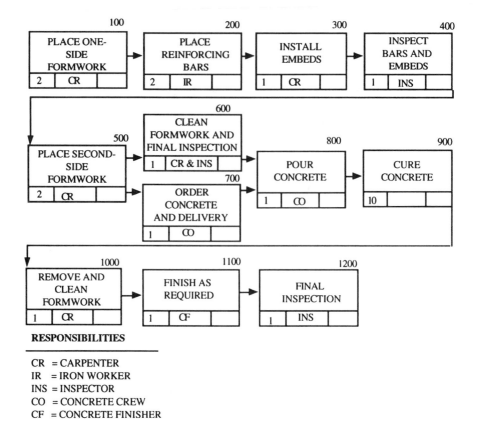

RESPONSIBILITIES

CR = CARPENTER
IR   = IRON WORKER
INS = INSPECTOR
CO = CONCRETE CREW
CF = CONCRETE FINISHER

Figure 1 Typical Concrete Pouring Fragnet

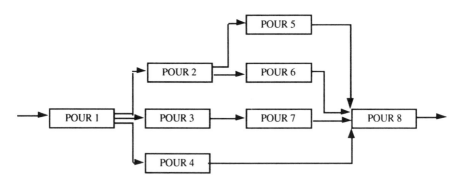

NOTE: EACH POUR CAN BE REPLACED WITH A DETAIL SEQUENCE OF OPERATIONS
AS INDICATED IN FIGURE 1, TYPICAL CONCRETE POURING FRAGNET

Figure 2 Concrete Pouring Sequence

FREE FLOAT (FF) OR FREE SLACK The amount of time by which the finish time of an activity may exceed its earliest finish time without increasing the earliest start time of any other activity immediately following.

Free float or free slack is the number of time units that allow an activity to be delayed without extending the early start of its immediate successors or completion of the project. It is equal to subtraction of the early finish time of the activity from the earliest early start time of its immediate successor(s). Free float can be equal to or less than total float, *(see Activity total float)* but it can never be greater than total float. Unlike total float, which belongs to all activities of the path, free float is an activity property that will not affect the schedule as long as the activity delay does not exceed its free float. Note that the free float of an activity can be positive only when the activity has more than one immediate successor; however, the reverse is not necessarily true.

The free float concept is not widely used nowadays since its meaning can be misleading in practice. As mentioned earlier, total float belongs to the path, whereas free float belongs to a particular activity. This can cause conflicts and can be demonstrated by Figures 1 and 2. From Figures 1 and 2, the activity Foundation has 2 days of free float and the path of the activities Excavation and Foundation has 2 days of total float. By applying the total float concept, the 2 days of total float will be shared between the two activities. If so, the number of float time units left for the activity Foundation will be less than 2 days which means that the activity Foundation no longer has 2 days as its free float.

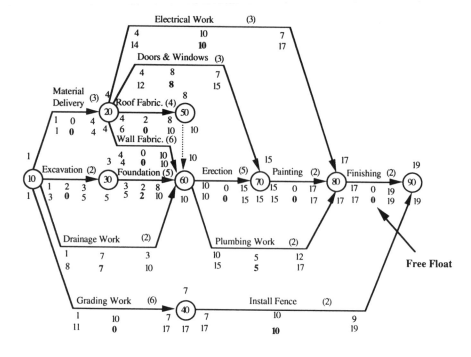

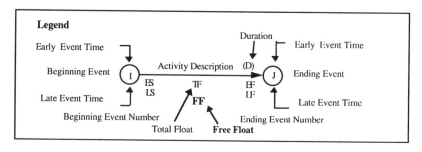

Legend

NOTE: The schedule computation in time units is based on the following assumption:

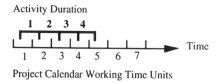

Activity Duration

Project Calendar Working Time Units

Activity with 4 time units starts at working time unit 1 and is finished at working time unit 5.

Figure 1 Activity Free Float Shown in Arrow Diagram

286

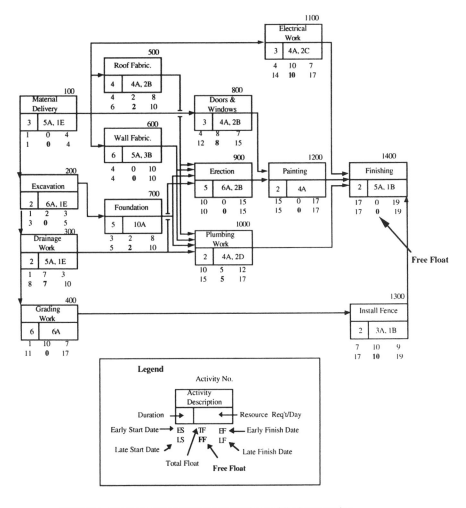

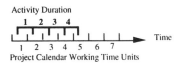

NOTE: The schedule computation in time units is based on the following assumption:

Activity with 4 time units starts at working time unit 1 and is finished at working time unit 5.

Figure 2 Activity Free Float Shown in Precedence Diagram

Figures 3 and 4 show activity free floats in schedule reports in arrow diagram and precedence diagram methods, respectively.

CIVIL ENGINEERING DEPARTMENT   PRIMAVERA PROJECT PLANNER

REPORT DATE 23NOV93 RUN NO.   9   PROJECT SCHEDULE REPORT   START DATE 17MAR93 FIN DATE 09APR93
            17:38
CLASSIC SCHEDULE REPORT   DATA DATE 17MAR93 PAGE NO.   1

ACTIVITY ID	ORIG DUR	REM DUR	CAL ID	%	CODE	ACTIVITY DESCRIPTION	EARLY START	EARLY FINISH	LATE START	LATE FINISH	TOTAL FLOAT	FREE FLOAT
10	20	3	3	1	0	MATERIAL DELIVERY	17MAR93	19MAR93	17MAR93	19MAR93	0	0
10	30	2	2	1	0	EXCAVATION	17MAR93	18MAR93	19MAR93	22MAR93	2	0
10	60	2	2	1	0	DRAINAGE WORK	17MAR93	18MAR93	26MAR93	29MAR93	7	7
10	40	6	6	1	0	GRADING WORK	17MAR93	24MAR93	31MAR93	07APR93	10	0
30	60	5	5	1	0	FOUNDATION	19MAR93	25MAR93	23MAR93	29MAR93	2	2
20	60	6	6	1	0	WALL FABRICATION	22MAR93	29MAR93	22MAR93	29MAR93	0	0
20	50	4	4	1	0	ROOF FABRICATION	22MAR93	25MAR93	24MAR93	29MAR93	2	0
20	70	3	3	1	0	DOORS AND WINDOWS	22MAR93	24MAR93	01APR93	05APR93	8	8
20	80	3	3	1	0	ELECTRICAL WORK	22MAR93	24MAR93	05APR93	07APR93	10	10
40	90	2	2	1	0	INSTALL FENCE	25MAR93	26MAR93	08APR93	09APR93	10	10
50	60	0	0	1	0	DUMMY 1	26MAR93	25MAR93	30MAR93	29MAR93	2	0
60	70	5	5	1	0	ERRECTION	30MAR93	05APR93	30MAR93	05APR93	0	0
60	80	2	2	1	0	PLUMBING WORK	30MAR93	31MAR93	06APR93	07APR93	5	5
70	80	2	2	1	0	PAINTING	06APR93	07APR93	06APR93	07APR93	0	0
80	90	2	2	1	0	FINISHING	08APR93	09APR93	08APR93	09APR93	0	0

Figure 3 Activity Free Float Shown in Tabular Format from Arrow Diagram

Note: Most existing CPM software packages calculate the early and late activity dates as follows:

- Early finish = early start + activity duration − 1 day.
- Early start of a successor = early finish of immediate predecessor + 1 day.
- Late start = late finish − activity duration + 1 day.
- Late finish of a predecessor = late start of direct successor − 1 day.
- First project day is day 1.

288

```
-------------------------------------------------------------------------------------------------
CIVIL ENGINEERING DEPARTMENT
                                        PRIMAVERA PROJECT PLANNER

REPORT DATE 23NOV93 RUN NO.   13
             17:32                      PROJECT SCHEDULE
                                                                      START DATE 13MAR93  FIN DATE 09APR93
CLASSIC SCHEDULE REPORT
                                                                      DATA DATE  17MAR93  PAGE NO.   1
-------------------------------------------------------------------------------------------------
```

ACTIVITY ID	ORIG DUR	REM DUR	CAL ID	%	CODE	ACTIVITY DESCRIPTION	EARLY START	EARLY FINISH	LATE START	LATE FINISH	TOTAL FLOAT	FREE FLOAT
100	3	3	1	0		MATERIAL DELIVERY	17MAR93	19MAR93	17MAR93	19MAR93	0	0
200	2	2	1	0		EXCAVATION	17MAR93	18MAR93	19MAR93	22MAR93	2	0
300	2	2	1	0		DRAINAGE WORK	17MAR93	18MAR93	26MAR93	29MAR93	7	7
400	6	6	1	0		GRADING WORK	17MAR93	24MAR93	31MAR93	07APR93	10	7
700	5	5	1	0		FOUNDATION	19MAR93	25MAR93	23MAR93	29MAR93	2	2
600	6	6	1	0		WALL FABRICATION	22MAR93	29MAR93	22MAR93	29MAR93	0	0
500	4	4	1	0		ROOF FABRICATION	22MAR93	25MAR93	24MAR93	29MAR93	2	2
800	3	3	1	0		DOORS & WINDOWS	22MAR93	24MAR93	01APR93	05APR93	8	8
1100	3	3	1	0		ELECTRICAL WORK	22MAR93	24MAR93	05APR93	07APR93	10	10
1300	2	2	1	0		INSTALL FENCE	25MAR93	26MAR93	08APR93	09APR93	10	10
900	5	5	1	0		ERECTION	30MAR93	05APR93	30MAR93	05APR93	0	0
1000	2	2	1	0		PLUMBING WORK	30MAR93	31MAR93	06APR93	07APR93	5	5
1200	2	2	1	0		PAINTING	06APR93	07APR93	06APR93	07APR93	0	0
1400	2	2	1	0		FINISHING	08APR93	09APR93	08APR93	09APR93	0	0

Figure 4 Activity Free Float Shown in Tabular Format from Precedence Diagram

Note: Most existing software packages calculate the early and late activity dates as follows:

- Early finish = early start + activity duration – 1 day.
- Early start of a successor = early finish of immediate predecessor + 1 day.
- Late start = late finish – activity duration + 1 day.
- Late finish of a predecessor = late start of direct successor – 1 day.
- First project day is day 1.

G

GANTT CHART Traditional bar chart scheduling popularized by Henry Gantt *(see Bar chart)*.

H

HAMMOCK ACTIVITY An aggregate or summary activity. A series or group of activities that spans the same events, is condensed as one summary activity, and is reported at the higher summary management level either for reporting purposes or for spreading resources or costs over a portion of or the entire project.

A hammock has no defined duration; its duration is calculated from the activities across which it spans. Figure 1 shows an arrow diagram containing two activities designated as hammock activities: Finishing Work, which spans events 20 and 90, and Structural Work, which spans events 10 and 80.

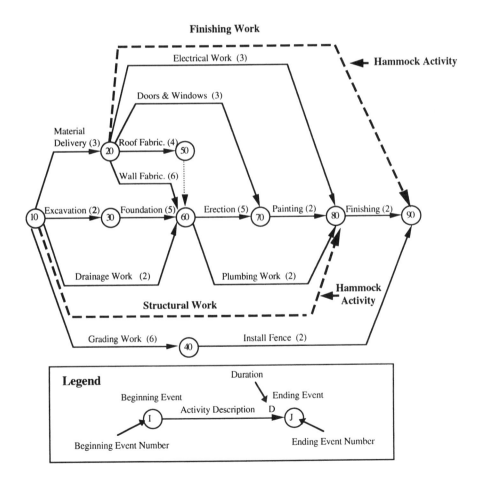

Figure 1 Two Hammock Activities in Arrow Diagram Network

The hammock activity does not affect the scheduled dates of activities across which it spans. Based on an algorithm for computing hammock activity duration, the hammock activity in an early start schedule is assumed to start at the earliest time of the preceding event and end at the earliest finish time of the ending event. The hammock activity in a late start schedule (scheduling all activities to start at the late start) is assumed to start at the latest time of the preceding event and to end at the latest finish time of the ending event.

The duration of a hammock activity is calculated after the dates of network activities and events have been determined. Figure 2 shows a computer-generated report for the arrow diagram containing hammock activities indicated in Figure 1. In a precedence diagram network (Figure 3), two hammock activities have been added to the network: activity 3000, Finishing Work, related to a start-to-start (SS) relation to activity 800, Doors and Windows, considered as a preceding activity and related to a finish-to-finish (FF) relation to activity 1400, Finishing, considered as a succeeding activity with a zero-time-unit delay.

PRED	SUCC	ORIG DUR	REM DUR	%	CODE	ACTIVITY DESCRIPTION	EARLY START	EARLY FINISH	LATE START	LATE FINISH	TOTAL FLOAT
10	20	3	3	0		MATERIAL DELIVERY	17MAR93	19MAR93	17MAR93	19MAR93	0
10	80	16*	16*	0		STRUCTURAL WORK	17MAR93	7APR93	17MAR93	7APR93	0
10	30	2	2	0		EXCAVATION	17MAR93	18MAR93	19MAR93	22MAR93	2
10	60	2	2	0		DRAINAGE WORK	17MAR93	18MAR93	26MAR93	29MAR93	7
10	40	6	6	0		GRADING WORK	17MAR93	24MAR93	31MAR93	7APR93	10
30	60	5	5	0		FOUNDATION	19MAR93	25MAR93	23MAR93	29MAR93	2
20	60	6	6	0		WALL FABRICATION	22MAR93	29MAR93	22MAR93	29MAR93	0
20	90	15*	15*	0		FINISHING WORK	22MAR93	9APR93	22MAR93	9APR93	0
20	50	4	4	0		ROOF FABRICATION	22MAR93	25MAR93	24MAR93	29MAR93*	2
20	70	3	3	0		DOORS AND WINDOWS	22MAR93	24MAR93	1APR93	5APR93	8
20	80	3	3	0		ELECTRICAL WORK	22MAR93	24MAR93	5APR93	7APR93	10
40	90	2	2	0		INSTALL FENCE	25MAR93	26MAR93	8APR93	9APR93	10
50	60	0	0	0		DUMMY 1	26MAR93	25MAR93	30MAR93	29MAR93	2
60	70	5	5	0		ERECTION	30MAR93	5APR93	30MAR93	5APR93*	0
60	80	2	2	0		PLUMBING WORK	30MAR93	31MAR93	6APR93	7APR93	5
70	80	2	2	0		PAINTING	6APR93	7APR93	6APR93	7APR93	0
80	90	2	2	0		FINISHING	8APR93	9APR93	8APR93	9APR93*	0

Figure 2 Tabular Schedule Report with Hammock Activities in Arrow Technique

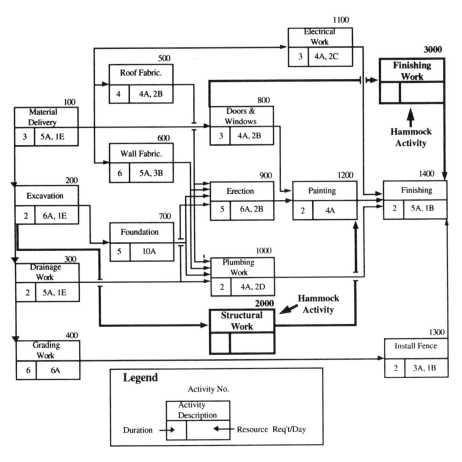

Figure 3 Two Hammock Activities in Precedence Diagram Network

Using the approach of designating a hammock activity, the early start of the preceding activity becomes the early start of the hammock activity and the early finish of the succeeding activity becomes the early finish of the hammock activity. Similarly, the late start of the preceding activity becomes the late start of the hammock activity and the late finish of the succeeding activity becomes the late finish of the hammock activity. Using this approach, a hammock activity with no preceding or succeeding activity is assumed to start at the data date *(see Data date)* or to end on the project early finish date.

Figure 4 shows a computer-generated report containing the hammock activities 2000 and 3000 and their calculated duration based upon early start of the preceding activity and early finish of the succeeding activity. Another approach to incorporating hammock activities into a precedence diagram is by indicating a start-to-start relation with no delay to all preceding activities and a finish-to-finish relation from the succeeding activity or activities. With this approach, it is not necessary for a hammock activity to have a succeeding activity.

CIVIL ENGINEERING DEPARTMENT   PRIMAVERA PROJECT PLANNER

REPORT DATE 24NOV93 RUN NO. 25   PROJECT SCHEDULE REPORT   START DATE 13MAR93 FIN DATE 9APR93
             2:53
CLASSIC SCHEDULE REPORT - SORT BY ES, TF   DATA DATE 17MAR93 PAGE NO. 1

ACTIVITY ID	ORIG DUR	REM DUR	%	CODE	ACTIVITY DESCRIPTION	EARLY START	EARLY FINISH	LATE START	LATE FINISH	TOTAL FLOAT
100	3	3	0		MATERIAL DELIVERY	17MAR93*	19MAR93	17MAR93*	19MAR93	0
2000	16*	16*	0		STRUCTURAL WORK	17MAR93	7APR93	19MAR93	7APR93	0
200	2	2	0		EXCAVATION	17MAR93	18MAR93	19MAR93	22MAR93	2
300	2	2	0		DRAINAGE WORK	17MAR93	18MAR93	26MAR93	29MAR93	7
400	6	6	0		GRADING WORK	17MAR93	24MAR93	31MAR93	7APR93	10
700	5	5	0		FOUNDATION	19MAR93	25MAR93	23MAR93	29MAR93	2
600	6	6	0		WALL FABRICATION	22MAR93	29MAR93	22MAR93	29MAR93	0
3000	15*	15*	0		FINISHING WORKS	22MAR93	9APR93	1APR93	9APR93	0
500	4	4	0		ROOF FABRICATION	22MAR93	25MAR93	24MAR93	29MAR93	2
800	3	3	0		DOORS & WINDOWS	22MAR93	24MAR93	1APR93	5APR93	8
1100	3	3	0		ELECTRICAL WORK	22MAR93	24MAR93	5APR93	7APR93	10
1300	2	2	0		INSTALL FENCE	25MAR93	26MAR93	8APR93	9APR93	10
900	5	5	0		ERECTION	30MAR93	5APR93	30MAR93	5APR93	0
1000	2	2	0		PLUMBING WORK	30MAR93	31MAR93	6APR93	7APR93	5
1200	2	2	0		PAINTING	6APR93	7APR93	6APR93	7APR93	0
1400	2	2	0		FINISHING	8APR93	9APR93	8APR93	9APR93	0

Figure 4 Tabular Schedule Report with
Hammock Activities in Precedence Technique

HANGER *(see Dangle)*

I

IMPACTED PROJECT SCHEDULE The schedule resulting from incorporating in a CPM schedule all project change orders, stop-work orders, disruptions, weather, and other impacts leading to project delay or acceleration. The revised schedule results from schedule modifications to incorporate change orders or other impacts leading to activity start or finish delay, project finish delay, duration changes, activity sequential changes, or earlier-than-planned activity start and completion dates.

During the planning stage, an as-planned project schedule *(see As-planned project schedule)* is developed to serve as a guideline for project execution. This schedule is, however, seldom followed exactly from start to completion of a project, for several reasons, including change orders, late material delivery, acts of God, and defective plans and specifications. As a result, the project schedule must be regularly updated to indicate the impact of unanticipated events in the overall project schedule. In the event of delay and acceleration claims, the impacted project schedule is important in identifying the party responsible for the delay.

When the as-planned project schedule incorporates all disturbing events, the revised schedule should identify the status of a project and its activities as being behind or ahead of schedule. The major cause of project delay is change orders, which normally prevent a project from completing on schedule. Change orders can also result in changes in work sequences or planned productivity for various trades. Identifying the impact of unanticipated events on a project schedule requires the integration of those events into the project schedule. For example, change order integration into a schedule requires the identification of change order time for preparation and direct work; then they can be added into a schedule to assess the impact.

Figure 1 shows the simple as-planned project schedule in a precedence diagram network as affected by change orders. The impact of two change orders on a project and on the activity start and completion dates is shown in Figure 2. The impact of change order 1 on activity 40 is to delay the start and completion dates for 4 days. Change order 2 causes a delay of 11 days in the start and completion dates of activity 50 and a delay of 6 days in the start and completion dates of activity 60, the project completion date.

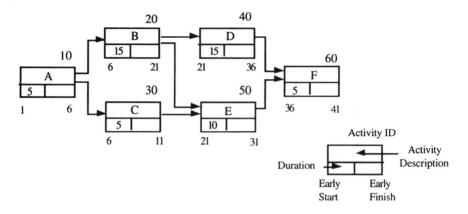

Figure 1 As-planned Project Schedule in Precedence Diagram

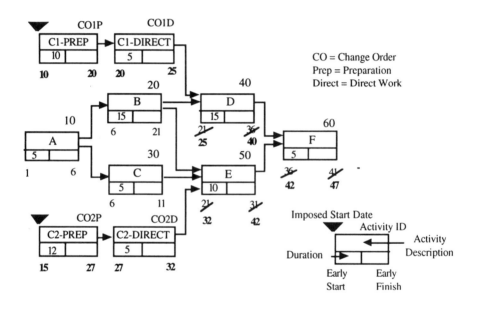

Figure 2 Impacted Project Schedule in Precedence Diagram

IMPOSED DATE *(see Contract date)*

INDEPENDENT FLOAT (INDF) The amount of delay that can be assigned to any one activity without delaying subsequent activities or restricting the scheduling of preceding activities.

INDF = earliest start of succeeding activity − latest finish of preceding activity − duration of subject activity

Independent float is the number of time units that allow an activity to be delayed and extend neither the completion of the project nor the early start of its immediate successors provided that all its immediate predecessors are finished on a late finish time. Independent float is equal to the earliest start of its immediate successor minus the latest finish of its immediate predecessor and its duration, as shown in the equation

independent float (Y) = early start date (Z) − late finish date (X) − duration (Y)

where activity X (X) immediately precedes activity Y (Y) and activity Y (Y) immediately precedes activity Z (Z).

The independent float cannot be greater than either the total float or the free float. Because of these limitations, the independent float, in many cases, results in a negative value. This situation is considered as having zero value, meaning that there is no independent float and it will not cause any delay in the project as negative values of total float.

Although the theory of independent float exists, many recent planning and scheduling publications do not mention this type of float. Furthermore, no existing scheduling software includes calculation of this type of float. Total float *(see Activity total float)* is generally covered and free float *(see Free float)* is mentioned occasionally.

IN-PROGRESS ACTIVITY An activity that has started but has not been completed by the data date (time now).

During project schedule updating *(see Monitoring)* an in-progress activity is recorded in the activity status report which collects the activity status from the site. An in-progress activity shows the actual start date before the data date *(see Data date)* and a remaining duration greater than zero. Another way to represent an in-progress activity is by the percent completion of that activity. Any activity with a percent completion greater than zero and less than 100 is considered as an in-progress activity.

Careful consideration must be given to an in-progress activity whose remaining duration plus its already spent duration is greater than its original duration, particularly an in-progress activity with lack of progress. There are many circumstances when the remaining duration (RD) of an in-progress activity has to be adjusted, such as for unforeseen events, an incorrect estimate, or inadequate resources.

In-progress activities can be recorded in different ways as shown in the following example containing two in-progress activities: Excavation, started on 03/17/93 with a duration to completion of 3 days, and Grading Work, started on 03/17/93 with a duration to completion of 4 days.

1. *Network diagram.* Figures 1 and 2 show arrow and precedence diagram networks with Excavation and Grading Work as in-progress activities.

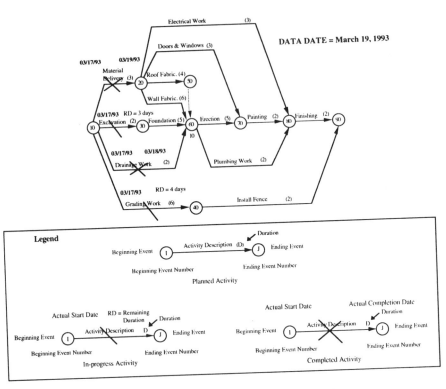

Figure 1 Updated Arrow Diagram Network

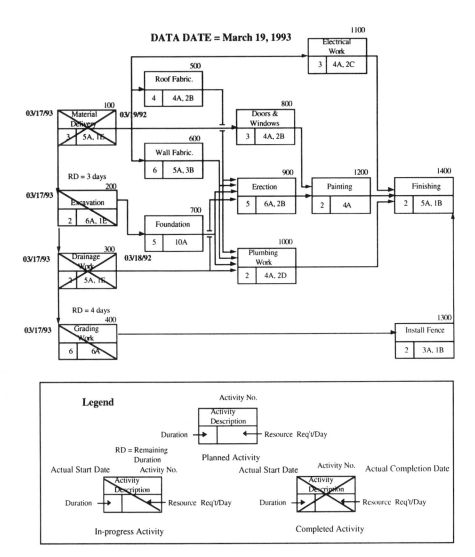

Figure 2 Updated Precedence Diagram Network

299

2. *Schedule report.* Figure 3 is a schedule report in tabular format with Excavation and Grading Work as in-progress activities.

CIVIL ENGINEERING DEPARTMENT   PRIMAVERA PROJECT PLANNER

REPORT DATE 23NOV93 RUN NO.   4   PROJECT SCHEDULE REPORT   START DATE 13MAR93 FIN DATE 12APR93
                21:54
CLASSIC SCHEDULE REPORT - SORT BY ES, TF   DATA DATE 19MAR93 PAGE NO.   1

ACTIVITY ID	ORIG DUR	REM DUR	%	CODE	ACTIVITY DESCRIPTION	EARLY START	EARLY FINISH	LATE START	LATE FINISH	TOTAL FLOAT
100	3	0	100		MATERIAL DELIVERY	17MAR93A	19MAR93A			
300	2	0	100		DRAINAGE WORK	17MAR93A	18MAR93A			
200	2	3	0		EXCAVATION	17MAR93A	23MAR93		23MAR93	0
400	6	4	33		GRADING WORK	17MAR93A	24MAR93		8APR93	11
500	7	7	0		ROOF FABRICATION	19MAR93	29MAR93	22MAR93	30MAR93	1
600	6	6	0		WALL FABRICATION	19MAR93	26MAR93	23MAR93	30MAR93	2
800	3	3	0		DOORS & WINDOWS	19MAR93	23MAR93	2APR93	6APR93	10
1100	3	3	0		ELECTRICAL WORK	19MAR93	23MAR93	6APR93	8APR93	12
700	5	5	0		FOUNDATION	24MAR93	30MAR93	24MAR93	30MAR93	0
1300	2	2	0		INSTALL FENCE	25MAR93	26MAR93	9APR93	12APR93	11
900	5	5	0		ERECTION	31MAR93	6APR93	31MAR93	6APR93	0
1000	2	2	0		PLUMBING WORK	31MAR93	1APR93	7APR93	8APR93	5
1200	2	2	0		PAINTING	7APR93	8APR93	7APR93	8APR93	0
1400	2	2	0		FINISHING	9APR93	12APR93	9APR93	12APR93	0

Figure 3 Updated Schedule Report (Precedence Diagram Network)

3. *Bar chart.* Figure 4 is a bar chart with Excavation and Grading Work as in-progress activities.

```
----------------------------------------------------------------------------------------------------
CIVIL ENGINEERING DEPARTMENT          PRIMAVERA PROJECT PLANNER

REPORT DATE 23NOV93  RUN NO.   5       PROJECT SCHEDULE REPORT                START DATE 13MAR93  FIN DATE 12APR93
       21:58
BAR CHART BY ES, EF, TF                                                      DATA DATE 19MAR93  PAGE NO.    1

                                                                                DAILY-TIME PER.    1
----------------------------------------------------------------------------------------------------
...........ACTIVITY DESCRIPTION.............      15    22    29    05    12    19    26    03    10    17
ACTIVITY ID  OD   RD  PCT   CODES   FLOAT  SCHEDULE  MAR   MAR   MAR   APR   APR   APR   APR   MAY   MAY   MAY
-----------  ----  ----  ---  ------------  -----    --------  93    93    93    93    93    93    93    93    93    93
----------------------------------------------------------------------------------------------------

DRAINAGE WORK                        CURRENT   . AA* .     .     .     .     .     .     .     .     .
       300    2   0 100              PLANNED   . EE* .     .     .     .     .     .     .     .     .

MATERIAL DELIVERY                    CURRENT   . AA* .     .     .     .     .     .     .     .     .
       100    3   0 100              PLANNED   . EEE .     .     .     .     .     .     .     .     .

EXCAVATION                           CURRENT   . AAE..EE    .     .     .     .     .     .     .     .
       200    2   3   0         0    PLANNED   . EE* .     .     .     .     .     .     .     .     .

GRADING WORK                         CURRENT   . AAE..EEE   .     .     .     .     .     .     .     .
       400    6   4  33        11    PLANNED   . EEE..EEE   .     .     .     .     .     .     .     .

DOORS & WINDOWS                      CURRENT   . E..EE    .     .     .     .     .     .     .     .
       800    3   3   0        10    PLANNED   . * EEE    .     .     .     .     .     .     .     .

ELECTRICAL WORK                      CURRENT   . E..EE    .     .     .     .     .     .     .     .
      1100    3   3   0        12    PLANNED   . * EEE    .     .     .     .     .     .     .     .

WALL FABRICATION                     CURRENT   . E..EEEEE  .     .     .     .     .     .     .     .
       600    6   6   0         2    PLANNED   . * EEEEE..E     .     .     .     .     .     .     .

ROOF FABRICATION                     CURRENT   . E..EEEE..E    .     .     .     .     .     .     .
       500    7   7   0         1    PLANNED   . * EEEE    .     .     .     .     .     .     .     .

FOUNDATION                           CURRENT   . *  . EEE..EE    .     .     .     .     .     .     .
       700    5   5   0         0    PLANNED   . E..EEEE   .     .     .     .     .     .     .     .

INSTALL FENCE                        CURRENT   . *  . EE   .     .     .     .     .     .     .     .
      1300    2   2   0        11    PLANNED   . *  . EE   .     .     .     .     .     .     .     .

PLUMBING WORK                        CURRENT   . *  .   . EE    .     .     .     .     .     .     .
      1000    2   2   0         5    PLANNED   . *  .   .EE    .     .     .     .     .     .     .

ERECTION                             CURRENT   . *  .   . EEE..EE    .     .     .     .     .     .
       900    5   5   0         0    PLANNED   . *  .   .EEEE..E    .     .     .     .     .     .

PAINTING                             CURRENT   . *  .     .  . EE   .     .     .     .     .     .
      1200    2   2   0         0    PLANNED   . *  .     .  .EE   .     .     .     .     .     .

FINISHING                            CURRENT   . *  .     .   . E..E   .     .     .     .     .
      1400    2   2   0         0    PLANNED   . *  .     .   . EE   .     .     .     .     .
```

Figure 4 Updated Bar Chart Report (Precedence Diagram Network)

301

INTERFACE ACTIVITY An activity connecting one subnetwork with another subnetwork, representing a logical or imposed interdependence.

A project may be comprised of many subprojects with various types of work, such as electrical or mechanical. In this circumstance, a master project schedule must contain several subproject schedules, and a logical constraint between subprojects can be represented by an interface activity. In construction a general contractor always has many specialty subcontractors working on one project; therefore, subcontractor schedules have to be incorporated with the general contractor schedule or master schedule. Another instance that requires an interface activity is when a project requires multiple prime contractors whose schedules must be tied together to represent an overall project schedule.

An interface activity linked between subprojects normally serves two purposes: (1) allowing an individual subproject to share common resources such as equipment and crews, and (2) allowing a project to schedule work sequencing between subprojects. An interface activity may be a dummy activity *(see Dummy activity)* with zero duration as shown in Figure 1, which imposes only logical constraints, or an actual activity as shown in Figure 2, such as moving a crawler crane or the transfer of crews from one subproject to another.

A network analysis of each subproject or project *(see Network analysis)* can be performed to recognize or ignore the constraint imposed by an interface activity. Networks can include one or more interface activities (Figures 3 and 4).

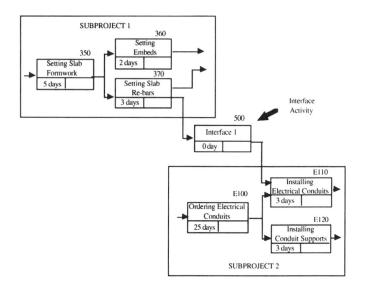

Figure 1 Interface Activity as Dummy Activity

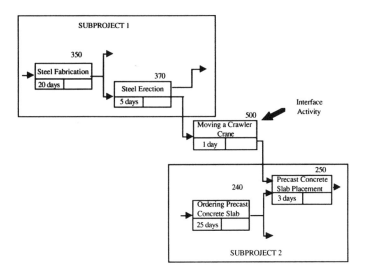

Figure 2 Interface Activity as Real Activity

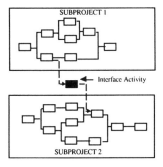

Figure 3 Networks with One Interface Activity

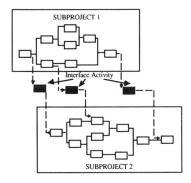

Figure 4 Networks with More Interface Activities

INTERFACE NODE (INTERFACE EVENT) A common node between two or more subnetworks representing a logical or imposed interdependence.

An interface node or event marks a definable point in time existing in more than one network or subnetwork. An interface node is required in an arrow diagram *(see Arrow diagramming method)* when a master project schedule contains several sub-project schedules from different subcontractors, such as electrical or mechanical. These subproject schedules need to be incorporated or tied together to represent a master project schedule. Another situation that requires an interface node between two or more networks is when a project requires multiple prime contractors whose schedules need to be coordinated. Figures 1 and 2 show two subprojects with one and more interface nodes.

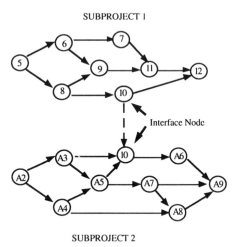

Figure 1 Two Subprojects (Networks) with One Interface Node

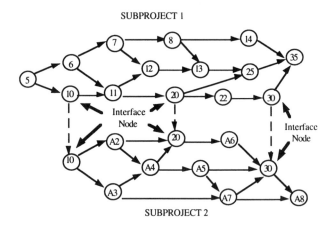

Figure 2 Two Subprojects (Networks) with Three Interface Nodes

304

An example of a work situation that requires an interface node is one in which the general contractor shows the activity of setting slab formwork (Figure 3). Then placement of electrical conduits by an electrical subcontractor is scheduled to be performed after the slab formwork is set. In this circumstance, an interface node is required to tie the general contractor's schedule into the electrical subcontractor's schedule.

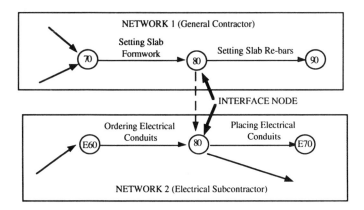

Figure 3 Sample of a Common Interface Node in Two Subprojects

INTERFERING FLOAT (IF) The amount of time by which the finish time of an activity will exceed its earliest finish time and its immediate successor(s) early start time but will not extend the project completion time.

Interfering float is the number of time units that allow an activity to be delayed without extending the completion of the project but will delay the early start of its immediate successors. It is equal to the subtraction of free float from total float (TF – FF) (Figure 1). The interfering float can be less than or equal to total float but never be greater than the total float. If an activity's interfering float is zero, it means that as soon as there is delay in the activity's immediate successor(s), it will absolutely lead to extension of the project completion time. If an activity's interfering float is equal to total float, which means that the activity's free float is zero, then as soon as the activity delays, its immediate successor(s) will also be delayed and the amount of interfering float indicates the maximum number of time units of delay that is permissible without extending the project completion time.

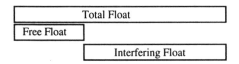

Figure 1 Interfering Float in Comparison with Free Float and Total Float

Theoretically, interfering float concept still exists; however, it is used very rarely in construction scheduling. Almost all scheduling software available in the current market does not include it as a calculation feature since the amount of the interfering float is obtained from the total float *(see Activity total float),* which is a property of all activities along the same path, and free float *(see Free float),* which is a property of an individual activity. Therefore, when the project is in progress, use of the interfering float can easily cause confusion. Figures 2 and 3 show the interfering float in arrow diagram and precedence diagram networks, respectively.

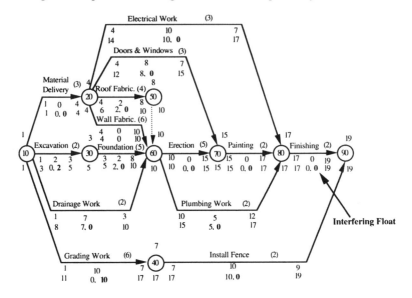

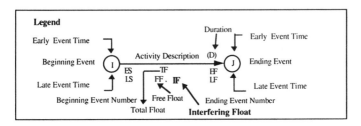

NOTE: The schedule computation in time units is based on the following assumption:

Project Calendar Working Time Units

Activity with 4 time units starts at working time unit 1 and is finished at working time unit 5.

Figure 2 Activity Interfering Float Shown in Arrow Diagram

306

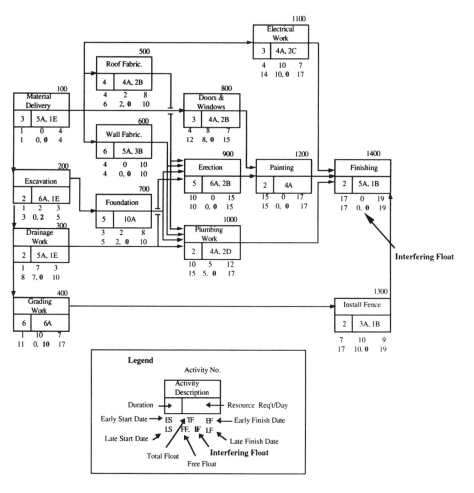

Legend

Activity No.

Activity Description

Duration → ← Resource Req't/Day

Early Start Date → ES TF EF ← Early Finish Date

LS FF, IF LF

Late Start Date ↗ ↘ Late Finish Date

Total Float **Interfering Float**

Free Float

Interfering Float

NOTE: The schedule computation in time units is based on the following assumption:

Activity Duration

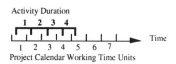

Time

1 2 3 4 5 6 7
Project Calendar Working Time Units

Activity with 4 time units starts at working time unit 1 and is finished at working time unit 5.

Figure 3 Activity Interfering Float Shown in Precedence Diagram

K

KEY ACTIVITY An activity occupying a place of strategic importance in the project network, such as an activity with numerous succeeding activities or a noncritical activity succeeding a critical activity. Key activities are not necessarily on the project critical path; however, they require careful attention during project execution.

After the project network is developed, key activities are normally defined to focus management attention during the project execution. A point in time when key activities start or are in-progress requires careful attention from management to closely monitor the progress status of those activities that usually are critical or nearly critical. When a key activity does not progress as planned, the project completion date has a tendency to be delayed or the original critical path is changed. If this situation arises, management has to take corrective action to accelerate the key activity or eliminate problems impeding its progress. The sample of key activities in a network is shown in the form of arrow and precedence diagram networks in Figures 1 and 2. The designated key activities are Erection and Finishing.

There are many reasons for management to define an activity as a key activity. A key activity may have one or more of the following characteristics:

1. Requirement for resources that are difficult to obtain (resource limitations)
2. Requirement for a long fabrication time for structural components
3. Requirement for high work quality that is difficult to achieve
4. Requirement for a special construction method or technique
5. Requirement for special equipment, machines, or tools that need to be mobilized from another job site
6. Requirement for coordination from many project participants (A/E, owner, specialty subcontractors, etc.)
7. Tight environmental or safety regulations
8. Critical or nearly critical activity

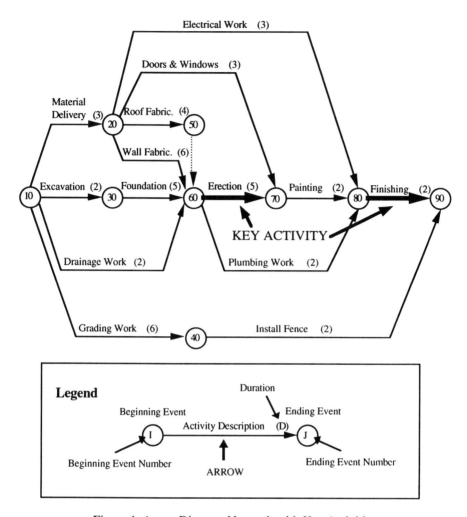

Figure 1 Arrow Diagram Network with Key Activities

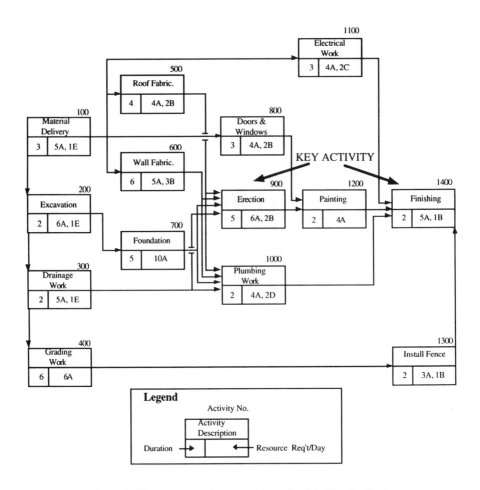

Figure 2 Precedence Diagram Network with Key Activities

KEY EVENT SCHEDULE *(see Milestone schedule)*

310

L

LADDERING A method of showing the logical relationship of a set of several parallel activities.

An example of a pipeline project showing the critical path in the arrow technique with several parallel activities using the laddering method shown in Figure 1. The activities Dig Trench, Lay Pipes (in trench), and Weld Pipes in Trench overlap at a given point in time. The relations between the start and completion of overlapping or parallel activities are shown with dummy activities, including time-unit delays, between the start or completion of the technologically related activities in the project. The laddering method can also be used to indicate the same managerial constraints in scheduling project activities in the arrow technique.

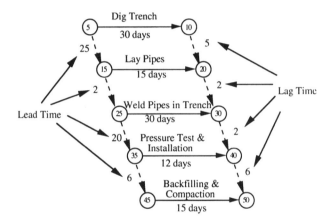

Figure 1 Pipeline Project (Arrow Technique)

LAG The specified time increment between the completion of an activity and the completion of its successor in a network diagram.

The lag time was used originally in the laddering technique *(see Laddering)* for imposing a technological or managerial delay between the finish of the predecessors and successors in the arrow diagram technique. Figure 1 illustrates the pipeline project with parallel activities in the laddering technique. The lag time between the finish of the activity Dig Trench (event 10) and the finish of the activity Lay Pipes (event 20) is shown as a dummy activity with a positive duration of 5 days. That means that the completion of the Lay Pipes activity cannot occur before at least 5 days elapsed after completion of the activity Dig Trench.

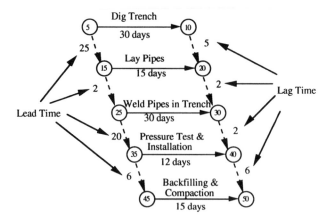

Figure 1 Pipeline Project (Arrow Technique)

In Figure 1, the following lag relations are indicated. Event 10–20 represents a 5-day delay, event 20–30 represents a 2-day delay, event 30–40 represents a 2-day delay, and event 40–50 represents a 6-day delay. These delays are the minimum estimated or imposed for managerial consideration. Every activity related to a lag relation is allowed to finish later than its predecessor's finish time plus lag time, but not earlier. The network computation is based on this consideration.

LATE COMPLETION SCHEDULE A schedule that indicates both an early finish date (project) and a late finish date (project) after the project finish date specified by the contract.

LATE DATE *(see Late start date* and *Late finish date)*

LATE EVENT TIME (LET) The latest time that an event may occur without increasing the project scheduled completion date.

In an arrow diagram network, a late event time, basically, can be used to identify an activity's late finish time *(see Late finish date)* except for activities with "not later than" contract dates *(see Contract date)*. The late event time is calculated through a backward pass calculation *(see Backward pass computation)* by subtracting the activity duration from the immediately succeeding event's late event time. In case there is more than one immediate succeeding event, the minimum number of the subtraction will be the event's late event time. In other words, the earliest late start time *(see Late start date)* of all direct successor activities will be the event's late event time.

Figure 1 illustrates the late event time calculation in the arrow diagram technique. In this example, event 60 is succeeded directly by the following activities:

60–70 with late start time = 15 – 5 = 10
60–80 with late start time = 17 – 2 = 15

The earlier late start time (10), which is from activity 60–70, will indicate the late event time for event 60.

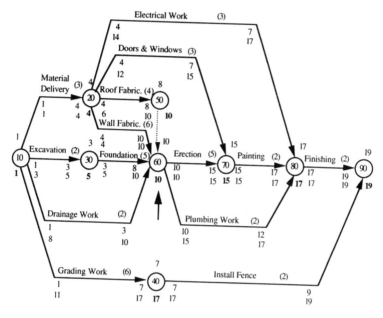

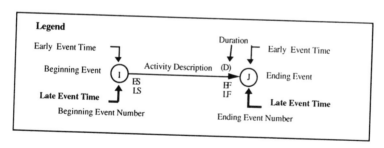

NOTE: The schedule computation in time units is based on the following assumption:

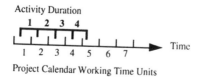

Project Calendar Working Time Units

Activity with 4 time units starts at working time unit 1 and is finished at working time unit 5.

Figure 1 Late Event Times Shown in Arrow Diagram Method

313

LATE FINISH DATE (ACTIVITY) (LF) The latest time at which an activity can be completed without affecting the project completion date.

The late finish date is computed from a backward pass calculation *(see Backward pass computation)* and is equal to the earliest late start date (LS) of immediately following activities. Also, the "not later than" contract date *(see Contract date)* and delay or lag of any activity must be considered for the late finish date computation. The latest allowable finish date for the end of the network is set equal to either an arbitrary scheduled completion time or the early finish date computed in the forward pass computation *(see Forward pass computation).*

Figures 1 to 8 show different situations for calculating the late finish date.

1. For an arrow diagram network, the late finish date of activity A is equal to the late event time *(see Late event time)* of event 50 (Figure 1).

$$LF(A) = LET \text{ of event } 50$$

For a precedence diagram network, the late finish date of activity A is equal to the late start date of activity B, the immediate successor (Figure 2).

$$LF(A) = LS(B) = 7$$

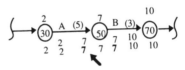

Figure 1 LF in Arrow Diagram

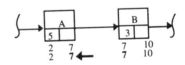

Figure 2 LF in Precedence Diagram

2. For an arrow diagram network, the late finish date of activity A is equal to the late event time *(see Late event time)* of event 30 as shown in Figure 3.

$$LF(A) = LET \text{ of event } 30 = 7$$

For a precedence diagram network, the late finish date of activity A is equal to the earliest late start date of immediate successors as shown in Figure 4. LF(A) is the minimum number among LS(B), LS(C), and LS(D).

$$LS(B) = 7; \quad LS(C) = 9; \quad LS(D) = 8$$

314

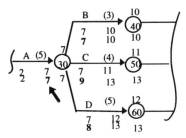

Figure 3 LF in Arrow Diagram

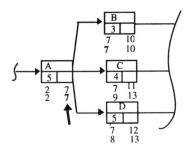

Figure 4 LF in Precedence Diagram

3. The late finish date of activity B is equal to 9, the "not later than" contract date, even if the LS of activity D, the immediate successor, is 12, as shown in Figures 5 and 6. LF(B) is the minimum between the "not later than" contract date for completing the activity and its direct successor's late finish date.

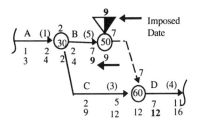

Figure 5 LF in Arrow Diagram

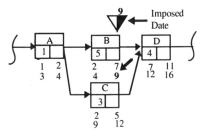

Figure 6 LF in Precedence Diagram

315

4. The late finish date of activity A is calculated from the dependency finish to finish *(see Dependency, finish to finish)*. Note that LF(A) is not the same as LS(B), although activity B is activity A's immediate successor (Figure 7).

$$LF(A) = LF(B) - lag = 10 - 2 = 8$$

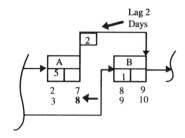

Figure 7 LF in Precedence Diagram

5. The late finish date of activity A will not be governed by the late start date of activity B because their relationship is start to start. Instead, the late finish date of activity A is equal to the late start date of activity C (Figure 8):

$$LF(A) = LS(C)$$

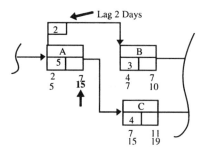

Figure 8 LF in Precedence Diagram

Figures 9 to 12 show the late finish dates of a sample construction network. Figure 9 shows an arrow diagram network, and Figure 10 is a computerized report related to the arrow diagram network. Figure 11 is a precedence diagram, and Figure 12 illustrates the computerized reports related to the precedence diagram.

316

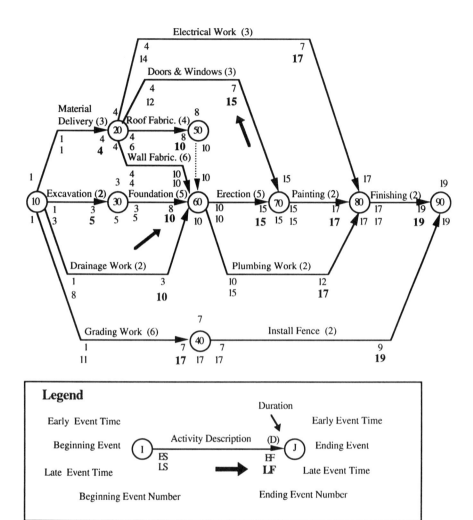

Electrical Work (3)

Doors & Windows (3)

Material Delivery (3)

Roof Fabric. (4)

Wall Fabric. (6)

Excavation (2)

Foundation (5)

Erection (5)

Painting (2)

Finishing (2)

Drainage Work (2)

Plumbing Work (2)

Grading Work (6)

Install Fence (2)

Legend

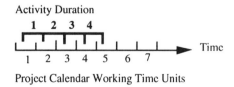

Duration

Early Event Time   Early Event Time

Beginning Event   Activity Description   (D)   Ending Event

Late Event Time   Late Event Time

Beginning Event Number   Ending Event Number

NOTE: The schedule computation in time units is based on the following assumption:

Activity Duration

Project Calendar Working Time Units

Activity with 4 time units starts at working time unit 1 and is finished at working time unit 5.

Figure 9 Late Finish Date in Arrow Diagram Network

CIVIL ENGINEERING DEPARTMENT   PRIMAVERA PROJECT PLANNER

REPORT DATE 23NOV93 RUN NO.   10   PROJECT SCHEDULE REPORT   START DATE 17MAR93 FIN DATE 9APR93
              17:51
CLASSIC SCHEDULE REPORT - SORT BY ES, TF   DATA DATE 17MAR93 PAGE NO.   1

PRED	SUCC	ORIG DUR	REM DUR	%	CODE	ACTIVITY DESCRIPTION	EARLY START	EARLY FINISH	LATE START	LATE FINISH	TOTAL FLOAT
10	20	3	3	0		MATERIAL DELIVERY	17MAR93*	19MAR93	17MAR93*	19MAR93	0
10	30	2	2	0		EXCAVATION	17MAR93	18MAR93	19MAR93	22MAR93	2
10	60	2	2	0		DRAINAGE WORK	17MAR93	18MAR93	26MAR93	29MAR93	7
10	40	6	6	0		GRADING WORK	17MAR93	24MAR93	31MAR93	7APR93	10
30	60	5	5	0		FOUNDATION	19MAR93	25MAR93	23MAR93	29MAR93	2
20	60	6	6	0		WALL FABRICATION	22MAR93	29MAR93	22MAR93	29MAR93	0
20	50	4	4	0		ROOF FABRICATION	22MAR93	25MAR93	24MAR93	29MAR93	2
20	70	3	3	0		DOORS AND WINDOWS	22MAR93	24MAR93	1APR93	5APR93	8
20	80	3	3	0		ELECTRICAL WORK	22MAR93	24MAR93	5APR93	7APR93	10
40	90	2	2	0		INSTALL FENCE	25MAR93	26MAR93	8APR93	9APR93	10
50	60	0	0	0		DUMMY 1	26MAR93	25MAR93	30MAR93	29MAR93	2
60	70	5	5	0		ERECTION	30MAR93	5APR93	30MAR93	5APR93	0
60	80	2	2	0		PLUMBING WORK	30MAR93	31MAR93	6APR93	7APR93	5
70	80	2	2	0		PAINTING	6APR93	7APR93	6APR93	7APR93	0
80	90	2	2	0		FINISHING	8APR93	9APR93	8APR93	9APR93	0

Figure 10 Late Finish Date in Tabular Format of Arrow Diagram Network

Note: Most existing software packages calculate the early and late activity dates as follows:

- Early finish = early start + activity duration – 1 day.
- Early start of a successor = early finish of immediate predecessor + 1 day.
- Late start = late finish – activity duration + 1 day.
- Late finish of a predecessor = late start of direct successor – 1 day.
- First project day is day 1.

318

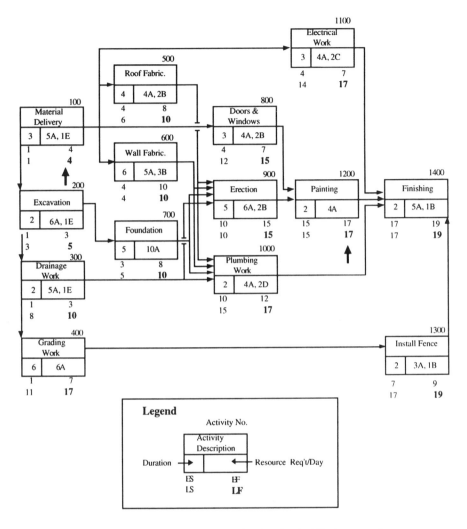

Legend

Activity No.

	Activity Description	
Duration →		← Resource Req't/Day
ES		EF
LS		**LF**

NOTE: The schedule computation in time units is based on the following assumption:

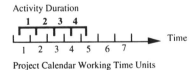

Activity Duration

1 2 3 4

Time

1 2 3 4 5 6 7

Project Calendar Working Time Units

Activity with 4 time units starts at working time unit 1 and is finished at working time unit 5.

Figure 11 Late Finish Date in Precedence Diagram Network

319

CIVIL ENGINEERING DEPARTMENT   PRIMAVERA PROJECT PLANNER

REPORT DATE 23NOV93 RUN NO.   14   PROJECT SCHEDULE REPORT   START DATE 13MAR93 FIN DATE 9APR93
           18:04
CLASSIC SCHEDULE REPORT - SORT BY ES, TF   DATA DATE 17MAR93 PAGE NO.   1

| ACTIVITY | ORIG | REM | | | ACTIVITY DESCRIPTION | EARLY | EARLY | LATE | LATE | TOTAL |
ID	DUR	DUR	%	CODE		START	FINISH	START	FINISH	FLOAT
100	3	3	0		MATERIAL DELIVERY	17MAR93*	19MAR93	17MAR93*	19MAR93	0
200	2	2	0		EXCAVATION	17MAR93	18MAR93	19MAR93	22MAR93	2
300	2	2	0		DRAINAGE WORK	17MAR93	18MAR93	26MAR93	29MAR93	7
400	6	6	0		GRADING WORK	17MAR93	24MAR93	31MAR93	7APR93	10
700	5	5	0		FOUNDATION	19MAR93	25MAR93	23MAR93	29MAR93	2
600	6	6	0		WALL FABRICATION	22MAR93	29MAR93	22MAR93	29MAR93	0
500	4	4	0		ROOF FABRICATION	22MAR93	25MAR93	24MAR93	29MAR93	2
800	3	3	0		DOORS & WINDOWS	22MAR93	24MAR93	1APR93	5APR93	8
1100	3	3	0		ELECTRICAL WORK	22MAR93	24MAR93	5APR93	7APR93	10
1300	2	2	0		INSTALL FENCE	25MAR93	26MAR93	8APR93	9APR93	10
900	5	5	0		ERECTION	30MAR93	5APR93	30MAR93	5APR93	0
1000	2	2	0		PLUMBING WORK	30MAR93	31MAR93	6APR93	7APR93	5
1200	2	2	0		PAINTING	6APR93	7APR93	6APR93	7APR93	0
1400	2	2	0		FINISHING	8APR93	9APR93	8APR93	9APR93	0

Figure 12 Late Finish Date in Tabular Format of Precedence Diagram Network

Note: Most existing CPM software packages calculate the early and late activity dates as follows:

- Early finish = early start + activity duration – 1 day.
- Early start of a successor = early finish of immediate predecessor + 1 day.
- Late start = late finish – activity duration + 1 day.
- Late finish of a predecessor = late start of direct successor – 1 day.
- First project day is day 1.

LATE FINISH DATE (PROJECT) (LFp) (1) The latest time at which a project is expected to be completed, based on a forward pass computation of the current schedule. (2) The completion date specified by the contract *(see Late completion schedule)*.

Normally, when a project does not have an imposed due date *(see Due date),* a late finish date of the project (LFp) is equal to the late finish date *(see Late finish date)* of end activity(s) *(see End activity)* of the project. Figures 1 and 2 illustrate a single-end-activity project network using the arrow diagram and precedence diagram techniques, respectively. Figures 3 and 4 show a computerized schedule report of a single-end-activity project as activity codes of the arrow diagram and precedence diagram techniques, respectively. In this case, the late finish date of the sample project is April 5, 1993, which is equal to the late finish date of the end activity (activity F).

320

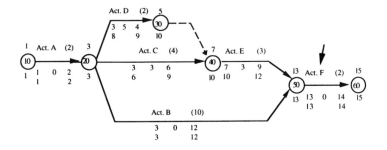

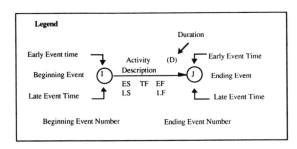

Figure 1 Arrow Diagram Network with Single End Activity

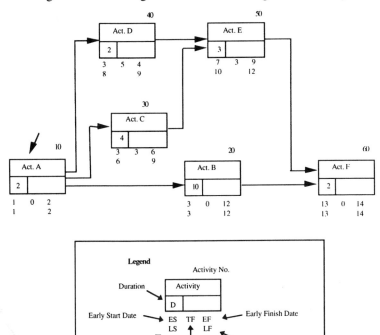

Figure 2 Precedence Diagram Network with Single End Activity

CIVIL ENGINEERING DEPARTMENT   PRIMAVERA PROJECT PLANNER

REPORT DATE 23NOV93 RUN NO.   4   PROJECT SCHEDULE REPORT   START DATE 17MAR93 FIN DATE 5APR93
              18:33
CLASSIC SCHEDULE REPORT - SORT BY ES, TF   DATA DATE 17MAR93 PAGE NO.   1

PRED	SUCC	ORIG DUR	REM DUR	%	CODE	ACTIVITY DESCRIPTION	EARLY START	EARLY FINISH	LATE START	LATE FINISH	TOTAL FLOAT
10	20	2	2	0		ACTIVITY A	17MAR93	18MAR93	17MAR93	18MAR93	0
20	50	10	10	0		ACTIVITY B	19MAR93	1APR93	19MAR93	1APR93	0
20	40	4	4	0		ACTIVITY C	19MAR93	24MAR93	24MAR93	29MAR93	3
20	30	2	2	0		ACTIVITY D	19MAR93	22MAR93	26MAR93	29MAR93	5
30	40	0	0	0		DUMMY	23MAR93	22MAR93	30MAR93	29MAR93	5
40	50	3	3	0		ACTIVITY E	25MAR93	29MAR93	30MAR93	1APR93	3
50	60	2	2	0		ACTIVITY F	2APR93	5APR93	2APR93	5APR93	0

Figure 3 Computerized Schedule Report (Single-End-Activity Project)
Sorted by Earliest Start, Arrow Diagram Technique

Note: Most existing CPM software packages calculate the early and late activity dates as follows:

- Early finish = early start + activity duration – 1 day.
- Early start of a successor = early finish of immediate predecessor + 1 day.
- Late start = late finish – activity duration + 1 day.
- Late finish of a predecessor = late start of direct successor – 1 day.
- First project day is day 1.

CIVIL ENGINEERING DEPARTMENT   PRIMAVERA PROJECT PLANNER

REPORT DATE 23NOV93 RUN NO.   2   PROJECT SCHEDULE REPORT   START DATE 17MAR93 FIN DATE 5APR93
              18:46
CLASSIC SCHEDULE REPORT - SORT BY ES, TF   DATA DATE 17MAR93 PAGE NO.   1

| ACTIVITY | ORIG | REM | | | ACTIVITY DESCRIPTION | EARLY | EARLY | LATE | LATE | TOTAL |
ID	DUR	DUR	%	CODE		START	FINISH	START	FINISH	FLOAT
10	2	2	0		ACTIVITY A	17MAR93	18MAR93	17MAR93	18MAR93	0
20	10	10	0		ACTIVITY B	19MAR93	1APR93	19MAR93	1APR93	0
30	4	4	0		ACTIVITY C	19MAR93	24MAR93	24MAR93	29MAR93	3
40	2	2	0		ACTIVITY D	19MAR93	22MAR93	26MAR93	29MAR93	5
50	3	3	0		ACTIVITY E	25MAR93	29MAR93	30MAR93	1APR93	3
60	2	2	0		ACTIVITY F	2APR93	5APR93	2APR93	5APR93	0

Figure 4 Computerized Schedule Report (Single-End-Activity Project)
Sorted by Earliest Start, Precedence Diagram Technique

Note: Most existing CPM software packages calculate the early and late activity dates as follows:

- Early finish = early start + activity duration – 1 day.
- Early start of a successor = early finish of immediate predecessor + 1 day.
- Late start = late finish – activity duration + 1 day.
- Late finish of a predecessor = late start of direct successor – 1 day.
- First project day is day 1.

For a project with multiple end activities without contract dates, their calculated late finish dates can be considered in two ways:

1. Each end activity's late finish date is equal to its early finish date.
2. All end activities' late finish dates will be the same and equal to the latest finish date.

Figures 5 and 6 illustrate a multiple-end-activity project network using the arrow diagram and precedence diagram techniques, respectively. The example project here has two end activities, Finishing and Install Fence, and the latest late finish date will be considered as the late finish date for all end activities (option 2). Figures 7 and 8 show computerized schedule reports of a multiple-end-activity project as activity codes of the arrow diagram and precedence diagram techniques, respectively. The late finish date of this sample project is April 9, 1993.

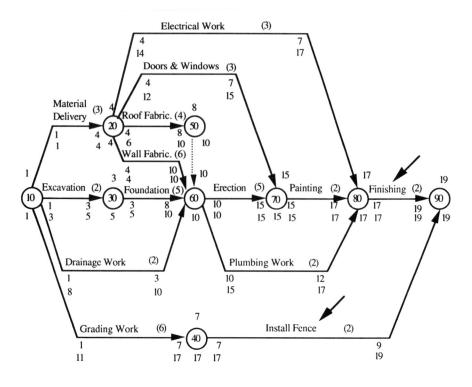

Electrical Work (3)

Doors & Windows (3)

Material Delivery (3)

Roof Fabric. (4)

Wall Fabric. (6)

Excavation (2)

Foundation (5)

Erection (5)

Painting (2)

Finishing (2)

Drainage Work (2)

Plumbing Work (2)

Grading Work (6)

Install Fence (2)

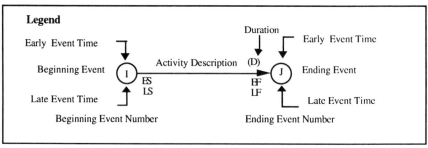

Legend

Duration

Early Event Time

Early Event Time

Beginning Event

Activity Description (D)

Ending Event

ES
LS

EF
LF

Late Event Time

Late Event Time

Beginning Event Number

Ending Event Number

NOTE: The schedule computation in time units is based on the following assumption:

Activity Duration

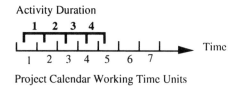

Time

Project Calendar Working Time Units

Activity with 4 time units starts at working time unit 1 and is finished at working time unit 5.

Figure 5 Arrow Diagram Network with Multiple End Activities

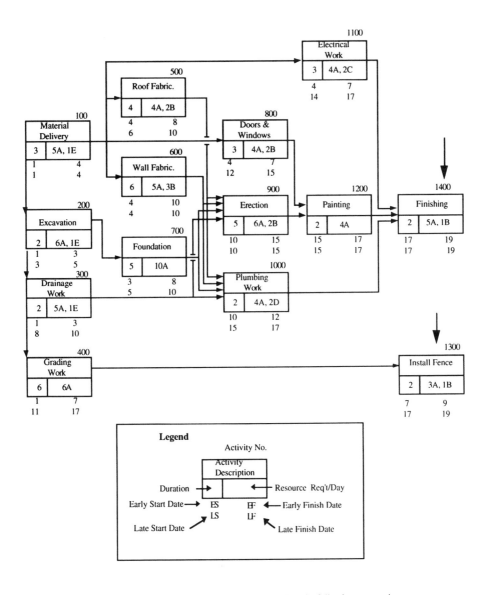

Legend

Activity No.

Activity
Description

Duration →

Early Start Date → ES   EF ← Early Finish Date
                    LS   LF
Late Start Date ↗   ↘ Late Finish Date

← Resource Req't/Day

NOTE: The schedule computation in time units is based on the following assumption:

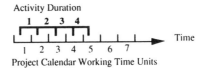

Activity Duration

1 2 3 4

Time

1 2 3 4 5 6 7
Project Calendar Working Time Units

Activity with 4 time units starts at working time unit 1 and is finished at working time unit 5.

Figure 6 Precedence Diagram Network with Multiple End Activities

PRED	SUCC	ORIG DUR	REM DUR	%	CODE	ACTIVITY DESCRIPTION	EARLY START	EARLY FINISH	LATE START	LATE FINISH	TOTAL FLOAT
10	20	3	3	0		MATERIAL DELIVERY	17MAR93*	19MAR93	17MAR93*	19MAR93	0
10	30	2	2	0		EXCAVATION	17MAR93	18MAR93	19MAR93	22MAR93	2
10	60	2	2	0		DRAINAGE WORK	17MAR93	18MAR93	26MAR93	29MAR93	7
10	40	6	6	0		GRADING WORK	17MAR93	24MAR93	31MAR93	7APR93	10
30	60	5	5	0		FOUNDATION	19MAR93	25MAR93	23MAR93	29MAR93	2
20	60	6	6	0		WALL FABRICATION	22MAR93	29MAR93	22MAR93	29MAR93	0
20	50	4	4	0		ROOF FABRICATION	22MAR93	25MAR93	24MAR93	29MAR93	2
20	70	3	3	0		DOORS AND WINDOWS	22MAR93	24MAR93	1APR93	5APR93	8
20	80	3	3	0		ELECTRICAL WORK	22MAR93	24MAR93	5APR93	7APR93	10
40	90	2	2	0		INSTALL FENCE	25MAR93	26MAR93	8APR93	9APR93	10
50	60	0	0	0		DUMMY 1	26MAR93	25MAR93	30MAR93	29MAR93	2
60	70	5	5	0		ERECTION	30MAR93	5APR93	30MAR93	5APR93	0
60	80	2	2	0		PLUMBING WORK	30MAR93	31MAR93	6APR93	7APR93	5
70	80	2	2	0		PAINTING	6APR93	7APR93	6APR93	7APR93	0
80	90	2	2	0		FINISHING	8APR93	9APR93	8APR93	9APR93	0

Figure 7 Computerized Schedule Report (Multiple-End-Activity Project) Sorted by Earliest Start, Arrow Diagram Technique

Note: Most existing CPM software packages calculate the early and late activity dates as follows:

- Early finish = early start + activity duration – 1 day.
- Early start of a successor = early finish of immediate predecessor + 1 day.
- Late start = late finish – activity duration + 1 day.
- Late finish of a predecessor = late start of direct successor – 1 day.
- First project day is day 1.

ACTIVITY ID	ORIG DUR	REM DUR	%	CODE	ACTIVITY DESCRIPTION	EARLY START	EARLY FINISH	LATE START	LATE FINISH	TOTAL FLOAT
100	3	3	0		MATERIAL DELIVERY	17MAR93*	19MAR93	17MAR93*	19MAR93	0
200	2	2	0		EXCAVATION	17MAR93	18MAR93	19MAR93	22MAR93	2
300	2	2	0		DRAINAGE WORK	17MAR93	18MAR93	26MAR93	29MAR93	7
400	6	6	0		GRADING WORK	17MAR93	24MAR93	31MAR93	7APR93	10
700	5	5	0		FOUNDATION	19MAR93	25MAR93	23MAR93	29MAR93	2
600	6	6	0		WALL FABRICATION	22MAR93	29MAR93	22MAR93	29MAR93	0
500	4	4	0		ROOF FABRICATION	22MAR93	25MAR93	24MAR93	29MAR93	2
800	3	3	0		DOORS & WINDOWS	22MAR93	24MAR93	1APR93	5APR93	8
1100	2	2	0		ELECTRICAL WORK	22MAR93	24MAR93	5APR93	7APR93	10
1300	2	2	0		INSTALL FENCE	25MAR93	26MAR93	8APR93	9APR93	10
900	5	5	0		ERECTION	30MAR93	5APR93	30MAR93	5APR93	0
1000	2	2	0		PLUMBING WORK	30MAR93	31MAR93	6APR93	7APR93	5
1200	2	2	0		PAINTING	6APR93	7APR93	6APR93	7APR93	0
1400	2	2	0		FINISHING	8APR93	9APR93	8APR93	9APR93	0

Figure 8 Computerized Schedule Report (Multiple-End-Activity Project) Sorted by Earliest Start, Precedence Diagram Technique

Note: Most existing CPM software packages calculate the early and late activity dates as follows:

- Early finish = early start + activity duration − 1 day.
- Early start of a successor = early finish of immediate predecessor + 1 day.
- Late start = late finish − activity duration + 1 day.
- Late finish of a predecessor = late start of direct successor − 1 day.
- First project day is day 1.

Figures 7 and 8 show the possible late completion time of the project on April 9, 1993. If the due date of the project is a "not later than" contract date *(see Contract date)* and is assigned to be later than the late completion time of the project (April 9, 1993). The late finish date of the project (LFp) will still be April 9, 1993. For instance, the scheduled report in Figure 9 shows that the due date is imposed on completion of the end activities—activity 1300, Install Fence, and activity 1400, Finishing—not later than April 13, 1993, later than the computed early finish dates. However, if, for example, the "not later than" contract due date is on April 7, 1993, the schedule will be recalculated and show activities with negative floats (Figure 10). This means that the as-planned project is very unlikely to be finished in time because even the early finish date of the end activity is later than the due date.

CIVIL ENGINEERING DEPARTMENT   PRIMAVERA PROJECT PLANNER

REPORT DATE 23NOV93   RUN NO.   4   PROJECT SCHEDULE REPORT   START DATE 17MAR93   FIN DATE 13APR93*

19:13

CLASSIC SCHEDULE REPORT - SORT BY ES, TF   DATA DATE   17MAR93   PAGE NO.   1

ACTIVITY ID	ORIG DUR	REM DUR	%	CODE	ACTIVITY DESCRIPTION	EARLY START	EARLY FINISH	LATE START	LATE FINISH	TOTAL FLOAT
100	3	3	0		MATERIAL DELIVERY	17MAR93*	19MAR93	17MAR93*	19MAR93	0
200	2	2	0		EXCAVATION	17MAR93	18MAR93	23MAR93	24MAR93	4
300	2	2	0		DRAINAGE WORK	17MAR93	18MAR93	30MAR93	31MAR93	9
400	6	6	0		GRADING WORK	17MAR93	24MAR93	2APR93	9APR93	12
700	5	5	0		FOUNDATION	19MAR93	25MAR93	25MAR93	31MAR93	4
600	6	6	0		WALL FABRICATION	22MAR93	29MAR93	24MAR93	31MAR93	2
500	4	4	0		ROOF FABRICATION	22MAR93	25MAR93	26MAR93	31MAR93	4
800	3	3	0		DOORS & WINDOWS	22MAR93	24MAR93	5APR93	7APR93	10
1100	3	3	0		ELECTRICAL WORK	22MAR93	24MAR93	7APR93	9APR93	12
1300	2	2	0		INSTALL FENCE	25MAR93	26MAR93	12APR93	13APR93	12
900	5	5	0		ERECTION	30MAR93	5APR93	1APR93	7APR93	2
1000	2	2	0		PLUMBING WORK	30MAR93	31MAR93	8APR93	9APR93	7
1200	2	2	0		PAINTING	6APR93	7APR93	8APR93	9APR93	2
1400	2	2	0		FINISHING	8APR93	9APR93	12APR93	13APR93	2

Figure 9   Computerized Schedule Report (Multiple-End-Activity Project) with "No Later Than" Contract Date Later Than Late Finish Date of End Activities, Sorted by Earliest Start, Precedence Diagram Technique

Note: Most existing CPM software packages calculate the early and late activity dates as follows:

- Early finish = early start + activity duration − 1 day.

327

- Early start of a successor = early finish of immediate predecessor + 1 day.
- Late start = late finish – activity duration + 1 day.
- Late finish of a predecessor = late start of direct successor – 1 day.
- First project day is day 1.

CIVIL ENGINEERING DEPARTMENT   PRIMAVERA PROJECT PLANNER

REPORT DATE 23NOV93 RUN NO.   6   PROJECT SCHEDULE REPORT   START DATE 17MAR93 FIN DATE 7APR93*
            19:17
CLASSIC SCHEDULE REPORT - SORT BY ES, TF   DATA DATE 17MAR93 PAGE NO.   1

| ACTIVITY | ORIG | REM | | | ACTIVITY DESCRIPTION | EARLY | EARLY | LATE | LATE | TOTAL |
ID	DUR	DUR	%	CODE		START	FINISH	START	FINISH	FLOAT
100	3	3	0		MATERIAL DELIVERY	17MAR93*	19MAR93	15MAR93*	17MAR93	-2
200	2	2	0		EXCAVATION	17MAR93	18MAR93	17MAR93	18MAR93	0
300	2	2	0		DRAINAGE WORK	17MAR93	18MAR93	24MAR93	25MAR93	5
400	6	6	0		GRADING WORK	17MAR93	24MAR93	29MAR93	5APR93	8
700	5	5	0		FOUNDATION	19MAR93	25MAR93	19MAR93	25MAR93	0
600	6	6	0		WALL FABRICATION	22MAR93	29MAR93	18MAR93	25MAR93	-2
500	4	4	0		ROOF FABRICATION	22MAR93	25MAR93	22MAR93	25MAR93	0
800	3	3	0		DOORS & WINDOWS	22MAR93	24MAR93	30MAR93	1APR93	6
1100	3	3	0		ELECTRICAL WORK	22MAR93	24MAR93	1APR93	5APR93	8
1300	2	2	0		INSTALL FENCE	25MAR93	26MAR93	6APR93	7APR93	8
900	5	5	0		ERECTION	30MAR93	5APR93	26MAR93	1APR93	-2
1000	2	2	0		PLUMBING WORK	30MAR93	31MAR93	2APR93	5APR93	3
1200	2	2	0		PAINTING	6APR93	7APR93	2APR93	5APR93	-2
1400	2	2	0		FINISHING	8APR93	9APR93	6APR93	7APR93	-2

Figure 10 Computerized Schedule Report (Multiple-End-Activity Project) with "No Later Than" Contract Date Earlier Than Late Finish Date of End Activities, Sorted by Earliest Start, Precedence Diagram Technique

Note: Most existing CPM software packages calculate the early and late activity dates as follows:

- Early finish = early start + activity duration – 1 day.
- Early start of a successor = early finish of immediate predecessor + 1 day.
- Late start = late finish – activity duration + 1 day.
- Late finish of a predecessor = late start of direct successor – 1 day.
- First project day is day 1.

LATE START DATE (ACTIVITY) (LS) The latest time at which an activity can start without lengthening the project:

Late start date = late finish date – activity duration

The late start date is computed by backward pass calculation *(see Backward pass computation)* and is equal to the late finish date *(see Late finish date)* minus the activity duration. Also, the "not later than" contract date *(see Contract date)* and the delay or lag of any activity must be considered for the late start date computation. For any activity the late start date cannot be earlier than the early start date.

Figures 1 to 7 show various situations for calculating the late start dates.

1. The late start date of activity B is equal to the late finish date minus its duration, as shown in Figures 1 and 2.

$$LS(B) = LF(B) - \text{duration}$$
$$= 10 - 3 = 7$$

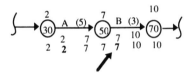

Figure 1 LS in Arrow Diagram

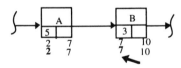

Figure 2 LS in Precedence Diagram

2. Activities B, C, and D have their own late start dates, obtained by deducting their respective durations from their late finish dates (Figures 3 and 4).

$$LS(B) = LF(B) - \text{duration} = 10 - 3 = 7$$
$$LS(C) = LF(C) - \text{duration} = 13 - 4 = 9$$
$$LS(D) = LF(D) - \text{duration} = 13 - 5 = 8$$

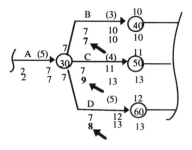

Figure 3 LS in Arrow Diagram

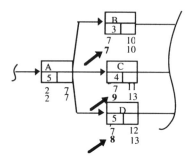

Figure 4 LS in Precedence Diagram

3. The late start date of activity B is 6 (the "not later than" contract date), even though subtraction of the duration from the late finish date is 7 (Figures 5 and 6).

Subtraction: LS(B) = LF(B) – duration = 12 – 5 = 7
"not later than" contract date = 6

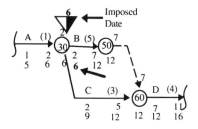

Figure 5 LS in Arrow Diagram

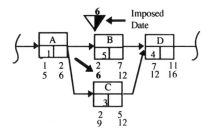

Figure 6 LS in Precedence Diagram

4. The late start date of activity B is 7 [LS(B) = LF(B) – duration = 10 – 3 = 7]. The late start date of activity A is the smaller number of the following:

$$LS(A) = LS(B) - lag = 7 - 2 = 5$$

or

$$LS(A) = LF(A) - duration = 15 - 5 = 10$$

Hence the late start date of activity A is 5, as shown in Figure 7.

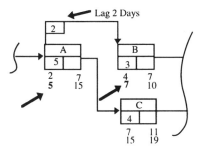

Figure 7 LS in Precedence Diagram

Figures 8 to 11 show the late start dates of sample construction networks. Figure 8 is an arrow diagram network, and Figure 9 shows a computerized report related to the arrow diagram. Figure 10 illustrates a precedence diagram network, and Figure 11 is a computerized report related to the precedence diagram.

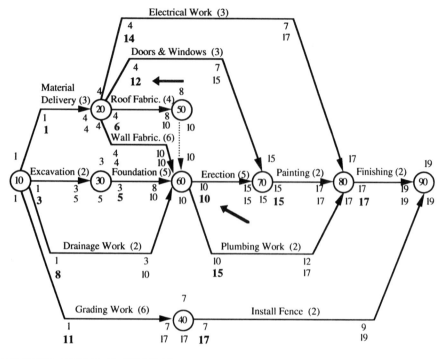

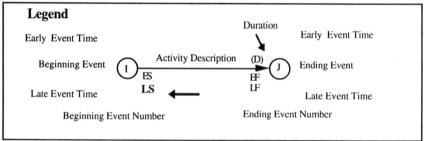

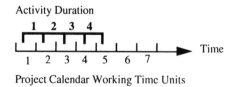

Legend

Early Event Time

Beginning Event   (I) — Activity Description — (D) (J)   Ending Event

ES   EF

Late Event Time   **LS** ←   LF   Late Event Time

Duration

Early Event Time

Beginning Event Number   Ending Event Number

NOTE: The schedule computation in time units is based on the following assumption:

Activity Duration

1 2 3 4

Time

1 2 3 4 5 6 7

Project Calendar Working Time Units

Activity with 4 time units starts at working time unit 1 and is finished at working time unit 5.

Figure 8 Late Start Date in Arrow Diagram Network

CIVIL ENGINEERING DEPARTMENT   PRIMAVERA PROJECT PLANNER

REPORT DATE 23NOV93 RUN NO.   10   PROJECT SCHEDULE REPORT   START DATE 17MAR93 FIN DATE 9APR93
     17:51
CLASSIC SCHEDULE REPORT - SORT BY ES, TF   DATA DATE 17MAR93 PAGE NO.   1

PRED	SUCC	ORIG DUR	REM DUR	%	CODE	ACTIVITY DESCRIPTION	EARLY START	EARLY FINISH	LATE START	LATE FINISH	TOTAL FLOAT
10	20	3	3	0		MATERIAL DELIVERY	17MAR93*	19MAR93	17MAR93*	19MAR93	0
10	30	2	2	0		EXCAVATION	17MAR93	18MAR93	19MAR93	22MAR93	2
10	60	2	2	0		DRAINAGE WORK	17MAR93	18MAR93	26MAR93	29MAR93	7
10	40	6	6	0		GRADING WORK	17MAR93	24MAR93	31MAR93	7APR93	10
30	60	5	5	0		FOUNDATION	19MAR93	25MAR93	23MAR93	29MAR93	2
20	60	6	6	0		WALL FABRICATION	22MAR93	29MAR93	22MAR93	29MAR93	0
20	50	4	4	0		ROOF FABRICATION	22MAR93	25MAR93	24MAR93	29MAR93	2
20	70	3	3	0		DOORS AND WINDOWS	22MAR93	24MAR93	1APR93	5APR93	8
20	80	3	3	0		ELECTRICAL WORK	22MAR93	24MAR93	5APR93	7APR93	10
40	90	2	2	0		INSTALL FENCE	25MAR93	26MAR93	8APR93	9APR93	10
50	60	0	0	0		DUMMY 1	26MAR93	25MAR93	30MAR93	29MAR93	2
60	70	5	5	0		ERECTION	30MAR93	5APR93	30MAR93	5APR93	0
60	80	2	2	0		PLUMBING WORK	30MAR93	31MAR93	6APR93	7APR93	5
70	80	2	2	0		PAINTING	6APR93	7APR93	6APR93	7APR93	0
80	90	2	2	0		FINISHING	8APR93	9APR93	8APR93	9APR93	0

Figure 9 Late Start Date in Tabular Format, Arrow Diagram Method

Note: Most existing CPM software packages calculate the early and late activity dates as follows:

- Early finish = early start + activity duration – 1 day.
- Early start of a successor = early finish of immediate predecessor + 1 day.
- Late start = late finish – activity duration + 1 day.
- Late finish of a predecessor = late start of direct successor – 1 day.
- First project day is day 1.

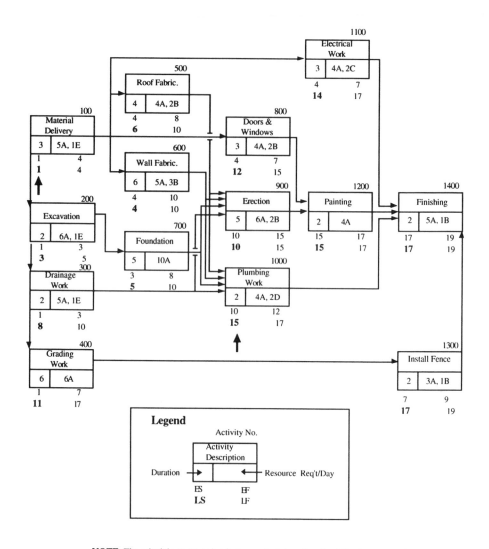

Legend

Activity No.

	Activity Description	
Duration →		← Resource Req't/Day
ES	EF	
LS	**LF**	

NOTE: The schedule computation in time units is based on the following assumption:

Activity Duration

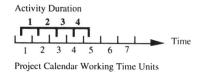

Project Calendar Working Time Units

Activity with 4 time units starts at working time unit 1 and is finished at working time unit 5.

Figure 10 Late Start Date in Precedence Diagram Network

CIVIL ENGINEERING DEPARTMENT PRIMAVERA PROJECT PLANNER

REPORT DATE 23NOV93 RUN NO. 14 PROJECT SCHEDULE REPORT START DATE 13MAR93 FIN DATE 9APR93
18:04

CLASSIC SCHEDULE REPORT - SORT BY ES, TF DATA DATE 17MAR93 PAGE NO. 1

ACTIVITY ID	ORIG DUR	REM DUR	%	CODE	ACTIVITY DESCRIPTION	EARLY START	EARLY FINISH	LATE START	LATE FINISH	TOTAL FLOAT
100	3	3	0		MATERIAL DELIVERY	17MAR93*	19MAR93	17MAR93*	19MAR93	0
200	2	2	0		EXCAVATION	17MAR93	18MAR93	19MAR93	22MAR93	2
300	2	2	0		DRAINAGE WORK	17MAR93	18MAR93	26MAR93	29MAR93	7
400	6	6	0		GRADING WORK	17MAR93	24MAR93	31MAR93	7APR93	10
700	5	5	0		FOUNDATION	19MAR93	25MAR93	23MAR93	29MAR93	2
600	6	6	0		WALL FABRICATION	22MAR93	29MAR93	22MAR93	29MAR93	0
500	4	4	0		ROOF FABRICATION	22MAR93	25MAR93	24MAR93	29MAR93	2
800	3	3	0		DOORS & WINDOWS	22MAR93	24MAR93	1APR93	5APR93	8
1100	3	3	0		ELECTRICAL WORK	22MAR93	24MAR93	5APR93	7APR93	10
1300	2	2	0		INSTALL FENCE	25MAR93	26MAR93	8APR93	9APR93	10
900	5	5	0		ERECTION	30MAR93	5APR93	30MAR93	5APR93	0
1000	2	2	0		PLUMBING WORK	30MAR93	31MAR93	6APR93	7APR93	5
1200	2	2	0		PAINTING	6APR93	7APR93	6APR93	7APR93	0
1400	2	2	0		FINISHING	8APR93	9APR93	8APR93	9APR93	0

Figure 11 Late Start Date in Tabular Format, Precedence Diagram Method

Note: Most existing CPM software packages calculate the early and late activity dates as follows:

- Early finish = early start + activity duration – 1 day.
- Early start of a successor = early finish of immediate predecessor + 1 day.
- Late start = late finish – activity duration + 1 day.
- Late finish of a predecessor = late start of direct successor – 1 day.
- First project day is day 1.

LATE START DATE (PROJECT) (LS_p) The latest time at which a project can be started, based on a forward pass calculation of the as-planned project schedule, and be completed within any project duration imposed by contract.

Since a project can have more than one start activity *(see Start activity)*, it is not necessary for the all start activities to start at the same start dates. A late start date of a project (LSp) will be equal to the earliest latest start date of the start activity(s) of the project.

Figures 1 and 2 illustrate a single-start-activity project network using the arrow diagram and precedence diagram techniques, respectively. Figures 3 and 4 show computerized schedule reports of a single-start-activity project as activity codes of the arrow diagram and precedence diagram techniques, respectively. The late start date of the project is March 17, 1993.

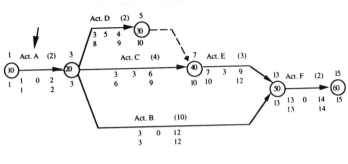

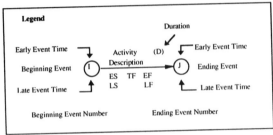

Figure 1 Arrow Diagram Network with Single Start Activity

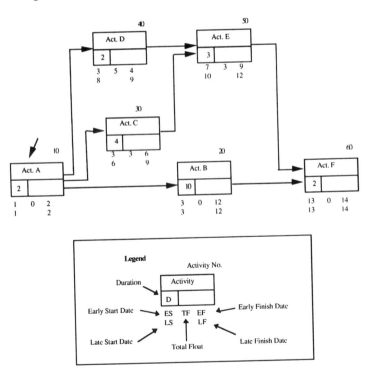

Figure 2 Precedence Diagram Network with Single Start Activity

336

REPORT DATE 23NOV93 RUN NO.   4   PROJECT SCHEDULE REPORT   START DATE 17MAR93   FIN DATE 5APR93
             18:33
CLASSIC SCHEDULE REPORT - SORT BY ES, TF   DATA DATE 17MAR93 PAGE NO.   1

		ORIG	REM			ACTIVITY DESCRIPTION	EARLY	EARLY	LATE	LATE	TOTAL
PRED	SUCC	DUR	DUR	%	CODE		START	FINISH	START	FINISH	FLOAT
10	20	2	2	0		ACTIVITY A	17MAR93	18MAR93	17MAR93	18MAR93	0
20	50	10	10	0		ACTIVITY B	19MAR93	1APR93	19MAR93	1APR93	0
20	40	4	4	0		ACTIVITY C	19MAR93	24MAR93	24MAR93	29MAR93	3
20	30	2	2	0		ACTIVITY D	19MAR93	22MAR93	26MAR93	29MAR93	5
30	40	0	0	0		DUMMY	23MAR93	22MAR93	30MAR93	29MAR93	5
40	50	3	3	0		ACTIVITY E	25MAR93	29MAR93	30MAR93	1APR93	3
50	60	2	2	0		ACTIVITY F	2APR93	5APR93	2APR93	5APR93	0

Figure 3 Computerized Schedule (Single-Start-Activity Project)
Sorted by Earliest Start, Arrow Diagram Technique

Note: Most existing CPM software packages calculate the early and late activity dates as follows:

- Early finish = early start + activity duration − 1 day.
- Early start of a successor = early finish of immediate predecessor + 1 day.
- Late start = late finish − activity duration + 1 day.
- Late finish of a predecessor = late start of direct successor − 1 day.
- First project day is day 1.

REPORT DATE 23NOV93 RUN NO.   2   PROJECT SCHEDULE REPORT   START DATE 17MAR93   FIN DATE 5APR93
             18:46
CLASSIC SCHEDULE REPORT - SORT BY ES, TF   DATA DATE 17MAR93 PAGE NO.   1

ACTIVITY	ORIG	REM			ACTIVITY DESCRIPTION	EARLY	EARLY	LATE	LATE	TOTAL
ID	DUR	DUR	%	CODE		START	FINISH	START	FINISH	FLOAT
10	2	2	0		ACTIVITY A	17MAR93	18MAR93	17MAR93	18MAR93	0
20	10	10	0		ACTIVITY B	19MAR93	1APR93	19MAR93	1APR93	0
30	4	4	0		ACTIVITY C	19MAR93	24MAR93	24MAR93	29MAR93	3
40	2	2	0		ACTIVITY D	19MAR93	22MAR93	26MAR93	29MAR93	5
50	3	3	0		ACTIVITY E	25MAR93	29MAR93	30MAR93	1APR93	3
60	2	2	0		ACTIVITY F	2APR93	5APR93	2APR93	5APR93	0

Figure 4 Computerized Schedule Report (Single-Start-Activity Project)
Sorted by Earliest Start, Precedence Diagram Technique

Note: Most existing CPM software packages calculate the early and late activity dates as follows:

- Early finish = early start + activity duration − 1 day.
- Early start of a successor = early finish of immediate predecessor + 1 day.
- Late start = late finish − activity duration + 1 day.
- Late finish of a predecessor = late start of direct successor −1 day.
- First project day is day 1.

Figures 5 and 6 illustrate a multiple-start-activity project network using the arrow diagram and precedence diagram techniques, respectively. The example project here has four start activities: Material Delivery, Excavation, Drainage Work, and Grading Work. Figures 7 and 8 show computerized schedule reports of a multiple-start-activity project as activity codes of the arrow diagram and precedence diagram techniques, respectively. Note that even though there are four start activities, the late start date of the activity Material Delivery is considered as the late start date of this sample project (March 17, 1993) because it is the earliest latest start date of the start activities.

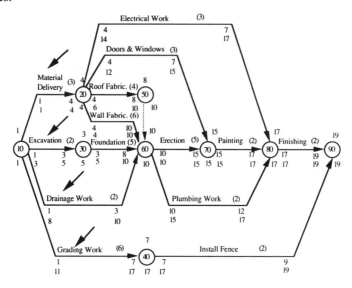

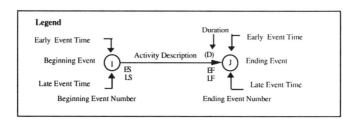

NOTE: The schedule computation in time units is based on the following assumption:

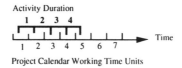

Project Calendar Working Time Units

Activity with 4 time units starts at working time unit 1 and is finished at working time unit 5.

Figure 5 Arrow Diagram Network with Multiple Start Activities

338

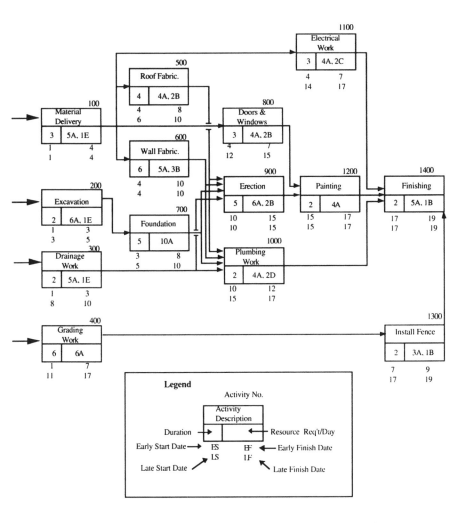

NOTE: The schedule computation in time units is based on the following assumption:

Activity Duration

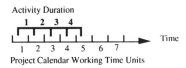

Activity with 4 time units starts at working time unit 1 and is finished at working time unit 5.

Figure 6 Precedence Diagram Network with Multiple Start Activities

339

CIVIL ENGINEERING DEPARTMENT   PRIMAVERA PROJECT PLANNER

REPORT DATE 23NOV93   RUN NO.   10   PROJECT SCHEDULE REPORT   START DATE 17MAR93 FIN DATE 9APR93
                      17:51
CLASSIC SCHEDULE REPORT - SORT BY ES, TF   DATA DATE 17MAR93 PAGE NO.   1

PRED	SUCC	ORIG DUR	REM DUR	%	CODE	ACTIVITY DESCRIPTION	EARLY START	EARLY FINISH	LATE START	LATE FINISH	TOTAL FLOAT
10	20	3	3	0		MATERIAL DELIVERY	17MAR93*	19MAR93	17MAR93*	19MAR93	0
10	30	2	2	0		EXCAVATION	17MAR93	18MAR93	19MAR93	22MAR93	2
10	60	2	2	0		DRAINAGE WORK	17MAR93	18MAR93	26MAR93	29MAR93	7
10	40	6	6	0		GRADING WORK	17MAR93	24MAR93	31MAR93	7APR93	10
30	60	5	5	0		FOUNDATION	19MAR93	25MAR93	23MAR93	29MAR93	2
20	60	6	6	0		WALL FABRICATION	22MAR93	29MAR93	22MAR93	29MAR93	0
20	50	4	4	0		ROOF FABRICATION	22MAR93	25MAR93	24MAR93	29MAR93	2
20	70	3	3	0		DOORS AND WINDOWS	22MAR93	24MAR93	1APR93	5APR93	8
20	80	3	3	0		ELECTRICAL WORK	22MAR93	24MAR93	5APR93	7APR93	10
40	90	2	2	0		INSTALL FENCE	25MAR93	26MAR93	8APR93	9APR93	10
50	60	0	0	0		DUMMY 1	26MAR93	25MAR93	30MAR93	29MAR93	2
60	70	5	5	0		ERECTION	30MAR93	5APR93	30MAR93	5APR93	0
60	80	2	2	0		PLUMBING WORK	30MAR93	31MAR93	6APR93	7APR93	5
70	80	2	2	0		PAINTING	6APR93	7APR93	6APR93	7APR93	0
80	90	2	2	0		FINISHING	8APR93	9APR93	8APR93	9APR93	0

Figure 7 Computerized Schedule Report (Multiple-Start-Activity Project)
Sorted by Earliest Start Date, Arrow Diagram Technique

Note: Most existing CPM software packages calculate the early and late activity dates as follows:

- Early finish = early start + activity duration – 1 day.
- Early start of a successor = early finish of immediate predecessor + 1 day.
- Late start = late finish – activity duration + 1 day.
- Late finish of a predecessor = late start of direct successor – 1 day.
- First project day is day 1.

340

CIVIL ENGINEERING DEPARTMENT   PRIMAVERA PROJECT PLANNER

REPORT DATE 23NOV93 RUN NO.   14   PROJECT SCHEDULE REPORT   START DATE 13MAR93 FIN DATE 9APR93
          18:04
CLASSIC SCHEDULE REPORT - SORT BY ES, TF   DATA DATE 17MAR93 PAGE NO.   1

ACTIVITY ID	ORIG DUR	REM DUR	%	CODE	ACTIVITY DESCRIPTION	EARLY START	EARLY FINISH	LATE START	LATE FINISH	TOTAL FLOAT
100	3	3	0		MATERIAL DELIVERY	17MAR93*	19MAR93	17MAR93*	19MAR93	0
200	2	2	0		EXCAVATION	17MAR93	18MAR93	19MAR93	22MAR93	2
300	2	2	0		DRAINAGE WORK	17MAR93	18MAR93	26MAR93	29MAR93	7
400	6	6	0		GRADING WORK	17MAR93	24MAR93	31MAR93	7APR93	10
700	5	5	0		FOUNDATION	19MAR93	25MAR93	23MAR93	29MAR93	2
600	6	6	0		WALL FABRICATION	22MAR93	29MAR93	22MAR93	29MAR93	0
500	4	4	0		ROOF FABRICATION	22MAR93	25MAR93	24MAR93	29MAR93	2
800	3	3	0		DOORS & WINDOWS	22MAR93	24MAR93	1APR93	5APR93	8
1100	3	3	0		ELECTRICAL WORK	22MAR93	24MAR93	5APR93	7APR93	10
1300	2	2	0		INSTALL FENCE	25MAR93	26MAR93	8APR93	9APR93	10
900	5	5	0		ERECTION	30MAR93	5APR93	30MAR93	5APR93	0
1000	2	2	0		PLUMBING WORK	30MAR93	31MAR93	6APR93	7APR93	5
1200	2	2	0		PAINTING	6APR93	7APR93	6APR93	7APR93	0
1400	2	2	0		FINISHING	8APR93	9APR93	8APR93	9APR93	0

Figure 8 Computerized Schedule Report (Multiple-Start-Activity Project)
Sorted by Earliest Start in Precedence Diagram Technique

Note: Most existing CPM software packages calculate the early and late activity dates as follows:

- Early finish = early start + activity duration – 1 day.
- Early start of a successor = early finish of immediate predecessor + 1 day.
- Late start = late finish – activity duration + 1 day.
- Late finish of a predecessor = late start of direct successor – 1 day.
- First project day is day 1.

From Figures 7 and 8, if the late start date of the project (LSp) is imposed as not later than March 18, 1993, which is later than the late start date of the activity Material Delivery, the late start date of the project will still be on March 17, 1993. On the other hand, if the late start date of the project is not to be later than March 16, 1993, the schedule should be revised and the early start date of the start activities should also be specified as not to be later than March 16, 1993 in the beginning.

LATEST REVISED ESTIMATE (LRE) The sum of the actual incurred costs plus the latest estimate to complete for a work package or summary item as currently reviewed and revised, or both (including applicable overhead where direct costs are specified).

The latest revised estimate is calculated by multiplying the actual quantities in place by the actual recorded up to the data date plus the remaining quantities to go by the forecasted unit price after the data date *(see Estimate at completion)*. The latest revised estimate is shown in Figure 1 using the project breakdown structure (PBS).

Data Date: Dec. 31, 1993

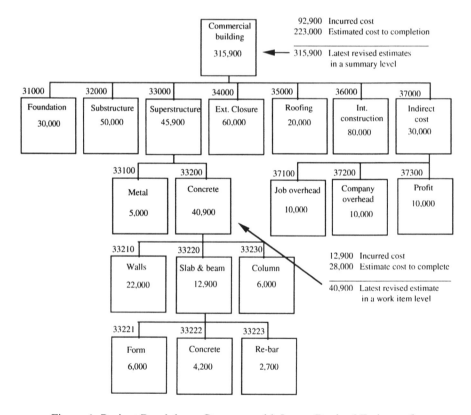

Figure 1 Project Breakdown Structure with Latest Revised Estimate for Work Items

LEAD (1) The specified time increment between the start of an activity and the start of its successor in a network diagram. (2) Nonnetwork activities, such as ordering, shop drawing preparation and review, custom manufacturing or fabrication, or shipping, which must occur prior to the start of a project activity.

Lead time is used in the laddering technique *(see Laddering)* for imposing a technological or managerial delay between the start of the predecessors and successors in the arrow diagram technique. Figure 1 illustrates the pipeline project with parallel activities in the laddering technique.

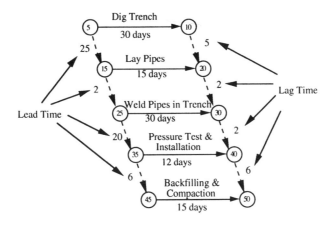

Figure 1 Pipeline Project (Arrow Technique)

The lead time between the start of the activity Dig Trench (event 5) and the start of the activity Lay Pipes (event 15) has been set at 25 time units (days). That means that activity Lay Pipes is not allowed to start earlier than 25 days after the start of the activity Dig Trench. This, in fact, will dictate when the Lay Pipes equipment and specialized crew are needed for the job. This represents the minimum time delay between the start of the two parallel or concurrent operations. If the project environment dictates that Lay Pipes start later than 15 days after the start of Dig Trench, this will reflect the progress report based on the CPM arrow technique.

It is possible that one activity starts to have lead-type relations and various lead times with several parallel activities. Originally, lead time could only be positive. However, present CPM software can handle both positive and negative lead times. Negative values are difficult to explain and monitor in real-life situations. In the precedence technique this type of relation is shown as a start-to-start relation with delay.

LEVEL The number of the level in the project breakdown structure at which a charge or summary number is assigned.

Level is a component of PBS *(see Project breakdown structure)*. The level refers to the management scope, which divides a project into clearly defined elements. For example, level 0 is for the end product of a project. Level 1 is for manageable elements of the project manager. Level 2 is for manageable elements of superintendents. Level 3 is for manageable elements of project engineers. Level 4 is for manageable elements of foremen. A maximum of seven levels is recommended to use in PBS. Also, a more detailed PBS is possible. Figure 1 indicates a PBS and its four levels for a hypothetical commercial building.

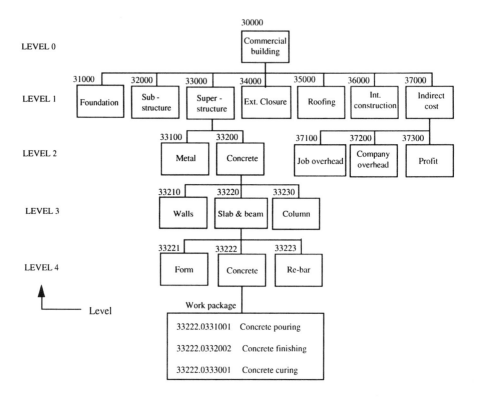

Figure 1 PBS and Its Levels for Commercial Building

344

LEVELING (RESOURCE) (1) The method of scheduling activities within their available float so as to minimize fluctuations in day-to-day resource requirements. (2) The procedure for scheduling project activities in CPM technique with no limit imposed on available resource(s) and to minimize resource(s) requirement fluctuations over a project's duration within the constraints imposed on project completion dates or interim milestone-imposed dates.

A systematic procedure for resource leveling has been developed by A. R. Burgess and J. B. Killebrew. This procedure is based on activity total float *(see Activity total float)* to shift (postpone) activities to reduce resource demand. The measure of effectiveness of the leveling procedure is to minimize the objective function selected, which is, in this example, to minimize the sum of the squared daily resource requirements. This leveling procedure is based on the following rules:

1. An early project completion date will not be postponed.
2. An activity started will continue until completion (no activity splitting).
3. Once an activity satisfies the maximum optimization function for a given project time unit, it will not be rescheduled (serial procedures).
4. Only one resource will be considered for leveling.

Figure 1 shows the Burgess–Killebrew flowchart algorithm for leveling resource requirements considering only one resource in the process.

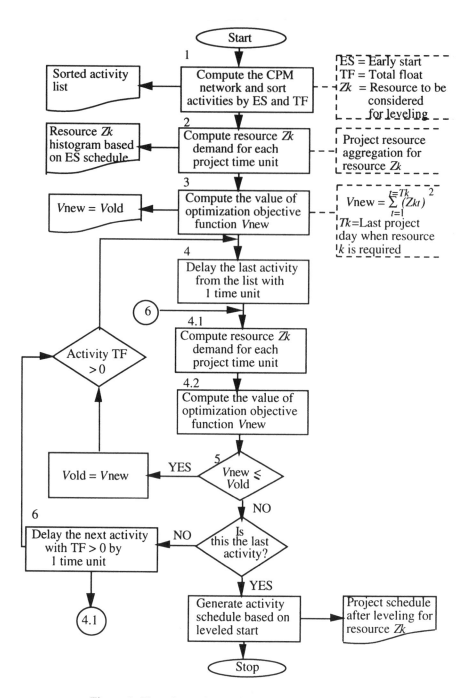

Figure 1 Flowchart of Algorithm for Resource Leveling

Step 1: Compute the CPM network and sort CPM activities in ascending order of early start and total float. (Resources required by project activities have been estimated.)

Step 2: Based on an early start schedule, compute the resource demand for each project time unit (in construction days).

Step 3: Compute the optimization objective function value selected for a resource histogram (e.g., sum of the squared resource requirements/time unit).

Step 4: Start with the last activity from the list (with total float) by shifting (delay) it one time unit and recalculate the value of the optimization function value.

Step 5: Compare the new value with the old value. If the new value < the old value and if the activity's total float > 0, delay the activity one more time unit and repeat the cycle until one or both conditions are not met.

Step 6: Move upward to the next-to-last activity from the list and repeat the process until all activities have been evaluated and/or no more improvements can be made in the objective function value.

LIBRARY NETWORK (STANDARD NETWORK DIAGRAM) A standard project or subproject network that can be amended separately or repeatedly to be included in a master network.

A library or standard network is a network diagram *(see Network)* that is created and recorded in a computer memory or hard copy. This type of network is intended to be used repetitively with some modifications in the network regarding project or activity data, such as duration, description, resources, and cost. A library network can be in a form of a subproject *(see Fragnet)* or a project.

Most of the time a general contractor or subcontractor performs the same type of work; as a result, a standard network should be developed and kept in a generic form such that it can be reused or customarily modified to meet the specific nature of a project. Each portion of the standard network can normally be extracted, allowing the extracted activities to be incorporated into a particular network. Most scheduling software allows a standard network to be extracted and modified globally to suit project requirements, such as changing activity duration or codes.

A sample of a standard network for a small house is shown in Figure 1. The contractual requirement is to build two houses. In this circumstance, the standard network is used to create a network for building two houses with interfacing relationships between them, as shown in Figure 2.

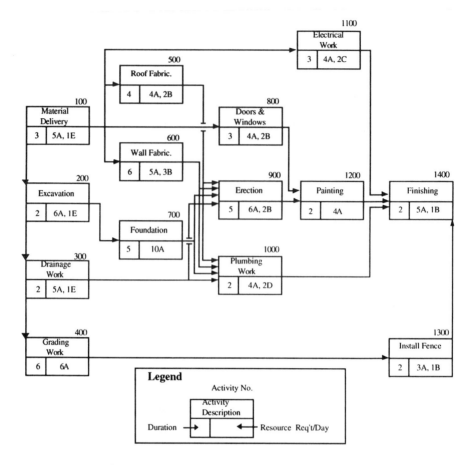

Figure 1 Standard Network in Precedence Diagram

348

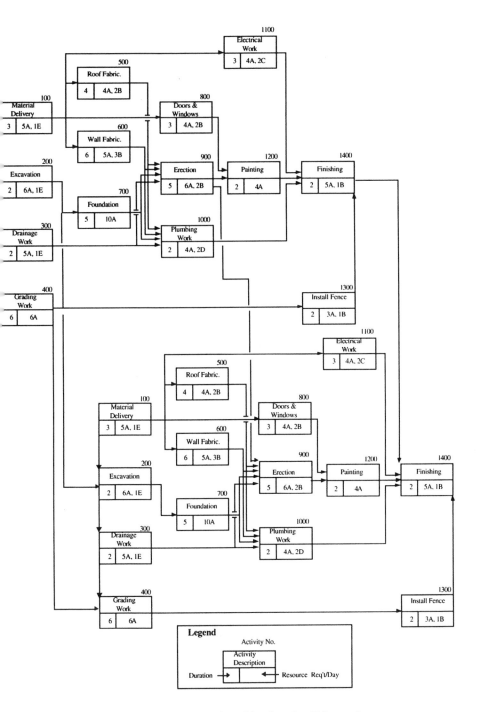

Figure 2 Network Developed by Standard Network

349

LINE OF BALANCE TECHNIQUE (L.O.B.) A precursor of the network analysis technique, which relates programs to forecast targets to determine how well a project is progressing. It is applicable to many similar repetitive subprojects (e.g., housing construction).

The line of balance technique (L.O.B.) was developed by the U.S. Navy in early 1960 to maintain control over production contracts. Line of balance was normally used for repetitive work units such as housing units in a large development, structural and finishing operations in a multistory building, and highway construction—in general for work that is linear in structure and repetitive in nature.
The line of balance technique uses four charts:

1. *Setback or sequence chart.* This chart shows the operations involved in a repetitive project in their proper order. The logic of the operations is developed on an end-of-time-scale basis. Figure 1 shows the dry-in operations over 16 working days for a low-cost wood-framed house.

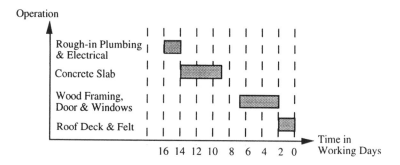

Figure 1 Setback Chart for Small Wood-Framed House Dry-in Operations

2. *Objective chart.* An objective chart is used to schedule the work. The rate of progress of similar units should meet contract requirements. Figure 2 shows the objective chart for completing 40 low-cost housing units over time (months) as a cumulative curve. The contract calls for completion of 5 houses by the end of the first month, 15 by the end of the second month, and 30 by the end of the third month.
On the same chart, the actual progress should be plotted, in a different color or pattern. Assume that at the end of the third month (data date), dry-in has been completed for only 20 houses. It is clear from Figure 2 that the project is behind schedule. The questions that can be addressed and answered with this technique are: What has caused this? What operation from the setback chart (Figure 1) needs more attention or acceleration to get back on schedule (original)?

350

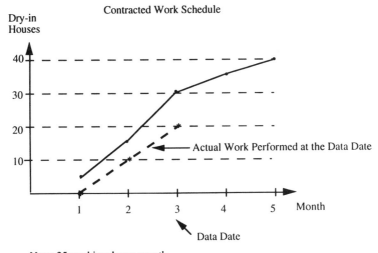

Contracted Work Schedule

Dry-in Houses

Actual Work Performed at the Data Date

Month

Data Date

Note: 25 working days a month

Figure 2 Objective Chart (Contract Requirements)

3. *Progress Chart.* This is a vertical bar chart indicating the original schedule for all activities indicated in the setback chart and actual progress at the data date (the third month, as indicated in Figure 2). Figure 3 shows the progress schedule for all activities in the low-cost housing project.

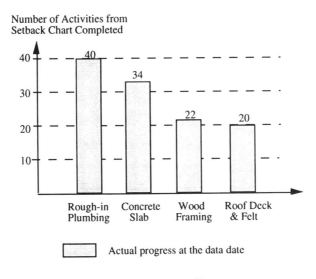

Number of Activities from Setback Chart Completed

Rough-in Plumbing   Concrete Slab   Wood Framing   Roof Deck & Felt

Actual progress at the data date

Figure 3 Progress Chart

351

4. *Line of balance.* An L.O.B. is drawn in stepped form on the progress chart shown in Figure 3. The level of the step for each activity is determined by using two charts: a setback chart (Figure 1), which provides the lead time to complete the project for each activity, and an objective chart (Figure 2), by measuring forward from the data date the lead time for each activity and using the vertical line until it intersects the cumulative contracted work schedule. From Figure 2 the Y axis represents activities to be completed at the data date to meet the schedule. The operation is repeated for all activities contained in the setback chart. Figure 4 shows the progress chart and the line of balance combined.

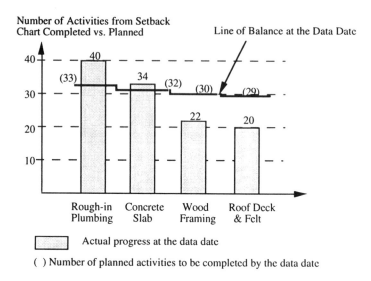

Figure 4 Progress Chart and Line of Balance Combined

A separate line of balance has to be drawn for each activity and every month. From the combination progress chart and line of balance for a given data date, the scheduled and actual dates can now identify activities progressing faster than schedule and those that are behind and require acceleration.

LINK *(see Dependency)*

LOGICAL RESTRAINT *(see Constraint)*

LOOP A circuitous and erroneous logic statement in CPM whereby an activity is shown to precede itself.

One of the most common network errors is when the sequences of work activities are arranged so that a continuous circle work sequence exists. This situation, called a loop, occurs when an activity imposes a logical constraint on an earlier activity, as shown in Figures 1 and 2. If a loop exists in a network, CPM-oriented software cannot schedule the project activities and an error and loop location will be detected which allows a planner to revise the network and break the loop.

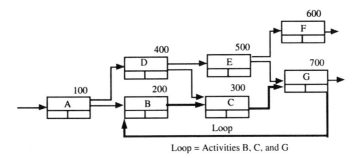

Loop = Activities B, C, and G

Figure 1 Loop in Arrow Diagram

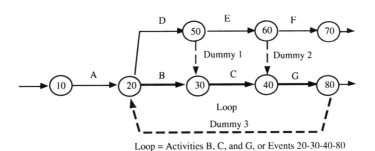

Loop = Activities B, C, and G, or Events 20-30-40-80

Figure 2 Loop in Precedence Diagram

A loop presents an illogical work sequence and prevents completion of the project schedule computation. Therefore, if detected, a loop must be eliminated. When a loop is presented in a network diagram whose activity data are entered in a computer, most scheduling software packages will stop the schedule computation and provide a report indicating a loop and the activities causing it, as shown in Figures 3 and 4.

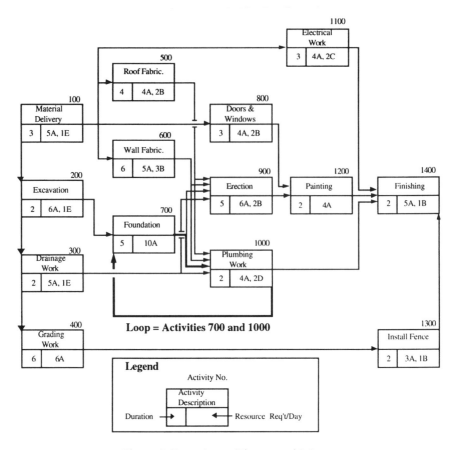

Figure 3 Precedence Diagram with Loop

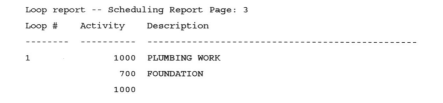

Figure 4 Computer Report Indicating Loop
(Primavera Project Planner 5.0)

Even though a loop is an obvious logical error in a network, when developing a complex network diagram, it may result in a loop or loops with inner loops. Minimizing a possibility of loop occurrence can be done by drawing a network diagram from far top left to bottom right and sequencing the activity identifiers or codes in ascending orders.

M

MASTER SCHEDULER (SCHEDULING ENGINEER) The job title of the person who manages the master project schedule. This person should be the best scheduler available as the consequences of planning and scheduling have a great impact on project performance. Ideally, the person should have substantial knowledge of the industry.

Planning and scheduling is one of the most critical factors for the success of a project. Therefore, a scheduling engineer should possess technical and managerial skills in order to generate the best possible plan and schedule for a project.
Duties of a scheduling engineer are:

1. Schedule development
   a. Develop a project plan and schedule that reflects the defined scope of work.
   b. Develop a quantity and labor expenditure rate for all project activities.
2. Progress monitoring
   a. Update a project schedule to assess planned progress against actual progress.
   b. Identify critical and nearly critical activities and report them to a project manager.
   c. Monitor schedule deviations and with appropriate project personnel, develop and recommend corrective actions.
3. Management coordination
   a. Present and answer questions on project plans and schedules at project meetings.
   b. Perform scheduling simulations as directed by project management.
   c. Work in concert with the project cost/scheduling team to ensure cost and schedule integration.

A scheduling engineer should have the following technical abilities:

1. Maintain a high level of technical expertise through continuing education programs.
2. Assist in the development and implementation of improved planning and scheduling techniques.
3. Prepare project planning and scheduling-related procedures and guidelines.
4. Assist in developing and recording historical scheduling information for use on future projects.
5. Understand networking and critical path analysis.
6. Know the CPM (critical path method) well.

A scheduling engineer should have the following managerial abilities:

1. Ask the right questions and accept the answer without bias.
2. Know how to ask questions.
3. Know how to listen, be objective, and be perceptive.
4. Be tenacious.
5. Be systematic and organized in collecting information and developing the logical sequence.
6. Understand the functions of project management levels and their needs in scheduling reports.

MILESTONE (KEY EVENT) (1) An event of major significance in achieving the program or project objectives. (2) A project activity or activities prior to the project end activity with a required completion date prior to the project late finish date.

These events may be designated as important delivery dates, a major phase of project completion or equipment installation, or a test for startup. These events may or may not be on the critical path. In Figure 1 the milestones are indicated. Usually, they have descriptions, alphanumeric or numeric codes similar to those of all other events on an arrow diagram and on contract dates (late or early times) reflected in contract documents.

1. Event 50 is declared a milestone, "Roof structure ready for erection."
2. Event 70 is marked as a milestone, "Building dry-in."
3. Event 90, "Project completed," is significant for project parties about the completion of the contractual requirements.

In arrow diagramming CPM it is recommended that the milestone be represented by a different symbol from that used for network events and that the milestone meaning (description) be shown. In the precedence technique (Figure 2), a milestone is represented by either the start or completion of an existing activity. Because the milestone is an event and because events do not have a duration, it is important to identify which event belongs to an activity to be declared as a milestone.

1. The end event of activity 500, Roof Fabrication, is declared a milestone, "Roof structure ready for erection."
2. The end event of activity 900, Erection, is declared a milestone, "Building dry-in."
3. The end event of activity 1400, Finishing, is declared a milestone, "Project completed."

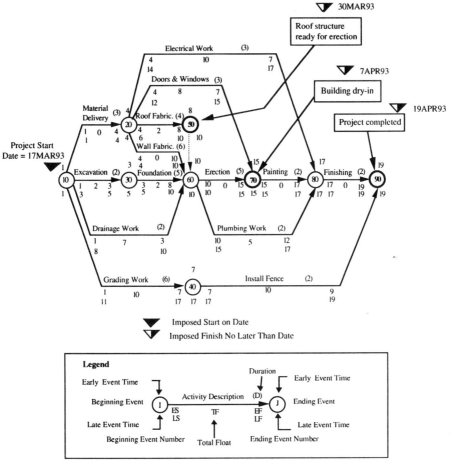

30MAR93

Roof structure
ready for erection

7APR93

Building dry-in

19APR93

Project completed

Imposed Start on Date
Imposed Finish No Later Than Date

Legend

Early Event Time

Beginning Event

Late Event Time

Beginning Event Number

Duration

Activity Description (D)

ES   TF   EF
LS   LF

Total Float

Early Event Time

Ending Event

Late Event Time

Ending Event Number

NOTE: The schedule computation in time units is based on the following assumption:

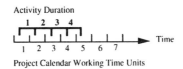

Activity Duration

1 2 3 4

Time

1 2 3 4 5 6 7

Project Calendar Working Time Units

Activity with 4 time units starts at working time unit 1 and is finished at working time unit 5.

Figure 1 Arrow Diagram Method with Milestones Flagged

357

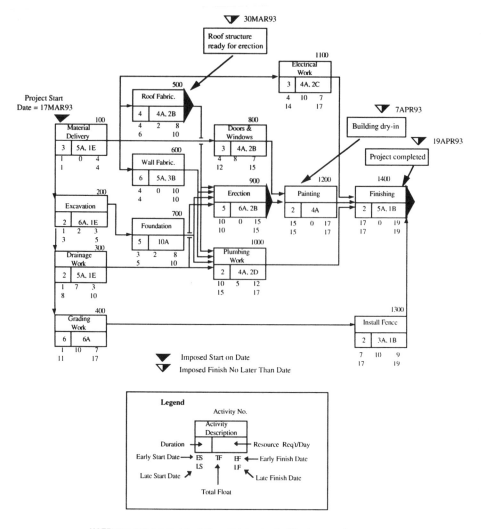

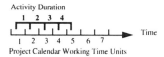

NOTE: The schedule computation in time units is based on the following assumption:

Activity Duration

Activity with 4 time units starts at working time unit 1 and is finished at working time unit 5.

Figure 2 Precedence Diagram Method with Milestones

The owner may include in the contract documents a reasonable number of milestones to be incorporated in the CPM schedule and against which to measure project status. The typical format of milestones required by the owner (client), with their contract completion dates *(see Contract date)*, is shown in Figure 3.

Milestone code	Milestone description	Imposed not later than date *
50	Roof structure ready for erection	10
70	Building dry-in	15
90	Project completed	19

* Note: Project working dates since notice to proceed.

Figure 3 Milestones Requested by Owner

The establishment of milestones as a contractual requirement helps ensure that the owner has a means of controlling the project's progress. The contract general conditions should include some flexibility to permit the owner or contractor to adjust milestone dates when necessitated by uncontrollable factors. Such an adjustment should be in writing with approval. The approval depends on the owner's project scope change and the contractor's ability to meet the new dates.

MILESTONE (LEVEL) The level of the project breakdown structure at which a particular event is considered to be a key event or milestone.

The scope of selected milestones to benchmark the project's progress and outlook is based on the project breakdown structure (PBS) *(see Project breakdown structure)*. Figure 1 shows a portion of a power plant PBS. In this example the power plant is the end item or level 1, the categories of work (e.g., civil, electrical, mechanical) are represented at level 4, and the work package is shown at level 7.

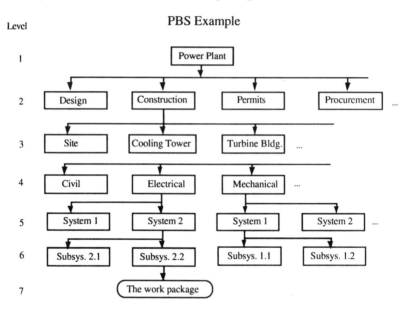

Figure 1 Power Plant PBS Example

The milestone is defined as the start or completion of a work item in a defined PBS. It is possible to have milestones defined at various levels of the PBS. Most of the time the owner defines milestones for project control at levels 3 to 6. At level 3 in this example, selected milestones indicate completion of the entire building, including all subsystems for operation (e.g., cooling towers completed, turbine building completed, etc.).

If the owner chooses to require level 4 milestones for monitoring project progress and outlook, various categories of the work belonging to a specific building could be reflected; such as that turbine building civil work is completed, turbine building electrical work is completed, and turbine building mechanical work is completed. These are typical milestone descriptions and meanings for the power plant PBS shown in Figure 1 at level 4 of the breakdown structure. The milestone computerized reports based on a CPM network could be sorted by level of detail if the level is included in the event coding system and may be beneficial for various levels of management in the project organizational structure.

360

MILESTONE REPORT A computer-generated report at a specified level of the project breakdown structure showing the milestone number (code), milestone description, responsibility, early date, late date, and float on a required completion date (date imposed). The report is generally sorted in ascending order of milestone code or milestone early date.

A computer-oriented scheduling technique incorporating a milestone *(see Milestone)* provides a condensed version or summary to help evaluate project status against a benchmark preset by the owner/contractor. It provides a "snapshot" of the project at a data date *(see Data date)* and presents a number of advantages:

1. The milestone listing in computer printout form provides a precise way in which the milestones, remaining total float, and their trends (increasing or decreasing) can be monitored and analyzed.
2. The milestone report can be sorted in ascending order of milestone codes or in ascending order of early times.
3. Many other sortings are possible, such as by project breakdown structure (PBS) level *[see Milestone (level)]*, by area, or by project phase, as required by the managers.

Figures 1 and 2 show arrow and precedence diagram networks, and the corresponding Figure 3 shows a computer-generated milestone report. Figure 3 illustrates a milestone report related to the Texas state capitol restoration project.

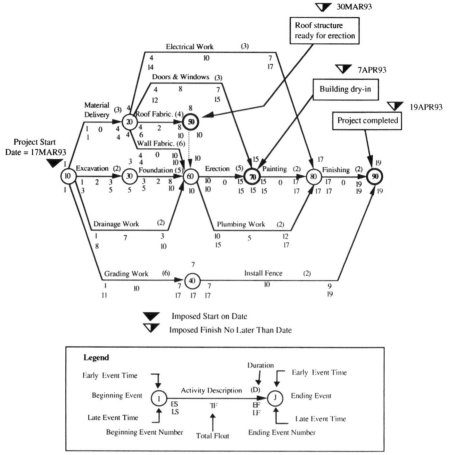

30MAR93

Roof structure
ready for erection

7APR93

Building dry-in

19APR93

Project completed

Imposed Start on Date
Imposed Finish No Later Than Date

Legend

Duration

Early Event Time | Early Event Time

Beginning Event — Activity Description (D) — Ending Event

ES TF EF
LS   LF

Late Event Time | Late Event Time

Beginning Event Number   Total Float   Ending Event Number

NOTE: The schedule computation in time units is based on the following assumption:

Activity Duration

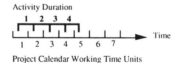

Time

Project Calendar Working Time Units

Activity with 4 time units starts at working time unit 1 and is finished at working time unit 5.

```
-----------------------------------------------------------------------------------------------------
ACME MOTORS                              PRIMAVERA PROJECT PLANNER          PLANT EXPANSION AND MODERNIZATION

REPORT DATE 24NOV93  RUN NO.   23          -----PROJECT SCHEDULE-----        START DATE 17MAR93  FIN DATE 09APR93
             02:30
CLASSIC SCHEDULE REPORT                                                     DATA DATE 17MAR93 PAGE NO.   1
-----------------------------------------------------------------------------------------------------
ACTIVITY    ORIG REM  CAL            ACTIVITY                     MILESTON MILESTON MILES
  ID        DUR  DUR  ID  %  CODE    DESCRIPTION                  LATE DAT IMPOSED  FLOAT
----------  ---- ---- --- --- -----  -----------                  -------- -------- -----
   20  50    4    4   1   0          ROOF FABRICATION             29MAR93  30MAR93     2
   60  70    5    5   1   0          ERECTION                     05APR93  07APR93     0
   80  90    2    2   1   0          FINISHING                    09APR93  19APR93     0
```

Figure 1 Arrow Diagramming with Milestones and Computer-Generated Report

362

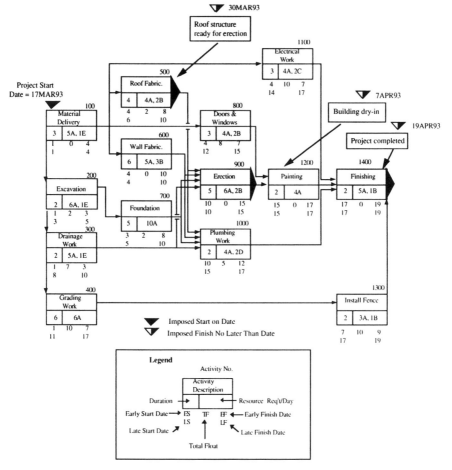

NOTE: The schedule computation in time units is based on the following assumption:

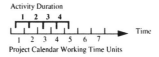

Activity Duration

Project Calendar Working Time Units

Activity with 4 time units starts at working time unit 1 and is finished at working time unit 5.

```
---------------------------------------------------------------------------------
ACME MOTORS                        PRIMAVERA PROJECT PLANNER         PLANT EXPANSION AND MODERNIZATION

REPORT DATE 24NOV93  RUN NO.   8        -----PROJECT SCHEDULE-----         START DATE 13MAR93  FIN DATE 16APR93
             02:04
CLASSIC SCHEDULE REPORT                                                    DATA DATE  17MAR93  PAGE NO.    1
---------------------------------------------------------------------------------
ACTIVITY   ORIG REM  CAL                 ACTIVITY                    MILESTON MILESTON MILES
  ID       DUR  DUR  ID  \   CODE        DESCRIPTION                 LATE DAT IMPOSED  FLOAT
--------- ---- ---- --- ---  ----------  -----------------------     -------- -------- -----
     500    4    4   1    0              ROOF FABRICATION            29MAR93  30MAR93    2
     900    5    5   1    0              ERECTION                    05APR93  07APR94    0
    1400    7    7   1    0              FINISHING                   16APR93  19APR94    0
```

Figure 2 Precedence Diagramming with Milestones and Computer-Generated Report

363

ACTIVITY ID	ORIG DUR	REM DUR	%	ACTIVITY DESCRIPTION	EARLY FINISH	LATE FINISH	TOTAL FLOAT
NORTH PORTICO, HANDICAP RAMPS, PROT WALKS							
"A" COMPL	0	0	0	"A" MILESTONE COMPLETE	17MAR94	30MAR94*	9
NORTH WING: LEV 5,4,3,2,1,G, ELEV/STAIRS/PED TUN							
"B" COMPL	0	0	0	4th Fl Nor, Att, B1/B2 Tun, Elev 1&3 ("B" Mile)	15MAR94	15MAR94*	0
NORT3/999	0	0	0	3rd Floor North Complete ("B" Milestone)	30MAR94	30MAR94*	0
"B1" COMPL	0	0	0	2nd Fl North (Library) Complete ("B1"Milestone)	28FEB94	1MAR94*	1
NORT1/999	0	0	0	1st Floor North Complete ("D" Milestone)	18APR94	29APR94*	9
NORTG/999	0	0	0	Ground Floor North Compl ("D" Milestone)	24MAY94	15JUN94*	15
EAST WING: LEVELS 4,3,2,1,G, E. TUNN/ W. TUNN							
"C" COMPL	0	0	0	"C" MILESTONE COMPLETE	1APR94	1APR94*	0
SOUTH WING: LEVELS 5, 4, 3, 2, 1, G							
"D" COMPL	0	0	0	"D" MILESTONE COMPLETE	28JUN94	30JUN94*	2
WEST WING: LEVELS 4, 3, 2, 1, G							
"E" COMPL	0	0	0	"E" MILESTONE COMPLETE	21NOV94	29NOV94*	5
ZZZZZZZZZZ	0	0	0	PROJECT COMPLETE	21NOV94	29NOV94*	5

Figure 3 Example of Milestone Report

MILESTONE SCHEDULE A graphical presentation of key events or milestones selected as a result of coordination between the client's and the contractor's project management and used as a basis to monitor overall project performance. The format may be a time-scale network, bar chart, or tabular format at a highly summarized project breakdown structure level.

The milestone approach is used for reporting project status in summary form to a higher manager, as it essentially summarizes the status of major events. The milestone schedule offers a "snapshot" of the project, as well as a number of advantages, such as:

1. The milestone computer-generated report *(see Milestone report)* provides a precise form in which program progress can be monitored.
2. Additional milestones can be added during project implementation for better project control or to reflect changes in the original planned project scope and schedule.

If the milestone schedule is specified by the owner, it should be based on a CPM master network which should be provided at the invitation to bid. Milestone-imposed dates should reflect at least the computed late event time to conserve the total float for project benefit. The number of milestones selected should be in a range that will not create an obstruction in managing project implementation and should be distributed over the total project duration and not be concentrated in the last 10% of project duration.

Milestones are points of significant accomplishments in the progress of a project. These points can be recognized by all parties in the early stage of project planning. They are physically verifiable and can represent checkpoints for evaluating progress against a preestablished plan.

The milestone schedule in graphical form is simply a refinement of a bar chart (bar Gantt) *(see Bar chart)*. To have better control of program status, milestones are indicated within individual bars. Figure 1 shows a bar chart in which the milestones have been indicated as a triangle.

For example, milestones 12, 13, and 14 represent the concrete finishes completed for A, B, and C. The milestone schedule in graphical format can also indicate the interdependence of the milestones, as indicated in Figure 2, by dashed arrows. Milestone 9, representing starting to pour a concrete slab on area A, depends on milestone 6, which is completion of the compacted fill base area A. Milestone 6 depends on milestone 2, completion of the subbase for area A.

The milestone schedule shown in Figure 2 represents a significant improvement over the traditional bar chart method. For complex projects with many milestones specified, it shows that the interdependence among milestones is of major importance for overall project control.

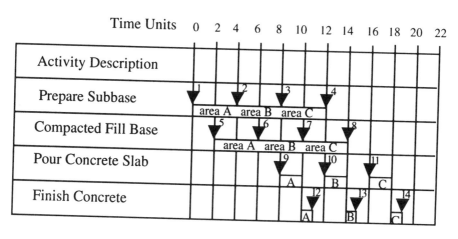

Figure 1 Milestone Schedule Combined with Bar Chart

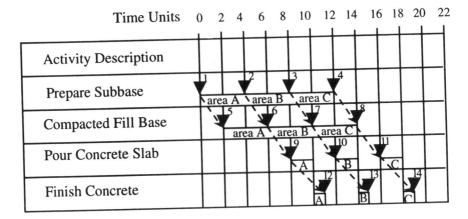

Time Units   0   2   4   6   8   10   12   14   16   18   20   22

Activity Description											
Prepare Subbase											
Compacted Fill Base											
Pour Concrete Slab											
Finish Concrete											

Figure 2   Milestone Schedule Indicating the
Interdependence Between Selected Milestones

MONITORING (UPDATING) Periodic gathering and validating of schedule data regarding project activity status and logical sequences. Monitoring involves the regular review, analysis, and recomputation of the project schedule.

Project monitoring is a very important function for a successful project and requires modification of the as-planned schedule *(see As-planned project schedule)* to reflect the current project progress status and to meet the project objectives. The monitoring process should account for unforeseen and unanticipated problems, change orders, adverse weather conditions, and new knowledge or technology. The result of project monitoring is feedback from project performance to management regarding a project completion date, delays, and problem areas impeding project progress.

After a project has been started, the schedule should be monitored and updated periodically so that comparison and evaluation of current and as-planned progress can be made. The information from this comparison is used to locate the causes of project delay. Monitoring a project schedule requires activity and network information to be amended or revised on the data date specified in the schedule specifications *(see Scheduling specifications)*. The following information should be considered for project monitoring:

1. An activity that has been added
2. An activity that has been deleted
3. Changes in logical constraints
4. Changes in activity duration
5. Changes in activity resources, such as labor, equipment, and materials
6. Estimated remaining duration and resources for in-progress activities
7. Actual start and completion dates of activities

The above-mentioned information should be marked and recorded on the network diagram *(see Network)* before it is input to a computer *(see Updated schedule)*. The schedule calculation *(see Network analysis)* for the updated schedule is performed based on the remaining incomplete activities at a given data date. All completed activities have no impact on the schedule computation. An updated schedule is then compared with the as-planned schedule and can be shown in the form of a network diagram, bar chart, or tabular report, as shown in Figures 1 to 6. Comparison reveals that the project is to be finished 1 day behind the planned completion date.

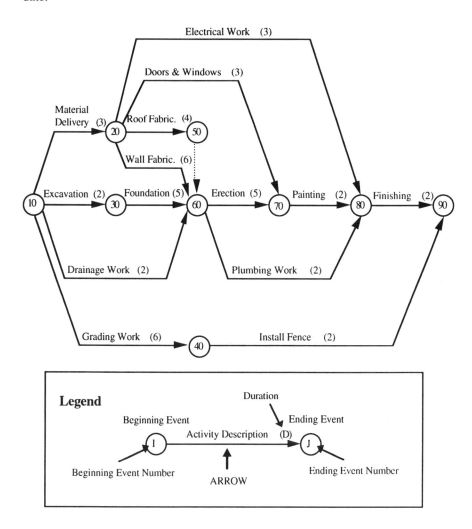

Figure 1 As-planned Arrow Diagram Network

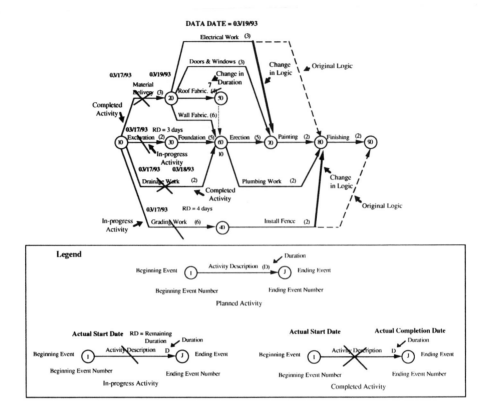

DATA DATE = 03/19/93

Figure 2 Updated Arrow Diagram Network

368

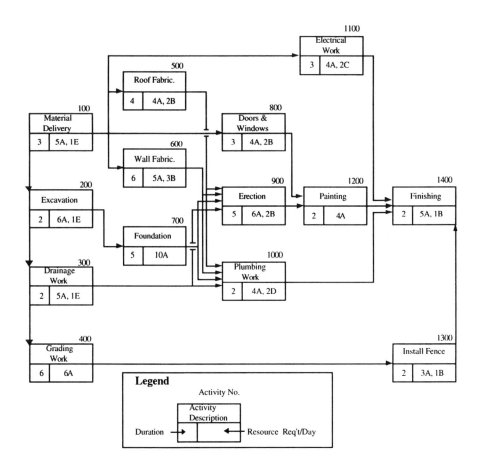

Figure 3 As-planned Precedence Diagram Network

369

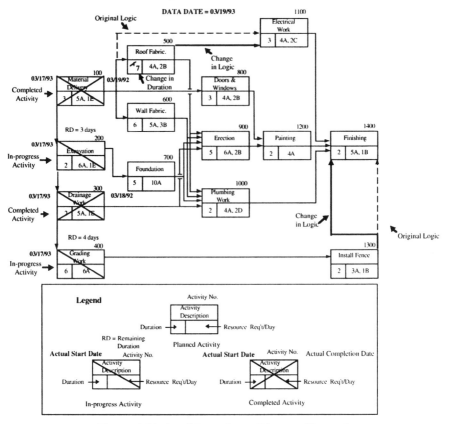

Figure 4 Updated Precedence Diagram Network

```
------------------------------------------------------------------------------------------
CIVIL ENGINEERING DEPARTMENT              PRIMAVERA PROJECT PLANNER

REPORT DATE 23NOV93  RUN NO.    7         PROJECT SCHEDULE REPORT          START DATE 13MAR93  FIN DATE 12APR93
           22:06
CLASSIC SCHEDULE REPORT - SORT BY ES, TF                                  DATA DATE  19MAR93  PAGE NO.    1
```

ACTIVITY ID	ORIG DUR	REM DUR	%	CODE	ACTIVITY DESCRIPTION	EARLY START	EARLY FINISH	LATE START	LATE FINISH	TOTAL FLOAT
100	3	0	100		MATERIAL DELIVERY	17MAR93A	19MAR93A			
300	2	0	100		DRAINAGE WORK	17MAR93A	18MAR93A			
200	2	3	0		EXCAVATION	17MAR93A	23MAR93		23MAR93	0
400	6	4	33		GRADING WORK	17MAR93A	24MAR93		8APR93	11
600	6	6	0		WALL FABRICATION	19MAR93	26MAR93	23MAR93	30MAR93	2
500	4	4	0		ROOF FABRICATION	19MAR93	24MAR93	25MAR93	30MAR93	4
800	3	3	0		DOORS & WINDOWS	19MAR93	23MAR93	2APR93	6APR93	10
1100	3	3	0		ELECTRICAL WORK	19MAR93	23MAR93	6APR93	8APR93	12
700	5	5	0		FOUNDATION	24MAR93	30MAR93	24MAR93	30MAR93	0
1300	2	2	0		INSTALL FENCE	25MAR93	26MAR93	9APR93	12APR93	11
900	5	5	0		ERECTION	31MAR93	6APR93	31MAR93	6APR93	0
1000	2	2	0		PLUMBING WORK	31MAR93	1APR93	7APR93	8APR93	5
1200	2	2	0		PAINTING	7APR93	8APR93	7APR93	8APR93	0
1400	2	2	0		FINISHING	9APR93	12APR93	9APR93	12APR93	0

Figure 5 Updated vs. As-planned Schedule Tabular Report (Precedence Diagram)

```
--------------------------------------------------------------------------------------------------------------------
CIVIL ENGINEERING DEPARTMENT                    PRIMAVERA PROJECT PLANNER

REPORT DATE 23NOV93  RUN NO.    8              PROJECT SCHEDULE REPORT                    START DATE 13MAR93  FIN DATE 12APR93
            22:08
BAR CHART BY ES, EF, TF                                                                  DATA DATE 19MAR93   PAGE NO.    1

                                                                                                            DAILY-TIME PER.   1
--------------------------------------------------------------------------------------------------------------------
.............ACTIVITY DESCRIPTION.............     15     22     29     05     12     19     26     03     10     17
ACTIVITY ID  OD   RD  PCT   CODES    FLOAT     SCHEDULE   MAR    MAR    MAR    APR    APR    APR    APR    MAY    MAY    MAY
-----------  ----  ---- ---- ------------- -----    --------   93     93     93     93     93     93     93     93     93     93
                                                    -------------------------------------------------------------------

DRAINAGE WORK                                 CURRENT   . AA* .      .      .      .      .      .      .      .      .
        300   2   0 100                       PLANNED   . EE* .      .`     .      .      .      .      .      .      .

MATERIAL DELIVERY                             CURRENT   . AA* .      .      .      .      .      .      .      .      .
        100   3   0 100                       PLANNED   . EEE  .      .      .      .      .      .      .      .      .

EXCAVATION                                    CURRENT   . AAE..EE     .      .      .      .      .      .      .      .
        200   2   3   0                    0  PLANNED   . EE* .      .      .      .      .      .      .      .      .

GRADING WORK                                  CURRENT   . AAE..EEE    .      .      .      .      .      .      .      .
        400   6   4  33                   11  PLANNED   . EEE..EEE    .      .      .      .      .      .      .      .

DOORS & WINDOWS                               CURRENT   .  E..EE   .      .      .      .      .      .      .      .
        800   3   3   0                    10  PLANNED   .  *  EEE   .      .      .      .      .      .      .      .

ELECTRICAL WORK                               CURRENT   .  E..EE   .      .      .      .      .      .      .      .
       1100   3   3   0                    12  PLANNED   .  *  EEE   .      .      .      .      .      .      .      .

ROOF FABRICATION                              CURRENT   .  E..EEE   .      .      .      .      .      .      .      .
        500   4   4   0                     4  PLANNED   .  *  EEEE   .      .      .      .      .      .      .      .

WALL FABRICATION                              CURRENT   .  E..EEEEE .      .      .      .      .      .      .      .
        600   6   6   0                     2  PLANNED   .  *  EEEEE..E     .      .      .      .      .      .      .

FOUNDATION                                    CURRENT   .  *  . EEE..EE    .      .      .      .      .      .      .
        700   5   5   0                     0  PLANNED   .  E..EEEE   .      .      .      .      .      .      .      .

INSTALL FENCE                                 CURRENT   .  *  . EE  .      .      .      .      .      .      .      .
       1300   2   2   0                    11  PLANNED   .  *  . EE  .      .      .      .      .      .      .      .

PLUMBING WORK                                 CURRENT   .  *  .      . EE  .      .      .      .      .      .      .
       1000   2   2   0                     5  PLANNED   .  *  .      .EE  .      .      .      .      .      .      .

ERECTION                                      CURRENT   .  *  .      . EEE..EE    .      .      .      .      .      .
        900   5   5   0                     0  PLANNED   .  *  .      .EEEE..E    .      .      .      .      .      .

PAINTING                                      CURRENT   .  *  .      .      . EE  .      .      .      .      .      .
       1200   2   2   0                     0  PLANNED   .  *  .      .      .EE  .      .      .      .      .      .

FINISHING                                     CURRENT   .  *  .      .      .  E..E     .      .      .      .      .
       1400   2   2   0                     0  PLANNED   .  *  .      .      . EE  .      .      .      .      .      .
```

Figure 6 Updated vs. As-planned Bar Chart Report (Precedence Diagram)

MOST CRITICAL TOTAL FLOAT In PDM scheduling, a convention of the forward and backward pass algorithms that looks at the float at the start and finish of a PDM activity and assigns the lesser float to the activity as its single total float value. In PDM scheduling, it is possible for activities to have a different float at the start of the activity than at the finish of a single activity due to the use of SS and FF relationships with lags.

MOST LIKELY TIME ESTIMATE   The most likely time for an activity to be completed *[see Expected duration (activity)]*.

A most likely time estimate *(Tm)* is the estimated performance time for an activity that would most frequently be most achieved or be most probable to happen under a number of repeated similar conditions regarding activity completion. In other words, it is the mode value of the performance time distribution. In the critical path method (CPM), the most likely time estimate is generally considered to be the activity duration.

Note that since PERT uses the central limit theorem to calculate the activity duration, the three time estimates, which are *Ta (see Optimistic time estimate), Tb (see Pessimistic time estimate),* and *Tm,* should be estimated independently based on each activity performance for each project. Also, the estimate should not be influenced by the time available to complete the project. It is not easy to determine the three time estimates since they need to include unexpected uncertainties of an activity in different conditions. The greater uncertainty an activity encounters, the wider the range of estimated time will become.

The general curve of *Tb* is shown in Figure 1. Table 1 shows *Tm* in the column Time Estimates, and Figure 2 shows an example of PERT diagram with *Tm* shown for each activity.

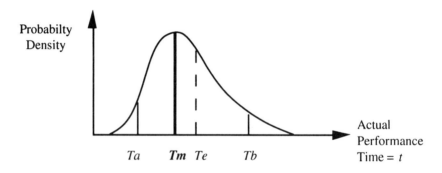

Figure 1   Most Likely Time Estimate Shown on Distribution of Performance Time

Table 1   Most Likely Time Estimates

		Time Estimates			$Te =$	$S(Te) =$	
Node	Activity	Ta	Tm	Tb	$(Ta + 4 Tm + Tb)/6$	$(Tb\text{-}Ta)/6$	$S(Te)\ x\ S(Te)$
1-2	A	4	4	4	4	0	0
1-3	B	1	3	5	3	0.67	0.44
2-4	C	2	4	12	5	1.67	2.79
2-5	D	9	12	33	15	4	16
3-4	E	4	7	22	9	3	9
4-5	F	2	4	8	4	1	1
5-6	G	5	5	5	5	0	0

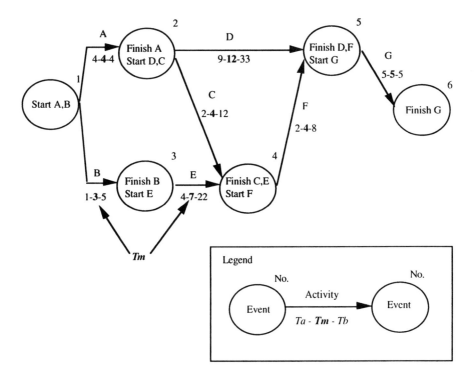

Figure 2 Example of PERT Diagram with *Tb* Indicated

MULTIPLE FINISH NETWORK A network that has more than one finish activity.

A network with many finish activities is used when parts of a project may need to be completed at different times. This type of the network creates a more realistic plan for a long-range project consisting of several subprojects requiring a finish activity for each subproject. For example, the development of multicomplex office buildings may require different facilities to be occupied at different times, so the completion of each building will be set according to the date required. The following computation can occur or be considered for the network with multiple finish activities:

1. *All late project end event times (all finish activities) = greatest early project end event time.* In this situation, the critical path will be the one with the greatest computed early project end event time and zero total float.
2. *All late project end event times (all finish activities) ≠ each end activity early project end event time.* In this situation, there will be multiple critical paths along which activities have zero total floats.
3. *All late project end event times (all finish activities) = each end activity early*

373

project end event time. In this situation, the late project end event times will be greater or smaller than each end activity early project end event time. There will be different paths with positive and/or negative total floats. The true critical path will be the one with the least positive total float or the most negative total floats.

An example of a network using the arrow and precedence techniques with multiple ending events is shown in Figures 1 and 2.

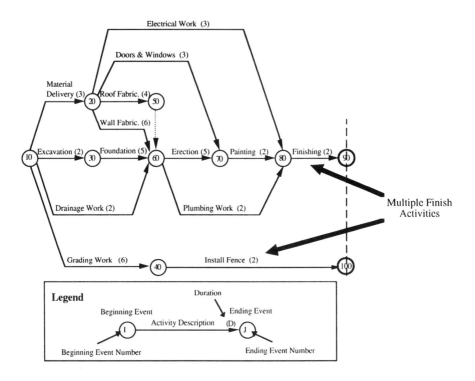

Figure 1 Multiple Finish Network in Arrow Diagram

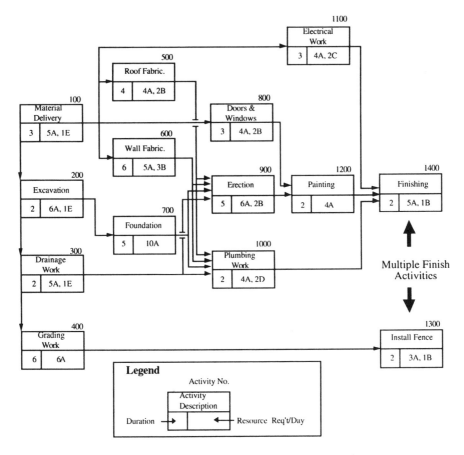

Figure 2 Multiple Finish Network in Precedence Diagram

375

MULTIPLE START NETWORK A network that has more than one start activity.

A multiple start network is used when a project has more than one starting point. In this case, the computations are inconvenient or sometimes incorrect if all beginning events are artificially brought together to a single initial event. Careful attention must be paid to establishing a starting date for different start activities because if the starting date for each activity is not specified, it is assumed that all start activities have the same start date. The multiple start network is suitable for the long-range project with different subprojects because some subprojects may not have to be started at the beginning of the project. For example, the development of the multicomplex office buildings, the main building may be started first; then other buildings may be started sometimes after the start of the main building. An example of a small network using the arrow and precedence techniques with multiple start activities is shown in Figures 1 and 2.

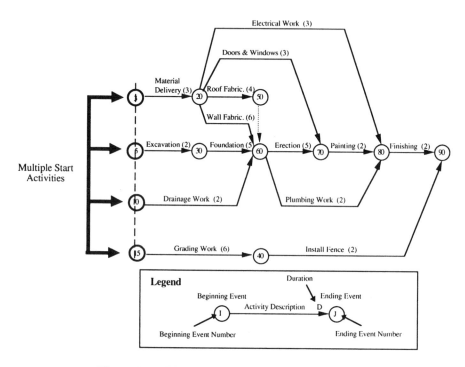

Figure 1 Multiple Start Network in Arrow Diagram

376

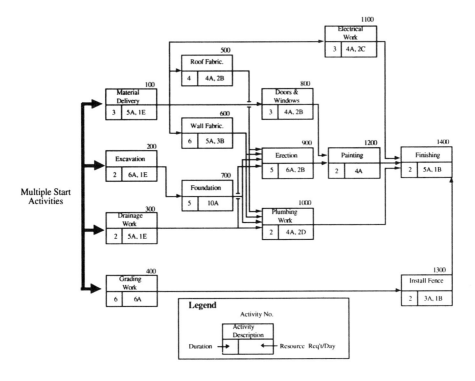

Figure 2 Multiple Start Network in Precedence Diagram

MULTIPROJECT SCHEDULE The result of having scheduled several projects or subprojects using a common pool of resources.

A project may be comprised of many subprojects whose schedules are developed independently, such as electrical and mechanical. In some situations a general contractor may be working on several projects using the same common resources, such as equipment and specialized crews. For both cases, a contractor or project manager has to develop a master schedule containing several projects or subprojects called a multiproject schedule.

Projects or subprojects in a multiproject schedule may or may not be interconnected. The connection between projects or subprojects is normally required when they use the same resources. If the connection is required to indicate the logical constraint between two or more activities of different projects, an interface node or activity is used to interconnect those activities (see *Interface activity* and *Interface node*).

The network analysis can be performed for a multiproject or each individual project schedule. In the event that a multiproject schedule is developed for the repetitive type of project such as multiple residential housings, the fragnet is mostly

377

used for the development of a multiproject schedule with different start dates for each project *(see Fragnet).*

When projects require the same resources, such as equipment, they must be interconnected and network analysis is performed to recognize the availability of the common resources *(see Resource-limited scheduling).* Figures 1 and 2 shows a multiproject precedence diagram network with one and more interface activities.

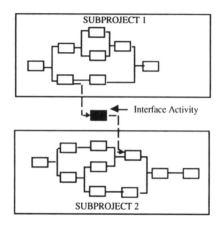

Figure 1 Multiproject Network with One Interface Activity

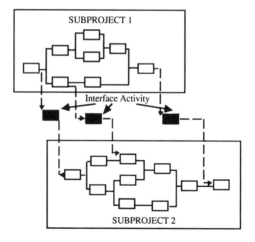

Figure 2 Multiproject Network with More Interface Activities

N

NEARLY CRITICAL ACTIVITY An activity with a relatively low amount of total float, thus indicating an increased possibility that changes or impacts on the project schedule could result in the activity becoming critical.

A nearly critical activity is a noncritical activity with less chance of delay than that of a critical activity *(see Critical activity)* but more likely than some other noncritical activities to cause delay. A nearly critical activity is determined from the amount of its total float *(see Activity total float),* which needs to be less than or equal to a specific small number set by a project manager. The purpose of determining a nearly critical activity is to let the project manager realize that these activities will possibly become critical as the project is undertaken so that he or she can prepare in advance to handle them.

It is important for the project manager to pay attention not only to critical activities but also to nearly critical activities since they have a great chance to become critical activities. In addition, for each updating, some noncritical activities can become nearly critical or critical activities later in a project. The project manager should be able to identify and manage them properly to avoid project delay.

Figures 1 and 2 show nearly critical activities of a simple network in the arrow diagram and precedence diagram techniques, respectively. Assuming that activities with a total float of less than 3 are considered nearly critical, the activities Excavation, Foundation, and Roof Fabrication are considered nearly critical activities.

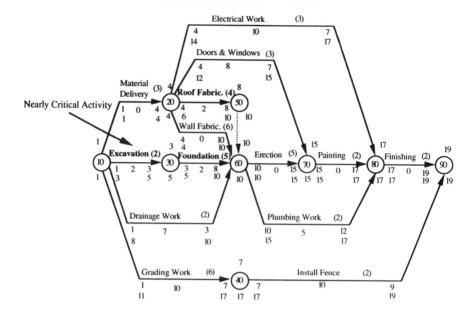

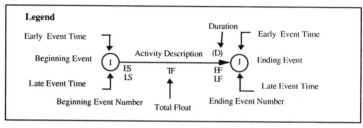

Legend

Early Event Time

Beginning Event

Late Event Time

Activity Description (D)

Duration

Early Event Time

Ending Event

Late Event Time

ES
LS

TF

EF
LF

Beginning Event Number

Total Float

Ending Event Number

NOTE: The schedule computation in time units is based on the following assumption:

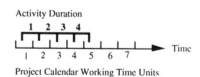

Activity Duration

Project Calendar Working Time Units

Activity with 4 time units starts at working time unit 1 and is finished at working time unit 5.

Figure 1 Nearly Critical Activities Shown in Arrow Diagram Network

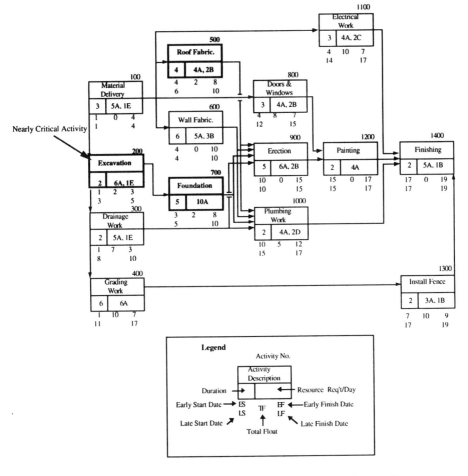

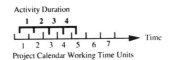

NOTE: The schedule computation in time units is based on the following assumption:

Activity Duration

Activity with 4 time units starts at working time unit 1 and is finished at working time unit 5.

Figure 2 Nearly Critical Activities Shown in Precedence Diagram Network

381

NEARLY CRITICAL PATH A series of connected activities with equal but relatively low total float, thus indicating an increased possibility that changes in or impacts on the project schedule could result in the path becoming critical.

A nearly critical path is a network path consisting of a nearly critical activity *(see Nearly critical activity)* or sequential nearly critical activities with the same amount of total float *(see Activity total float)* which have a tendency to become critical as the project gets under way. Normally, critical paths *(see Critical path)* of a project are much more focused than the nearly critical paths since activities of the critical paths have no or very little flexibility to be delayed. However, the nearly critical paths should not be totally ignored because although they contain float at present, they will, with high potential, become critical eventually. During the project update, some noncritical paths can become nearly critical or critical paths. It is necessary for the project manager to be aware of and be prepared to deal with them because those paths may not be realized before.

Figures 1 and 2 show nearly critical paths of a simple network using the arrow diagram and precedence diagram techniques, respectively. Assuming that paths with a total float of less than 3 are considered nearly critical, the path of the activities Excavation and Foundation and the path of the activity Roof Fabrication are considered nearly critical paths.

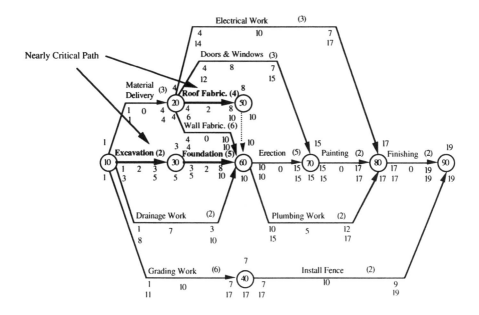

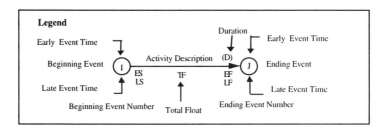

Legend

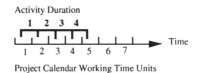

NOTE: The schedule computation in time units is based on the following assumption:

Activity Duration

Project Calendar Working Time Units

Activity with 4 time units starts at working time unit 1 and is finished at working time unit 5.

Figure 1 Nearly Critical Paths Shown in Arrow Diagram Network

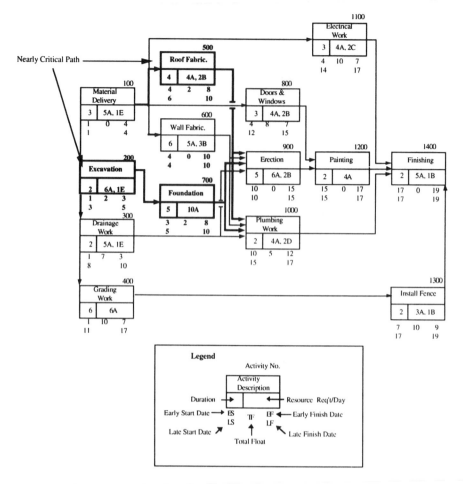

Nearly Critical Path

Legend

Activity No.

Activity Description

Duration →

Resource Req't/Day

Early Start Date → ES   TF   EF ← Early Finish Date
LS   LF
Late Start Date ↗   ↑   ↖ Late Finish Date
Total Float

NOTE: The schedule computation in time units is based on the following assumption:

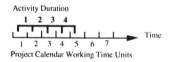

Activity Duration

Project Calendar Working Time Units

Activity with 4 time units starts at working time unit 1 and is finished at working time unit 5.

Figure 2 Nearly Critical Paths Shown in Precedence Diagram Network

384

NEGATIVE FLOAT (NEGATIVE ACTIVITY DURATION) An improper calculation of total float for an activity or path expressed as a negative number. Generally derived by establishing or imposing a project late finish date earlier than that indicated by a network forward pass computation and assigning activity float times to reflect completion on the earlier finish date.

NEGATIVE LAG A convention of PDM scheduling that allows for a negative integer (lag) to be applied to any of the four PDM restraint types (FS, SS, FF, SF) which causes the successor activity to start or finish n time units prior to the start or finish time of the predecessor.

NETWORK ANALYSIS A technique used to determine the critical path and float times for noncritical activities for a project consisting of a sequence of activities and their interrelationships within a network of activities.

Network analysis is a process by which activity and project start and completion dates are identified. Network analysis is performed after the network diagram *(see Network)* is developed and the duration is assigned to all activities. The critical path method *(see Critical path method)* is used as a means for network analysis. Network analysis using CPM involves forward and backward pass computation *(see Forward pass computation* and *Backward pass computation).*

The forward pass computation identifies the early start and finish times for each activity comprising a project in time units or calendar dates. The backward pass computation identifies the late start and finish times for each activity comprising a project in time units or calendar dates. Critical activities *(see Critical activity)* and activity floats *(see Activity total float* and *Free float)* are defined after the forward and backward pass computations are completed. Normally, network analysis is accomplished through the computerized system.

During the planning stage, network analysis enables management to determine when each activity must start and finish in order to complete the project within the specified completion date. After the start of a project start, network analysis must be performed periodically or regularly to update the activity information in the network so that any change or modification can be represented in the network. The project evaluation can then be performed to ensure that the project is progressing as planned. In the event that the project performance is unsatisfactory, the network analysis can identify the location of delay in a project. An example of network analysis is shown in Figures 1 to 4.

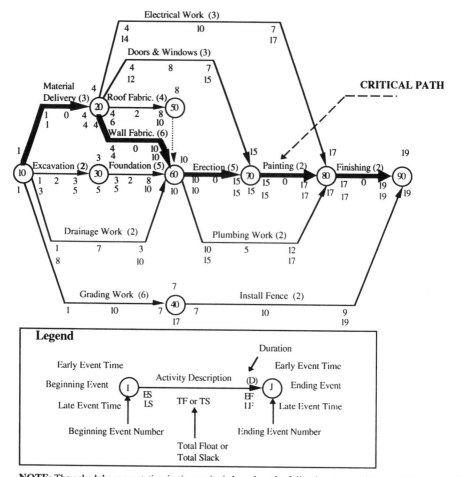

Electrical Work (3)

Doors & Windows (3)

Material Delivery (3)

Roof Fabric. (4)

Wall Fabric. (6)

Excavation (2)

Foundation (5)

Erection (5)

Painting (2)

Finishing (2)

Drainage Work (2)

Plumbing Work (2)

Grading Work (6)

Install Fence (2)

CRITICAL PATH

Legend

Early Event Time
Beginning Event
Late Event Time

Activity Description

Duration

Early Event Time
Ending Event
Late Event Time

Beginning Event Number

Total Float or
Total Slack

Ending Event Number

NOTE: The schedule computation in time units is based on the following assumption:

Activity Duration

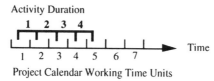

Project Calendar Working Time Units

Activity with 4 time units starts at working time unit 1 and is finished at working time unit 5.

Figure 1 Arrow Diagram Network with Schedule Computation

386

		ORIG	REM			ACTIVITY DESCRIPTION	EARLY	EARLY	LATE	LATE	TOTAL
PRED	SUCC	DUR	DUR	%	CODE		START	FINISH	START	FINISH	FLOAT
10	20	3	3	0		MATERIAL DELIVERY	17MAR93*	19MAR93	17MAR93*	19MAR93	0
10	30	2	2	0		EXCAVATION	17MAR93	18MAR93	19MAR93	22MAR93	2
10	60	2	2	0		DRAINAGE WORK	17MAR93	18MAR93	26MAR93	29MAR93	7
10	40	6	6	0		GRADING WORK	17MAR93	24MAR93	31MAR93	7APR93	10
30	60	5	5	0		FOUNDATION	19MAR93	25MAR93	23MAR93	29MAR93	2
20	60	6	6	0		WALL FABRICATION	22MAR93	29MAR93	22MAR93	29MAR93	0
20	50	4	4	0		ROOF FABRICATION	22MAR93	25MAR93	24MAR93	29MAR93	2
20	70	3	3	0		DOORS AND WINDOWS	22MAR93	24MAR93	1APR93	5APR93	8
20	80	3	3	0		ELECTRICAL WORK	22MAR93	24MAR93	5APR93	7APR93	10
40	90	2	2	0		INSTALL FENCE	25MAR93	26MAR93	8APR93	9APR93	10
50	60	0	0	0		DUMMY 1	26MAR93	25MAR93	30MAR93	29MAR93	2
60	70	5	5	0		ERECTION	30MAR93	5APR93	30MAR93	5APR93	0
60	80	2	2	0		PLUMBING WORK	30MAR93	31MAR93	6APR93	7APR93	5
70	80	2	2	0		PAINTING	6APR93	7APR93	6APR93	7APR93	0
80	90	2	2	0		FINISHING	8APR93	9APR93	8APR93	9APR93	0

Figure 2 Schedule Report, Arrow Diagram Network

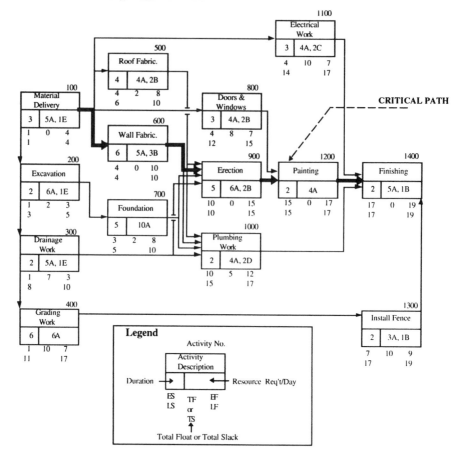

CRITICAL PATH

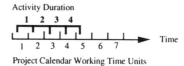

NOTE: The schedule computation in time units is based on the following assumption:

Activity Duration

Time

Project Calendar Working Time Units

Activity with 4 time units starts at working time unit 1 and is finished at working time unit 5.

Figure 3 Precedence Diagram Network with Schedule Computation

388

CIVIL ENGINEERING DEPARTMENT   PRIMAVERA PROJECT PLANNER

REPORT DATE 23NOV93 RUN NO.   20   PROJECT SCHEDULE REPORT   START DATE 13MAR93 FIN DATE 9APR93
          22:13
CLASSIC SCHEDULE REPORT - SORT BY ES, TF   DATA DATE 17MAR93 PAGE NO.   1

ACTIVITY ID	ORIG DUR	REM DUR	%	CODE	ACTIVITY DESCRIPTION	EARLY START	EARLY FINISH	LATE START	LATE FINISH	TOTAL FLOAT
100	3	3	0		MATERIAL DELIVERY	17MAR93*	19MAR93	17MAR93*	19MAR93	0
200	2	2	0		EXCAVATION	17MAR93	18MAR93	19MAR93	22MAR93	2
300	2	2	0		DRAINAGE WORK	17MAR93	18MAR93	26MAR93	29MAR93	7
400	6	6	0		GRADING WORK	17MAR93	24MAR93	31MAR93	7APR93	10
700	5	5	0		FOUNDATION	19MAR93	25MAR93	23MAR93	29MAR93	2
600	6	6	0		WALL FABRICATION	22MAR93	29MAR93	22MAR93	29MAR93	0
500	4	4	0		ROOF FABRICATION	22MAR93	25MAR93	24MAR93	29MAR93	2
800	3	3	0		DOORS & WINDOWS	22MAR93	24MAR93	1APR93	5APR93	8
1100	3	3	0		ELECTRICAL WORK	22MAR93	24MAR93	5APR93	7APR93	10
1300	2	2	0		INSTALL FENCE	25MAR93	26MAR93	8APR93	9APR93	10
900	5	5	0		ERECTION	30MAR93	5APR93	30MAR93	5APR93	0
1000	2	2	0		PLUMBING WORK	30MAR93	31MAR93	6APR93	7APR93	5
1200	2	2	0		PAINTING	6APR93	7APR93	6APR93	7APR93	0
1400	2	2	0		FINISHING	8APR93	9APR93	8APR93	9APR93	0

Figure 4 Schedule Report, Precedence Diagram Network

NETWORK (NETWORK DIAGRAM) A flow diagram or graphical representation consisting of the activities and events that must be accomplished to reach the project objectives. A network shows the sequence of work, interdependencies, and interrelationships among project activities.

A network diagram is an important aspect of using a critical path method *(see Critical path method)* as a tool for project planning, scheduling, and controlling. A network is normally used as a means for communicating a project plan and progress to various project participants. A network consists of activities that must be completed to accomplish project objectives and logical interrelationships among activities that represent work sequences. The graphical representation of a network can be in the form of an arrow or precedence diagram *(see Arrow diagramming method* and *Precedence diagramming method).* An activity in arrow and precedence diagram networks is represented by a node and an arrow, respectively, as shown in Figures 1 and 2.

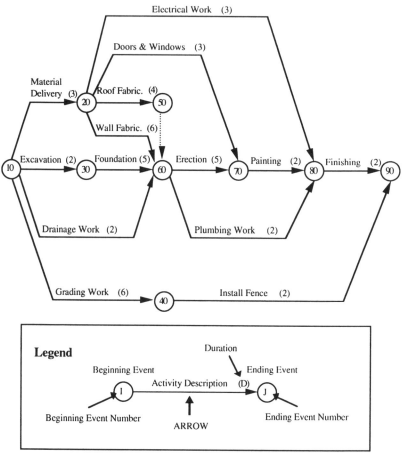

Figure 1 Arrow Diagram Network

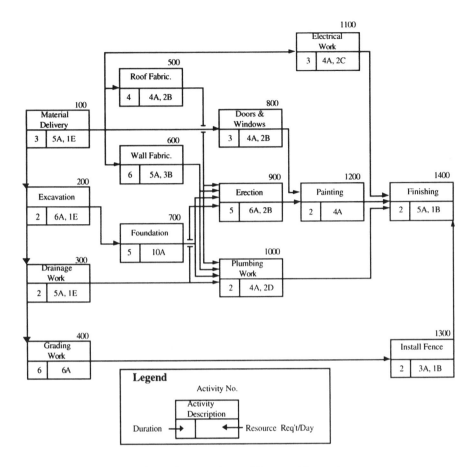

Figure 2 Precedence Diagram Network

Network development requires a project end objective to be broken down into smaller operations or activities by using a project breakdown structure *(see Project breakdown structure)*. The logical constraints representing the relationships among activities are then defined. The final product of network development is the graphical display of sequences in which activities are performed. Using a network diagram for project scheduling requires the realistic and accurate estimate of duration *(see Activity duration)* for each activity. After a duration is assigned to all activities, the CPM schedule computation *(see Network analysis)* can be performed to identify each activity start and finish dates.

For complex project control, the network allows cost and resources to be considered in each activity so that management can monitor the project cost and resources as well as schedule. During the project execution, a network must be updated to reflect the project current status *(see Monitoring)* and to allow modifications or changes associated with each activity as the project progresses.

NETWORK PLANNING A broad generic term for techniques used to plan a complex project. The most popular techniques are CPM (arrow and precedence diagramming techniques) and PERT (for a research and new product development project).

CPM and PERT are the most popular techniques that employ a network as a means to plan a project *(see Critical path method* and *PERT).* Network planning allows management to think logically from the start to the finish of a project in order to identify the most efficient way to carry out a project. Network planning starts with the selection of a planning technique based on planning and scheduling specifications *(see Scheduling specifications).* Work activities and relationships among them are then carefully identified. Network planning is shown in a logical diagram *(see Network)* representing the intended sequences of planned operations. The work schedule *(see Network analysis)* is calculated to indicate when each activity should start and finish in order to complete the project.

The critical elements of network planning that must be defined accurately are the estimated duration for each activity and the interrelationships among activities. Cost and resources can also be included in the network planning as required by management or specifications. Network planning may cover different project functions, such as design, procurement, and construction, and involve various project participants, such as the A/E, owners, general contractors, and subcontractors.

The most popular technique used for network planning is the critical path method (CPM). Arrow and precedence diagrams *(see Arrow diagramming method* and *Precedence diagramming method)* are commonly used for network planning, as shown in Figures 1 to 4.

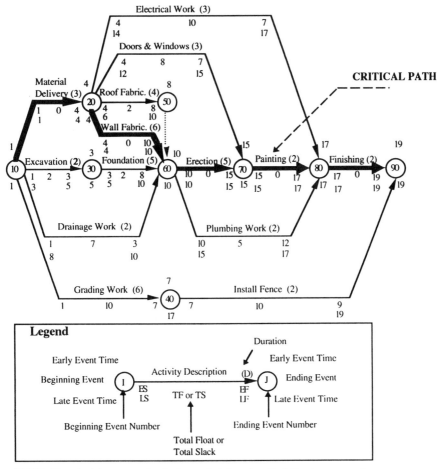

Electrical Work (3)

Doors & Windows (3)

CRITICAL PATH

Material
Delivery (3)

Roof Fabric. (4)

Wall Fabric. (6)

Excavation (2)

Foundation (5)

Erection (5)

Painting (2)

Finishing (2)

Drainage Work (2)

Plumbing Work (2)

Grading Work (6)

Install Fence (2)

Legend

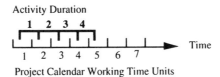

Duration

Early Event Time

Early Event Time

Beginning Event

Activity Description

(D)

Ending Event

ES
LS

TF or TS

EF
LF

Late Event Time

Late Event Time

Beginning Event Number

Ending Event Number

Total Float or
Total Slack

NOTE: The schedule computation in time units is based on the following assumption:

Activity Duration

1   2   3   4

Time

Project Calendar Working Time Units

Activity with 4 time units starts at working time unit 1 and is finished at working time unit 5.

Figure 1 Arrow Diagram Network with Schedule Computation

CIVIL ENGINEERING DEPARTMENT   PRIMAVERA PROJECT PLANNER

REPORT DATE 23NOV93 RUN NO.   12   PROJECT SCHEDULE REPORT   START DATE 17MAR93 FIN DATE 9APR93
            22:19
CLASSIC SCHEDULE REPORT - SORT BY ES, TF   DATA DATE 17MAR93 PAGE NO.   1

PRED	SUCC	ORIG DUR	REM DUR	%	CODE	ACTIVITY DESCRIPTION	EARLY START	EARLY FINISH	LATE START	LATE FINISH	TOTAL FLOAT
10	20	3	3	0		MATERIAL DELIVERY	17MAR93*	19MAR93	17MAR93*	19MAR93	0
10	30	2	2	0		EXCAVATION	17MAR93	18MAR93	19MAR93	22MAR93	2
10	60	2	2	0		DRAINAGE WORK	17MAR93	18MAR93	26MAR93	29MAR93	7
10	40	6	6	0		GRADING WORK	17MAR93	24MAR93	31MAR93	7APR93	10
30	60	5	5	0		FOUNDATION	19MAR93	25MAR93	23MAR93	29MAR93	2
20	60	6	6	0		WALL FABRICATION	22MAR93	29MAR93	22MAR93	29MAR93	0
20	50	4	4	0		ROOF FABRICATION	22MAR93	25MAR93	24MAR93	29MAR93	2
20	70	3	3	0		DOORS AND WINDOWS	22MAR93	24MAR93	1APR93	5APR93	8
20	80	3	3	0		ELECTRICAL WORK	22MAR93	24MAR93	5APR93	7APR93	10
40	90	2	2	0		INSTALL FENCE	25MAR93	26MAR93	8APR93	9APR93	10
50	60	0	0	0		DUMMY 1	26MAR93	25MAR93	30MAR93	29MAR93	2
60	70	5	5	0		ERECTION	30MAR93	5APR93	30MAR93	5APR93	0
60	80	2	2	0		PLUMBING WORK	30MAR93	31MAR93	6APR93	7APR93	5
70	80	2	2	0		PAINTING	6APR93	7APR93	6APR93	7APR93	0
80	90	2	2	0		FINISHING	8APR93	9APR93	8APR93	9APR93	0

Figure 2 Schedule Report, Arrow Diagram Network

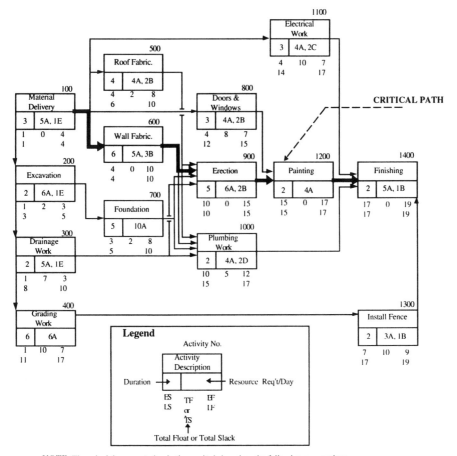

NOTE: The schedule computation in time units is based on the following assumption:

Activity Duration

Time

Project Calendar Working Time Units

Activity with 4 time units starts at working time unit 1 and is finished at working time unit 5.

Figure 3 Precedence Diagram Network with Schedule Computation

CIVIL ENGINEERING DEPARTMENT   PRIMAVERA PROJECT PLANNER

REPORT DATE 23NOV93 RUN NO.   20   PROJECT SCHEDULE REPORT   START DATE 13MAR93 FIN DATE 9APR93
         22:13
CLASSIC SCHEDULE REPORT - SORT BY ES, TF   DATA DATE 17MAR93 PAGE NO.   1

ACTIVITY ID	ORIG DUR	REM DUR	%	CODE	ACTIVITY DESCRIPTION	EARLY START	EARLY FINISH	LATE START	LATE FINISH	TOTAL FLOAT
100	3	3	0		MATERIAL DELIVERY	17MAR93*	19MAR93	17MAR93*	19MAR93	0
200	2	2	0		EXCAVATION	17MAR93	18MAR93	19MAR93	22MAR93	2
300	2	2	0		DRAINAGE WORK	17MAR93	18MAR93	26MAR93	29MAR93	7
400	6	6	0		GRADING WORK	17MAR93	24MAR93	31MAR93	7APR93	10
700	5	5	0		FOUNDATION	19MAR93	25MAR93	23MAR93	29MAR93	2
600	6	6	0		WALL FABRICATION	22MAR93	29MAR93	22MAR93	29MAR93	0
500	4	4	0		ROOF FABRICATION	22MAR93	25MAR93	24MAR93	29MAR93	2
800	3	3	0		DOORS & WINDOWS	22MAR93	24MAR93	1APR93	5APR93	8
1100	3	3	0		ELECTRICAL WORK	22MAR93	24MAR93	5APR93	7APR93	10
1300	2	2	0		INSTALL FENCE	25MAR93	26MAR93	8APR93	9APR93	10
900	5	5	0		ERECTION	30MAR93	5APR93	30MAR93	5APR93	0
1000	2	2	0		PLUMBING WORK	30MAR93	31MAR93	6APR93	7APR93	5
1200	2	2	0		PAINTING	6APR93	7APR93	6APR93	7APR93	0
1400	2	2	0		FINISHING	8APR93	9APR93	8APR93	9APR93	0

Figure 4 Schedule Report, Precedence Diagram Network

"NO DAMAGE FOR DELAY" (Contract Clause) Currently popular contract wording that attempts to strictly limit a contractor's remedy for otherwise compensatory critical path delays to an extension of time only, thus denying recovery by the contractor of extended field and home office performance costs even when the contractor can prove that critical path delays were solely the fault of the owner or his agents.

NODE *(see Event)*

NODE *i* A point in time representing the start of an activity (or activities) in an arrow diagram network.

Node *i* is a small circle with an integer number or a combination of numbers and letters inside and represents the start point of an activity in an arrow diagram network. Figure 1 shows the relationships of the node *i* (node 10), node *j* *(see Node j)* (node 20), and an arrow (activity A). The arrow is always started out from node *i* as activity A begins from node 10. Since an activity consists of two nodes, *i* and *j*, the smaller one will be referred to as node *i*. However, existing scheduling software packages on the market do not require node *i* to be smaller than node *j* for any activity.

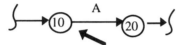

Figure 1 Node *i* (Node 10) of Activity A

It is noted that any node *i* of an activity can also be node *j* of its immediate predecessors except for the predecessors with the same node *i*. For instance, Figure 2 shows that node 30, activity C's node *i*, cannot be both activity A's and B's node *j* because they both have the same node *i*. Hence it implies that more than one activity can have the same node *i* as long as their node *j*'s are different. Figure 3 shows node *i* of an example arrow diagram network.

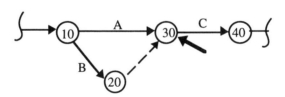

Figure 2 Node *i* (Node 30) of Activity C

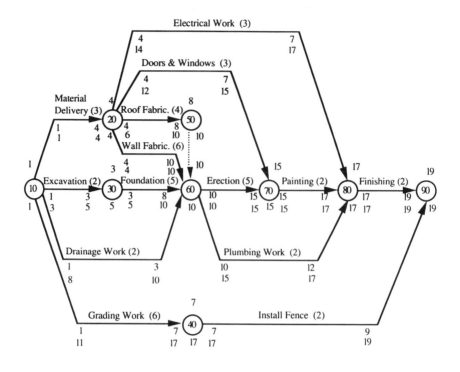

Electrical Work (3)

Doors & Windows (3)

Material Delivery (3)

Roof Fabric. (4)

Wall Fabric. (6)

Excavation (2)

Foundation (5)

Erection (5)

Painting (2)

Finishing (2)

Drainage Work (2)

Plumbing Work (2)

Grading Work (6)

Install Fence (2)

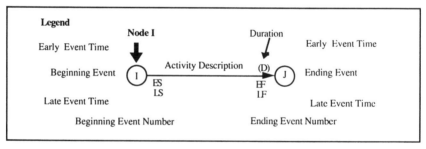

Legend

Node I

Duration

Early Event Time

Early Event Time

Beginning Event

Activity Description

(D)

Ending Event

ES
LS

EF
LF

Late Event Time

Late Event Time

Beginning Event Number

Ending Event Number

NOTE: The schedule computation in time units is based on the following assumption:

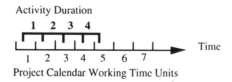

Activity Duration

Time

Project Calendar Working Time Units

Activity with 4 time units starts at working time unit 1 and is finished at working time unit 5.

Figure 3 Node i of Activities in Arrow Diagram Network

NODE *j* A point in time representing the completion of an activity (or activities) in an arrow diagram network.

Node *j* is a small circle with an integer number or a combination of numbers and letters inside and represents the completion point of an activity in an arrow diagram. Figure 1 shows the relationships of node *i* *(see Node i)* (node 10), node *j* (node 20), and an arrow (activity A). The arrow is always ended at node *j* as activity A is finished at node 20. Since an activity consists of two nodes, *i* and *j*, the larger one will be considered as node *j*. However, existing scheduling software packages on the market do not require node *j* to be larger than node *i* for any activity.

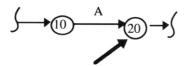

Figure 1 Node *j* (Node 20) of Activity A

It is noted that any node *j* of an activity's immediate predecessors can become node *i* of the activity. However, it is not necessary that node *j* for an activity's immediate predecessors must be the activity's node *i*. For example, Figure 2 shows that even though node 30, the activity C's node *i*, is only the activity A's node *j*, activities A and B are both activity C's immediate predecessors. Figure 3 shows Node *j* of an example arrow diagram network.

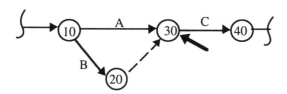

Figure 2 Node *j* (Node 30) of Activity A

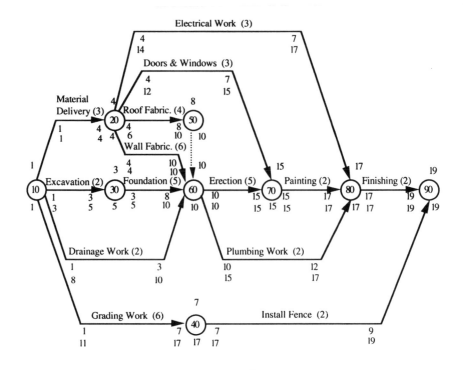

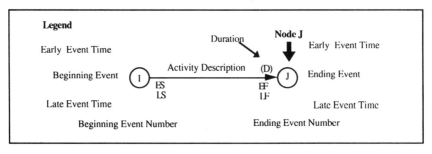

NOTE: The schedule computation in time units is based on the following assumption:

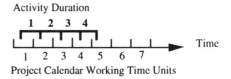

Activity with 4 time units starts at working time unit 1 and is finished at working time unit 5.

Figure 3 Node j of Activities in Arrow Diagram Network

NONWORK UNIT A calendar-specified time unit during which work cannot be scheduled or performed, such as weekends, legal holidays, or other dates selected by the project managing team.

Nonwork units are defined in project calendar information as time when project activities are not scheduled to be performed. Typically, nonwork units are weekends and holidays. When a project consists of many activities to be performed in various parts of the country or world, various calendars can be created with specific nonwork units and be assigned to each activity. For example, a project may require structural component fabrications from a country that has different holidays or nonwork days than those in the United States. Figure 1 shows a sample of nonwork units in project calendar information, indicating two nonworking days (July 4, 1993 and November 25, 1993) and one nonworking period (December 24, 1993 to January 1, 1994).

```
→
                          PROJECT CALENDAR                      TAR1
========================================================================

        Calendar ID: 1                      Title: DAILY CALENDAR


        Workdays:  X MON   X TUE   X WED   X THU   X FRI    SAT    SUN

_____
   Nonwork periods        Page:   1 |  Exceptions              Page:    1
                                    |
     Start            End           |    Start             End
   ~~~~~~~~~~~~~~~   ~~~~~~~~~~~~~~~ |  ~~~~~~~~~~~~~~~   ~~~~~~~~~~~~~~~
     04JUL93                        |
     25NOV93                        |                       .
     25DEC93                        |
                                    |
                                    |
                                    |
                                    |
                                    |
                                    |
       Scroll using Home, End       |    Scroll using PgUp, PgDn
========================================================================
Commands: Add Delete Edit Help More Next Print Return Transfer View Window
Windows : Calendar Global List
```

Figure 1 Project Calendar Information (Primavera Project Planner 5.0)

NORMAL ACTIVITY TIME COST POINT The absolute minimum of direct costs required to perform an activity and the corresponding activity duration, called normal duration *(see Activity duration).*

The normal activity time cost point is the basic information for accomplishing the time-cost trade-off analysis *(see Time-cost trade-off, Cost slope,* and *Crash activity time cost point).* Prager's mechanical analogy is used to explain the normal activity time cost point and crash activity time cost point (Figure 1). In Figure 1 a structural member represents each network activity *(i,j)*. It is composed of a rigid sleeve of a certain length (crash time) and a compressible rod whose natural length is *Dij* (normal time), like a piston. When the compressive force, *fij* (cost), is gradually increased to the member, the rod remains rigid until this force reaches intensity *Cij* (cost slope), which is referred to as the *yield limit* of the member *ij*. The point at which the compressible rod (normal time) stays without any additional compressible force (normal cost) is called the normal activity time cost point.

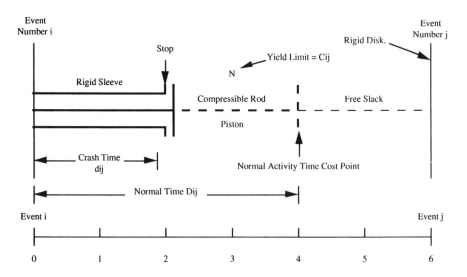

Figure 1 Normal Activity Time Cost Point Using Prager's Mechanical Analogy

A normal activity time cost point is shown in Figure 2. Here point *N* is indicated by coordinate *D* and *CD* on the activity duration time axis and activity direct cost axis respectively. Point *D* on the time axis expresses normal time, and point *CD* on the cost axis expresses normal cost. So point *N* is determined by normal time and normal cost and is called the normal activity time cost point.

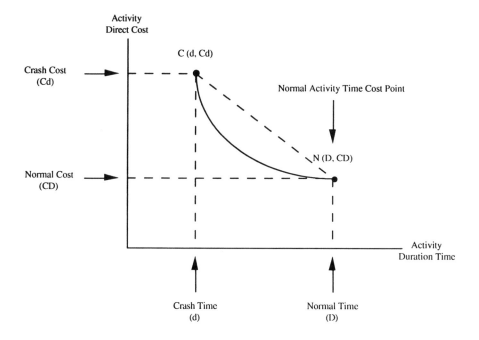

Figure 2 Normal Activity Time Cost Point on the Time-cost Curve

O

OPTIMISTIC TIME ESTIMATE In PERT scheduling techniques, the shortest time in which an activity is expected to be completed, based on a favorable outcome or resolution for each assumption or contingency associated with the estimate of activity duration.

The optimistic time estimate *(Ta)* is the minimum estimated time required for an activity to be performed repeatedly under similar conditions. It is considered less than 5 percentile of the distribution of performance time. The condition of performing the work is assumed to be exceptionally well planned. *Ta* is not considered as a crashed time unless the other two time estimates, *Tb (see Pessimistic time estimate)* and *Tm (see Most likely time estimate),* are considered as crashed time to complete the work.

Calculation of the three time estimates forces the project manager to analyze and evaluate carefully all possible difficulties involved in the activity, to come up with the estimates. The general curve of *Ta* is shown in Figure 1. Table 1 shows *Ta* appearing in the column Time Estimates and Figure 2 shows a PERT diagram with *Ta* shown for each activity.

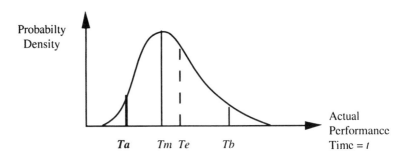

Figure 1 Optimistic Time Estimate Shown on the Distribution of Performance Time

Table 1 Optimistic Time Estimates

Node	Activity	Ta	Time Estimates Tm	Tb	Te = (Ta + 4Tm + Tb)/6	S(Te) = (Tb-Ta)/6	S(Te) x S(Te)
1-2	A	4	4	4	4	0	0
1-3	B	1	3	5	3	0.67	0.44
2-4	C	2	4	12	5	1.67	2.79
2-5	D	9	12	33	15	4	16
3-4	E	4	7	22	9	3	9
4-5	F	2	4	8	4	1	1
5-6	G	5	5	5	5	0	0

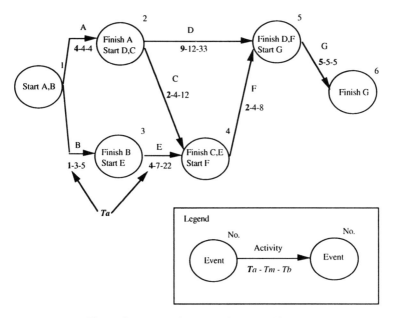

Figure 2 PERT Diagram with *Ta* Indicated

ORGANIZATIONAL CODES The numbers or alphabetized characters that are used to identify the responsibility for each work item in PBS.

Organizational codes are developed from OBS and specify the responsibility of each work item through the combination of PBS. For example, the OBS for the commercial building shown in Figure 1 demonstrates the use of organizational codes.

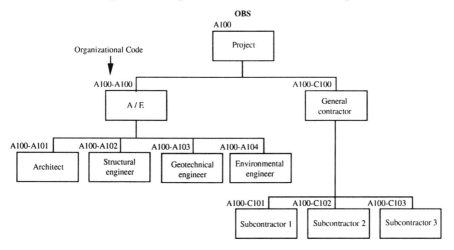

Figure 1 OBS and Organizational Code

ORGANIZATION BREAKDOWN STRUCTURE (OBS) The identification of participants and their hierarchical relationship.

OBS is developed to assign clear responsibilities to each project participant and show the relationships among their responsibilities. In a method of work packaging, a work package is created through the combination of PBS with OBS. So this process gives responsibility to each work item through the combination. For example, the OBS developed for a commercial building is shown in Figure 1.

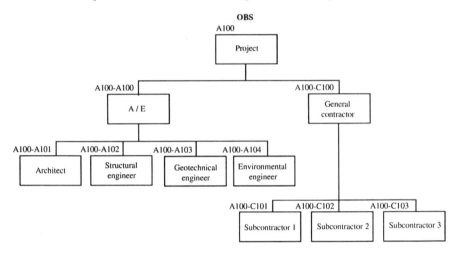

Figure 1 Organization Breakdown Structure for Commercial Building

ORIGINAL DURATION The estimated duration of an activity belonging to an as-planned approved CPM schedule.

The original duration is the amount of time an activity needs to be completed as a planned estimate and can be determined by experienced planners or schedulers. Also, it can be obtained from records or an as-built schedule of a similar project that has been completed. Original duration is used in planning as a baseline or target schedule for a project.

All activity durations *(see Activity duration)* initially estimated for the project target schedule are considered original duration. Original duration can be reassessed when the project is updated and the quantity of work, type of work, production rate, or environmental factors differ from what has been expected. The normal predictable unfavorable work conditions should be included in an original duration estimate. It is necessary to have a complete review of an activity's original duration to ensure accuracy.

The original duration is used as the base for establishing as-planned network computerization and for network analysis. Figures 1 and 2 illustrate the original

406

duration of activities to be used for project schedule calculation in the arrow diagram and precedence diagram techniques, respectively. Figures 3 and 4 show the original duration of activities from a schedule report in tabular format in the arrow diagram and precedence diagram techniques, respectively.

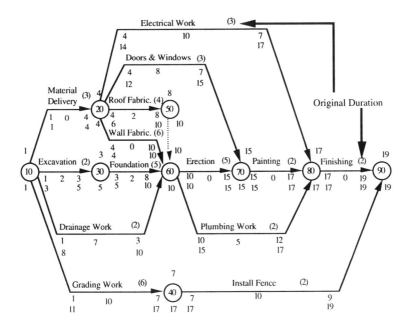

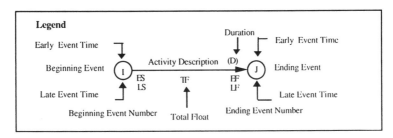

NOTE: The schedule computation in time units is based on the following assumption:

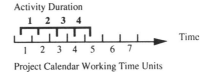

Activity Duration

Project Calendar Working Time Units

Activity with 4 time units starts at working time unit 1 and is finished at working time unit 5.

Figure 1   Original Duration of Activities in Arrow Diagram

407

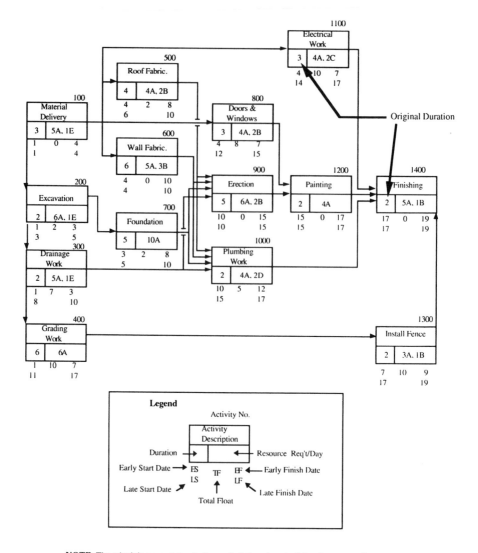

Legend

Activity No.

Activity Description

Duration → ← Resource Req't/Day

Early Start Date → ES TF EF ← Early Finish Date
LS LF
Late Start Date ↗ ↑ ↖ Late Finish Date
Total Float

NOTE: The schedule computation in time units is based on the following assumption:

Activity Duration

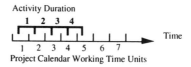

Time

Project Calendar Working Time Units

Activity with 4 time units starts at working time unit 1 and is finished at working time unit 5.

Figure 2 Original Duration of Activities in Precedence Diagram

408

CIVIL ENGINEERING DEPARTMENT   PRIMAVERA PROJECT PLANNER

REPORT DATE 23NOV93 RUN NO.   5   PROJECT SCHEDULE REPORT   START DATE 17MAR93 FIN DATE 9APR93
         16:58
CLASSIC SCHEDULE REPORT - SORT BY ES, TF   DATA DATE 17MAR93 PAGE NO.   1

PRED	SUCC	ORIG DUR	REM DUR	%	CODE	ACTIVITY DESCRIPTION	EARLY START	EARLY FINISH	LATE START	LATE FINISH	TOTAL FLOAT
10	20	3	3	0		MATERIAL DELIVERY	17MAR93*	19MAR93	17MAR93*	19MAR93	0
10	30	2	2	0		EXCAVATION	17MAR93	18MAR93	19MAR93	22MAR93	2
10	60	2	2	0		DRAINAGE WORK	17MAR93	18MAR93	26MAR93	29MAR93	7
10	40	6	6	0		GRADING WORK	17MAR93	24MAR93	31MAR93	7APR93	10
30	60	5	5	0		FOUNDATION	19MAR93	25MAR93	23MAR93	29MAR93	2
20	60	6	6	0		WALL FABRICATION	22MAR93	29MAR93	22MAR93	29MAR93	0
20	50	4	4	0		ROOF FABRICATION	22MAR93	25MAR93	24MAR93	29MAR93	2
20	70	3	3	0		DOORS AND WINDOWS	22MAR93	24MAR93	1APR93	5APR93	8
20	80	3	3	0		ELECTRICAL WORK	22MAR93	24MAR93	5APR93	7APR93	10
40	90	2	2	0		INSTALL FENCE	25MAR93	26MAR93	8APR93	9APR93	10
50	60	0	0	0		DUMMY 1	26MAR93	25MAR93	30MAR93	29MAR93	2
60	70	5	5	0		ERECTION	30MAR93	5APR93	30MAR93	5APR93	0
60	80	2	2	0		PLUMBING WORK	30MAR93	31MAR93	6APR93	7APR93	5
70	80	2	2	0		PAINTING	6APR93	7APR93	6APR93	7APR93	0
80	90	2	2	0		FINISHING	8APR93	9APR93	8APR93	9APR93	0

Figure 3 Original Duration of Activities in Tabular Schedule Report
in Arrow Diagram Technique

CIVIL ENGINEERING DEPARTMENT   PRIMAVERA PROJECT PLANNER

REPORT DATE 23NOV93 RUN NO.   12   PROJECT SCHEDULE REPORT   START DATE 13MAR93 FIN DATE 9APR93
         17:02
CLASSIC SCHEDULE REPORT - SORT BY ES, TF   DATA DATE 17MAR93 PAGE NO.   1

ACTIVITY ID	ORIG DUR	REM DUR	%	CODE	ACTIVITY DESCRIPTION	EARLY START	EARLY FINISH	LATE START	LATE FINISH	TOTAL FLOAT
100	3	3	0		MATERIAL DELIVERY	17MAR93*	19MAR93	17MAR93*	19MAR93	0
200	2	2	0		EXCAVATION	17MAR93	18MAR93	19MAR93	22MAR93	2
300	2	2	0		DRAINAGE WORK	17MAR93	18MAR93	26MAR93	29MAR93	7
400	6	6	0		GRADING WORK	17MAR93	24MAR93	31MAR93	7APR93	10
700	5	5	0		FOUNDATION	19MAR93	25MAR93	23MAR93	29MAR93	2
600	6	6	0		WALL FABRICATION	22MAR93	29MAR93	22MAR93	29MAR93	0
500	4	4	0		ROOF FABRICATION	22MAR93	25MAR93	24MAR93	29MAR93	2
800	3	3	0		DOORS & WINDOWS	22MAR93	24MAR93	1APR93	5APR93	8
1100	3	3	0		ELECTRICAL WORK	22MAR93	24MAR93	5APR93	7APR93	10
1300	2	2	0		INSTALL FENCE	25MAR93	26MAR93	8APR93	9APR93	10
900	5	5	0		ERECTION	30MAR93	5APR93	30MAR93	5APR93	0
1000	2	2	0		PLUMBING WORK	30MAR93	31MAR93	6APR93	7APR93	5
1200	2	2	0		PAINTING	6APR93	7APR93	6APR93	7APR93	0
1400	2	2	0		FINISHING	8APR93	9APR93	8APR93	9APR93	0

Figure 4 Original Duration of Activities in Tabular Schedule Report
in Precedence Diagram Technique

OUT-OF-SEQUENCE PROGRESS The work started or completed for an activity before it is logically scheduled to occur. For example, in a conventional relationship, an activity that starts before its predecessor is completed indicates out-of-sequence progress.

A network diagram is logically developed to represent a project execution plan. It is comprised of a series of activities connected by arrows representing their inter-relationships. Theoretically, tasks must be carried out in a specified logical order established in a network. In a specific circumstance, however, it may be possible to start and/or complete some tasks before the logical constraints from predecessors are satisfied. This situation does not follow the logical sequence of work scheduled and is called out-of-sequence progress. Any activity causing out-of-sequence progress is called an out-of-sequence activity.

When out-of-sequence progress occurs, it may happen only to the start of that activity. Normally, succeeding activities of an out-of-sequence activity are not able to start until all logical constraints are satisfied. Figure 1 shows a sample of an updated precedence diagram network with an out-of-sequence activity, Install Fence. Logically, this activity cannot be started until its preceding activity, Grading Work, is completed.

It is practical and recommended that the schedule computation be set to retain the original logic; otherwise, the out-of-sequence progress may decrease the project duration and cause an earlier-than-expected project completion date. Figure 2 shows an updated project schedule in tabular format and a computer message detecting the out-of-sequence activity.

410

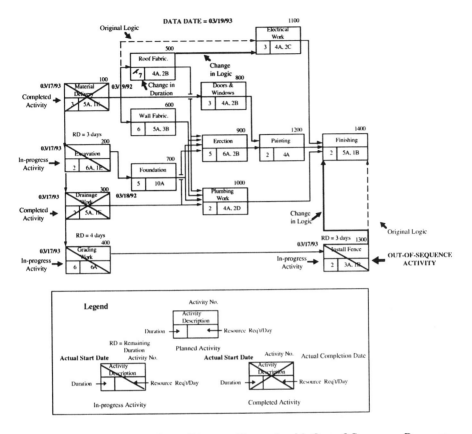

Figure 1 Updated Precedence Diagram Network with Out-of-Sequence Progress

411

Activity	Predecessor	Rel	Lag	Description
1300	400	FS	0	Activity started, predecessor has not finished.

CIVIL ENGINEERING DEPARTMENT   PRIMAVERA PROJECT PLANNER

REPORT DATE 24NOV93 RUN NO.   2   PROJECT SCHEDULE REPORT   START DATE 13MAR93 FIN DATE 13APR93
                1:31
CLASSIC SCHEDULE REPORT - SORT BY ES, TF   DATA DATE 20MAR93 PAGE NO.   1

ACTIVITY ID	ORIG DUR	REM DUR	%	CODE	ACTIVITY DESCRIPTION	EARLY START	EARLY FINISH	LATE START	LATE FINISH	TOTAL FLOAT
100	3	0	100		MATERIAL DELIVERY	17MAR93A	19MAR93A			
300	2	0	100		DRAINAGE WORK	17MAR93A	18MAR93A			
200	2	3	0		EXCAVATION	17MAR93A	24MAR93		24MAR93	0
400	6	4	33		GRADING WORK	17MAR93A	25MAR93		6APR93	8
1300	2	3	0		INSTALL FENCE	17MAR93A	30MAR93		9APR93	8
500	4	7	0		ROOF FABRICATION	22MAR93	30MAR93	23MAR93	31MAR93	1
600	6	6	0		WALL FABRICATION	22MAR93	29MAR93	24MAR93	31MAR93	2
800	3	3	0		DOORS & WINDOWS	22MAR93	24MAR93	5APR93	7APR93	10
700	5	5	0		FOUNDATION	25MAR93	31MAR93	25MAR93	31MAR93	0
1100	3	3	0		ELECTRICAL WORK	31MAR93	2APR93	7APR93	9APR93	5
900	5	5	0		ERECTION	1APR93	7APR93	1APR93	7APR93	0
1000	2	2	0		PLUMBING WORK	1APR93	2APR93	8APR93	9APR93	5
1200	2	2	0		PAINTING	8APR93	9APR93	8APR93	9APR93	0
1400	2	2	0		FINISHING	12APR93	13APR93	12APR93	13APR93	0

Figure 2 Updated Schedule Report in Tabular Format
with Out-of-Sequence Activity

OVERHEAD *(see Activity indirect costs and Burden)*

OVER PLAN (UNDER PLAN) The current planned or budgeted cost minus the latest revised estimate for work items in the project breakdown structure (PBS). When the planned cost exceeds the latest revised estimate, an under-plan condition exists. When the latest revised estimate exceeds the planned cost, an over-plan condition exists.

An over- or under-plan condition indicates the project cost status at completion based on the latest revised estimate *(see Latest revised estimate)*. Obtaining the cost status at completion of a project requires the determination of planned cost as originally estimated *(see Budget at completion)* and the latest revised estimate at completion. The difference between the planned or budgeted and latest revised estimate indicates an over- and under-plan condition expressed as at-completion cost variance (ACV) *(see At-completion variance)*. For example, at a given data date *(see Data date)* to complete a project, the planned or budgeted cost has allocated $1,000,000

412

and the latest revised estimate is $1,100,000. An over-plan condition of $100,000 then exists. In contrast, if the latest revised estimate is $900,000, an under-plan condition of $100,000 exists. Over- and under-plan conditions are identified at completion by a cost variance as follows:

$$ACV = BAC - EAC$$

where ACV is the at-completion cost variance, BAC is the budgeted cost at completion, and EAC is the latest revised estimate at completion.

Variance	–	0	+
ACV	Over Plan	On Cost	Under Plan

An over- or under-plan condition can also be represented by the project cost progress report as shown in Figures 1 and 2.

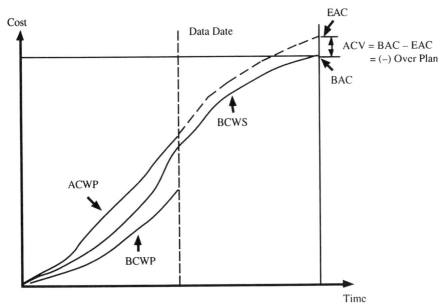

BAC = Budget Cost at Completion
EAC = Latest Revised Estimate at Completion
BCWP = Budget Cost of Work Performed
ACWP = Actual Cost of Work Performed
ACV = At-completion Variance

Figure 1 Project Cost Progress Report Showing Over-plan Status

413

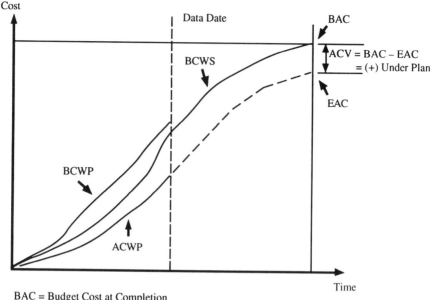

Cost

Data Date

BAC

BCWS

ACV = BAC – EAC
= (+) Under Plan

EAC

BCWP

ACWP

Time

BAC = Budget Cost at Completion
EAC = Latest Revised Estimate at Completion
BCWP = Budget Cost of Work Performed
ACWP = Actual Cost of Work Performed
ACV = At-completion Variance

Figure 2 Project Cost Progress Report Showing Under-plan Status

The following are possible over- or under-plan causations:

1. Change orders (over or under plan)
2. Inaccurate EAC or BAC (over or under plan)
3. Lower (over plan) or higher (under plan) productivity than expected
4. Unexpected increase (over plan) or decrease (under plan) in material, labor, or equipment costs

OVERRUN (UNDERRUN) (WORK PERFORMED TO DATE) The value of the worked performed to date minus the actual cost for the same work. When the value exceeds the actual cost, an underrun condition exists. When the actual cost exceeds the value, an overrun condition exists.

An overrun or underrun condition indicates the current project cost status. Determining the cost overrun and underrun requires measurement of the actual value (budget cost of work performed) and the actual cost of work performed to date *(see Budgeted cost of work performed and Actual cost of work performed)*. The differ-

414

ence between budgeted and actual cost of work performed can then be calculated and identified as current cost variance (CV) *(see Cost variance)*. For example, at a given data date *(see Data date)* the budget has allocated $25,000 for a certain amount of work and the actual cost for that amount of work has been $30,000. A cost overrun condition of $5000 then exists. In contrast, if the actual cost has been $20,000, a cost underrun condition of $5000 exists. The overrun and underrun conditions are identified by cost variance as follows:

$$CV = BCWP - ACWP$$

where CV is the cost variance, BCWP is the budgeted cost of work performed, and ACWP is the actual cost of work performed.

Variance	−	0	+
ACV	Overrun	On Cost	Underrun

The overrun or underrun can also be represented by the project cost progress report as shown in Figures 1 and 2.

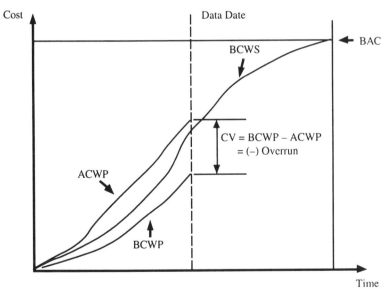

BAC = Budget Cost at Completion
BCWP = Budget Cost of Work Performed
ACWP = Actual Cost of Work Performed
CV = Cost Variance

Figure 1 Project Cost Progress Report Showing Overrun Status

415

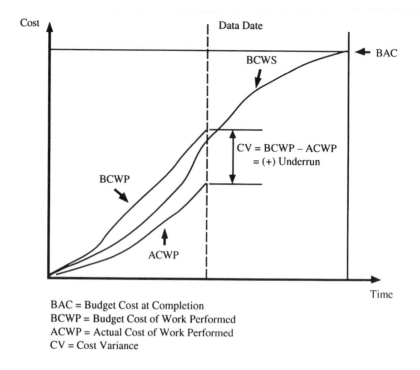

BAC = Budget Cost at Completion
BCWP = Budget Cost of Work Performed
ACWP = Actual Cost of Work Performed
CV = Cost Variance

Figure 2 Project Cost Progress Report Showing Underrun Status

The following are possible overrun or underrun causations:

1. Technical problems requiring the allocation of extra resources (overrun)
2. Inaccurate estimation of original work (overrun or underrun)
3. Lower (overrun) or higher (underrun) productivity than expected
4. Unexpected increase (overrun) or decrease (underrun) in material, labor, or equipment costs

416

P

PATH The physical continuous series of connected activities throughout a network.

In CPM *(see Critical path method),* a logical network diagram *(see Network)* shows activities and their interrelationships representing work sequences. A network diagram possesses the multiple continuous series of connected activities from the start to finish of a project. Each continuous series, called a path, can be shown only on a network diagram. After the schedule calculation *(see Network analysis),* management can identify the critical path that is the longest path in the network *(see Critical path).* Each activity in the critical path is called a critical activity *(see Critical activity).* Most of the time when project early and late completion dates are the same, the critical path and critical activities have zero total float *(see Activity total float).* Management may also be interested in a nearly critical path *(see Nearly critical path),* whose activities contain total floats close to critical activities.

In the event that management focus is in a specific time frame, a path may not be from the start to the finish of a project, but from the specified start to finish time frame. Figures 1 and 2 show the network in arrow and precedence diagrams with critical and nearly critical paths.

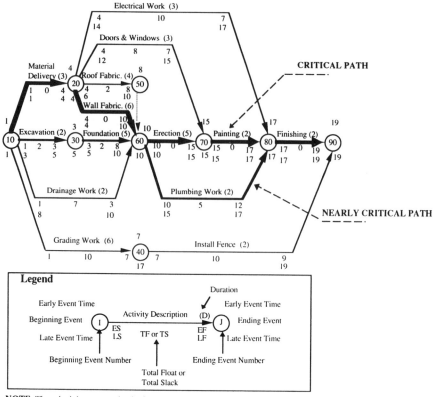

Electrical Work (3)
4   10   7
14   17

Doors & Windows (3)
4   8   7
12   15

Material
Delivery (3)   4   Roof Fabric. (4)   8
1   0   4   [20] 4   2   8   [50]
1   4   4   6   10
                    Wall Fabric. (6)
1   4   0   10
Excavation (2)   3   4   10   10
[10] 1   2   3   [30] 3   2   8   [60] 10
1   3   5   5   5   10   10

Erection (5)
15
10   0   15
15   15   15

Painting (2)
17
15   0   17
17   17

Finishing (2)
17   19
17   0   19
17   17   [80] 19   19 [90]

CRITICAL PATH

NEARLY CRITICAL PATH

Drainage Work (2)
1   7   3
8   10

Plumbing Work (2)
10   5   12
15   17

Grading Work (6)   7
1   10   7 [40] 7   Install Fence (2)
         17   7   10   9
                        19

Legend

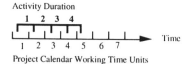

Duration
Early Event Time   Early Event Time
Beginning Event   [I] Activity Description (D) [J] Ending Event
                  ES   EF
Late Event Time   LS   TF or TS   LF   Late Event Time
Beginning Event Number   Ending Event Number
                  Total Float or
                  Total Slack

NOTE: The schedule computation in time units is based on the following assumption:

Activity Duration
1   2   3   4
|————|————|————|
1   2   3   4   5   6   7   Time
Project Calendar Working Time Units

Activity with 4 time units starts at working time unit 1 and is finished at working time unit 5.

Figure 1 Arrow Diagram Network

418

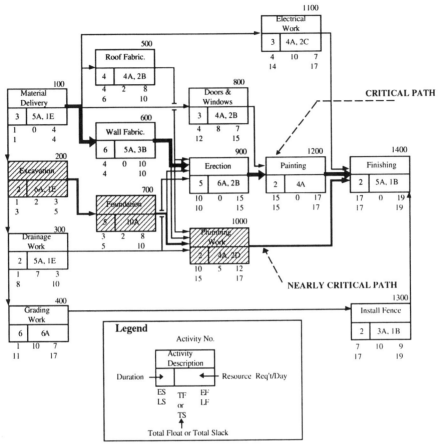

CRITICAL PATH

NEARLY CRITICAL PATH

Legend

Activity No.

Activity Description

Duration →

← Resource Req't/Day

ES TF EF
LS or LF
TS

Total Float or Total Slack

NOTE: The schedule computation in time units is based on the following assumption:

Activity Duration

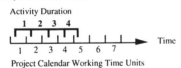

Time

Project Calendar Working Time Units

Activity with 4 time units starts at working time unit 1 and is finished at working time unit 5.

Figure 2 Precedence Diagram Network

419

PBS *(see Project breakdown structure)*

PDM *(see Precedence diagramming method)*

PDM FINISH-TO-FINISH RELATION *(see Dependency, finish to finish)*

PDM FINISH-TO-START RELATION *(see Dependency, finish to start)*

PDM START-TO-FINISH RELATION *(see Dependency, start to finish)*

PDM START-TO-START RELATION *(see Dependency, start to start)*

PERCENT COMPLETE (PC) The performance measurement of earned value to date in comparison with the total budget at completion in terms of cost.

The earned value to date is determined by multiplying actual work quantities by the original unit price. Percent complete cannot be greater than 100% and can be at either the work item or project level.

Percent complete (PC) is one of the C/SCSC elements *(see Cost and schedule control system criteria).*

The PC is calculated using the formula

$$PC = BCWP / BAC$$

(See Budgeted cost of work completed and Budget at completion.) Interpretation of the PC is as follows:

VARIANCE	Less Than 1.00	Equal To 1.00	More Than 1.00
PC	% Complete	Project Complete	Incorrect PC

For example, a PC of 0.90 (90%) indicates that if the total budgeted cost at completion were $100, a total of $90 worth of work would have been accomplished to date. That is, the project is 90% complete as of the report date.

To develop the PC in a project, the following steps are generally taken:

Step 1: Use the original PBS of a project.

Step 2: Use the original cost estimation of the project based on the PBS.

Step 3: Use the latest updated CPM schedule by the data date (time now) *(see Data date).*

Step 4: Calculate the BCWP of the activity and project level on the data date (time now).

Step 5: Calculate the BAC on the data date (time now).

Step 6: Calculate the PC by dividing the BCWP by the BAC. The results of steps 4 to 6 are shown in the monthly cost performance report.

The preceding six steps are explained using a residential house as an example.

1. *Step 1:* PBS

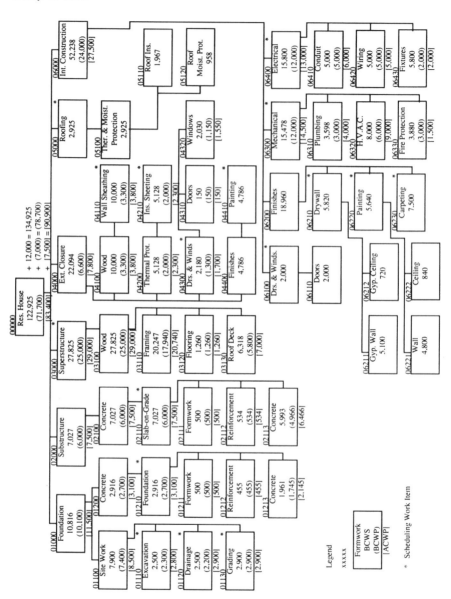

422

2. *Step 2:* Cost Estimation

DESCRIPTION	U/M	U/P	Q'TY	TOTAL
01 General Requirements				
Subtotal	L/S	12,000	1	12,000
02 Site Work				
01110 Excavation	C.Y.	1.89	1,323	2,500
01120 Drainage	L.F.	3.11	804	2,500
01130 Grading	C.Y.	1.06	2,736	2,900
Subtotal				7,900
03 Concrete				
01211 Foundation Form	SFCA	2.89	173	500
01212 Foundation Re-bar	Tons	910	0.5	455
01213 Foundation Concrete	C.Y.	103.2	19	1,961
02111 Slab-on-Grade Form	L.F.	1.38	362	500
02112 Slab-on-Grade Re-bar	Tons	890	0.6	534
02113 Slab-on-Grade Concrete	C.Y.	90.8	66	5,993
Subtotal				9,943
06 Wood				
03110 Superstructure Framing	M.B.F.	1,191	17	20,247
03120 Superstructure Flooring	S.F.	0.63	2,000	1,260
03130 Roof Decking	S.F.	5.01	1,261	6,318
04110 Wall Sheathing	S.F.	1.17	8,547	10,000
Subtotal				37,825
07 Thermal & Moisture Protection				
04210 Wall Insulating Sheeting	S.F.	0.6	8,547	5,128
05110 Roof Insulating	S.F.	1.56	1,261	1,967
05120 Roof Moisture Protection	S.F.	0.76	1,261	958
Subtotal				8,053
08 Doors & Windows				
04310 Ext. Doors	Ea.	150	1	150
04320 Ext. Windows	Ea.	203	10	2,030
06110 Int. Doors	Ea.	200	10	2,000
Subtotal				4,180
09 Finishes				
04410 Ext. Painting	S.F.	0.56	8,547	4,786
06211 Int. Gyp. Wallboard	S.F.	0.34	15,000	5,100
06212 Int. Gyp. Ceiling Board	S.F.	0.36	2,000	720
06311 Int. Wall Painting	S.F.	0.32	15,000	4,800
06312 Int. Ceiling Painting	S.F.	0.42	2,000	840
06230 Carpeting	S.Y.	5	1,500	7,500
Subtotal				23,746
15 Mechanical				
06310 Plumbing	L.F.	3.98	904	3,598
06320 H.V.A.C.	L/S	8,000	1	8,000
06330 Fire Protection	Ea.	776	5	3,880
Subtotal				15,478
16 Electrical				
06410 Electrical Conduit	L/S	5,000	1	5,000
06420 Electrical Wiring	L/S	5,000	1	5,000
06430 Electrical Fixtures	L/S	5,800	1	5,800
Subtotal				15,800
GRAND TOTAL				134,925

3. *Step 3:* Updated CPM Network Schedule

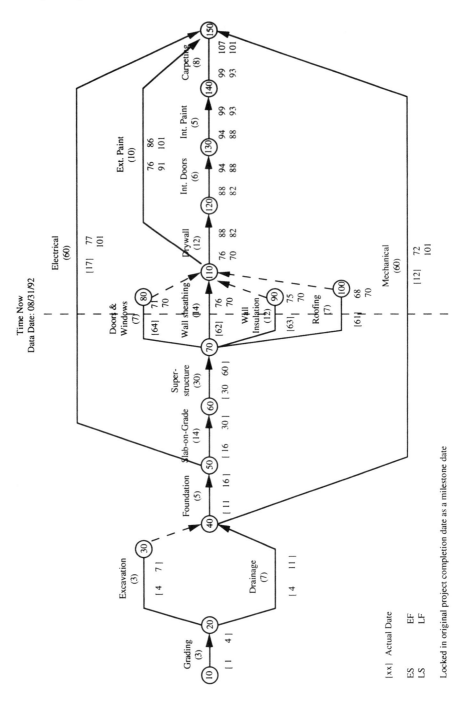

4. *Steps 4 to 6:* Monthly Cost Performance Report

MONTHLY COST PERFORMANCE REPORT: CPM ITEM

PROJECT : RESIDENTIAL HOUSE
REPORT DATE: 08/31/92

ITEM		CURRENT PERIOD									CUMULATIVE TO DATE									AT COMPLETION		
		BUDGET COST			VARIANCE			INDEX			BUDGET COST			VARIANCE			INDEX					
PBS	CPM	BCWS	BCWP	ACWP	SV	CV	AV	SPI	CPI	PC	BCWS	BCWP	ACWP	SV	CV	AV	SPI	CPI	PC	BAC	EAC	ACV
01130	10 - 20										2,900	2,900	2,900	0	0	0	1.00	1.00	1.00	2,900	2,900	0
01110	20 - 30										2,500	2,300	2,800	-200	-500	-300	0.92	0.82	0.92	2,500	2,800	-300
01120	20 - 40										2,500	2,200	2,800	-300	-600	-300	0.88	0.79	0.88	2,500	2,800	-300
01210	40 - 50										2,916	2,700	3,100	-216	-400	-184	0.93	0.87	0.93	2,916	3,100	-184
06300	40 - 150	5,160	5,000	5,200	-160	-200	-40	0.97	0.96	0.32	13,416	12,000	14,500	-1,416	-2,500	-1,084	0.89	0.83	0.78	15,478	16,562	-1,084
02110	50 - 60										7,027	6,000	7,500	-1,027	-1,500	-473	0.85	0.80	0.85	7,027	7,500	-473
03000	60 - 70	14,903	13,900	15,000	-1,003	-1,100	-97	0.93	0.93	0.50	27,825	25,000	29,000	-2,825	-4,000	-1,175	0.90	0.86	0.90	27,825	29,000	-1,175
06400	50 - 150	5,260	5,100	5,300	-160	-200	-40	0.97	0.96	0.32	12,361	12,000	13,000	-361	-1,000	-639	0.97	0.92	0.76	15,800	16,439	-639
04300	70 - 80	1,555	1,300	1,700	-255	-400	-145	0.84	0.76	0.60	1,555	1,300	1,700	-255	-400	-145	0.84	0.76	0.60	2,180	2,325	-145
04210	70 - 90	2,135	2,000	2,300	-135	-300	-165	0.94	0.87	0.39	2,135	2,000	2,300	-135	-300	-165	0.94	0.87	0.39	5,128	5,293	-165
04110	70 - 110	3,570	3,300	3,800	-270	-500	-230	0.92	0.87	0.33	3,570	3,300	3,800	-270	-500	-230	0.92	0.87	0.33	10,000	10,230	-230
05000	70 - 100																			2,925	2,925	0
06210	110-120																			5,820	5,820	0
04410	110-150																			4,786	4,786	0
06100	120-130																			2,000	2,000	0
06220	130-140																			5,640	5,640	0
06230	140-150																			7,500	7,500	0
SUBTOTAL		32,583	30,600	33,300	-1,983	-2,700	-717	0.94	0.92	0.25	78,705	71,700	83,400	-7,005	-11,700	-4,695	0.91	0.86	0.58	122,925	127,620	-4,695
GEN. REQ.		2,400	2,400	2,500		-100	-100	1.00	0.96	0.20	7,200	7,000	7,500	-200	-500	-300	0.97	0.93	0.58	12,000	12,300	-300
TOTAL		34,983	33,000	35,800	-1,983	-2,800	-817	0.94	0.92	0.24	85,905	78,700	90,900	-7,205	-12,200	-4,995	0.92	0.87	0.58	134,925	139,920	-4,995

Note: The project at the data date has 24% in the current period and 58% cumulatively as the percent complete (PC).
The PC is shown either by the work item or by the project level, and is calculated in terms of cost ($).

425

PERFORMANCE MEASUREMENT BASELINE (BASELINE PLAN) The time-phased budgeted plan against which contract performance is measured; it is formed by the budget assigned to project activities and/or work items and the applicable indirect budget.

After the contract is awarded and before the project is begun, a contractor has to develop a baseline plan for project execution. Typically, the baseline plan is based on detailed estimates of the quality of labor, materials, equipment, and other resources required to accomplish all project activities. The most common baseline is the planned or budgeted cost.

When the detailed estimate is complete and costs have been allocated for each project activity, the time-phased budget can be drawn through the computation provided in most scheduling softwares. The time-phased budget is also called a budgeted cost of work performed (BCWP) or S curve *(see Budgeted cost of work performed),* representing the baseline for the performance measurement as shown in Figure 1.

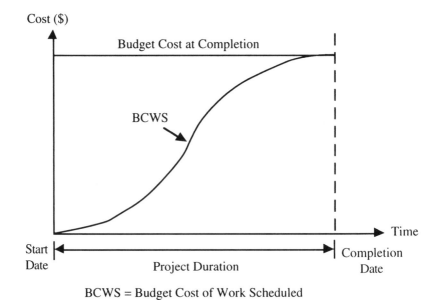

Figure 1 Performance Measurement Baseline

During the project execution phase, the actual cost of work performed (ACWP) and the budgeted cost of work performed (BCWP) *(see Actual cost of work performed* and *Budgeted cost of work performed)* are measured periodically and compared against the baseline plan or budgeted cost of work scheduled at the data date as shown in Figure 2. The comparison reveals if the work is over or under budget or behind or ahead of schedule *(see Overrun* and *Over plan).*

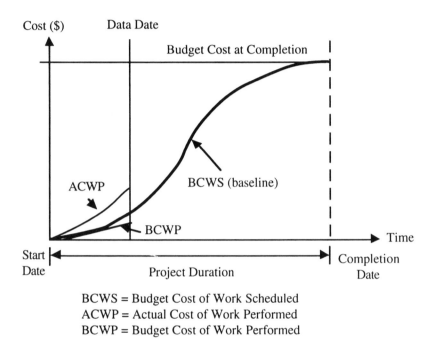

BCWS = Budget Cost of Work Scheduled
ACWP = Actual Cost of Work Performed
BCWP = Budget Cost of Work Performed

Figure 2 Comparison of Baseline Plan and Actual Project Performance

427

PERT Acronym for *performance evaluation and review technique*, or *program evaluation and review technique*. It is another mathematical method of analyzing a network of interdependent activities, offering more probability options than CPM.

PERT was devised originally as a tool to aid program managers in predicting the probability of reaching specific milestones and forecasting uncertain events in research and new product development programs. The development of PERT was intended to assist in planning when no historical cost or time information are available. It was developed by the U.S. Navy Special Project Office in conjunction with Booz-Allen and Hamilton, the management consultants, and the Lockheed Aircraft Corporation to control the Poraris missile system in 1958. At that time, the system was named the program evaluation and review technique.

The PERT system was, in the very beginning, adopted by the Air Force under the name *program evaluation procedure* (PEP). After that, PERT was modified and named differently as follows: PERT I, PERT II, PERT III, PERT/TIME, and PERT/COST. In June 1962, the Department of Defense and the National Aeronautics and Space Administration adopted a uniform approach of planning and controlling procedures on major weapon systems.

PERT, rarely used on today's construction projects, has the potential to be used on any construction project where quantities or productivity is uncertain. PERT is a statistical treatment of uncertain activity performance time and assumes a variability in activity durations based on a variability in production rates. The CPM assumes that activity durations do not vary or vary so little that they can be assumed to be deterministic. Like CPM scheduling, PERT uses logic diagrams to analyze performance times. PERT focuses on events (nodes) that must occur prior to successful completion of a project. The probability of achieving performance of all activities that define an event is the outcome of PERT calculation. Because of PERT's emphasis on events, PERT is considered event oriented. CPM, in contrast, focuses on completion of the individual activities that comprise the project. CPM is thus considered activity oriented.

PERT model uses arrows to represent activity durations while nodes or events refer to the status of the project at that point in time. Figure 1 shows three diagrams of CPM, in precedence, arrow, and PERT diagrams for comparison.

The primary use of PERT is for projects that have not been done before. PERT permits more information than with other deterministic methods. The PERT model requires three estimates for activity duration: the optimistic time *(Ta) (see Optimistic time estimate)*, the most likely time *(Tm) (see Most likely time estimates)*, and the pessimistic time *(Tb) (see Pessimistic time estimate)*, to compute the statistical limits that describe the activity duration and the distribution. The expected duration or expected mean time *(Te) (see Expected duration)* is derived from the equation

$$Te = \frac{Ta + 4 \times Tm + Tb}{6}$$

For every activity with three time estimates, a probability distribution for all possible activity durations is assumed to be distributed as shown in Figure 2.

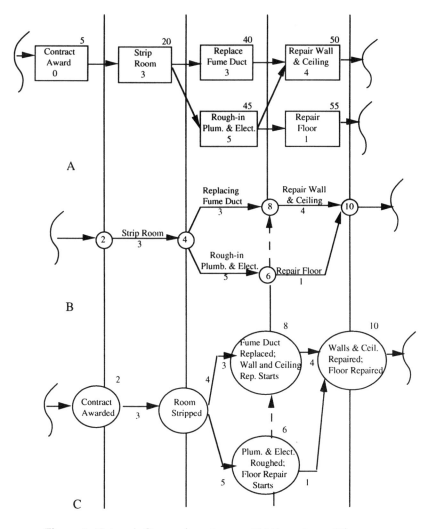

Figure 1 Network Comparison Among (A) Precedence Diagram,
(B) Arrow Diagram, and (C) PERT

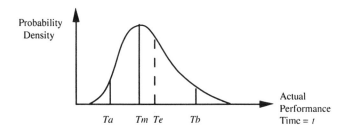

Figure 2 PERT System of Three Time Estimates

The standard deviation, which is a statistical measure of uncertainty, the spread of the distribution curve about its mean value, is given by

$$S(Te) = \frac{Tb - Ta}{6}$$

The variance is defined as the square of the standard deviation, so that

$$V(Te) = S^2(Te) = \left[\ \frac{Tb - Ta}{6}\ \right]^2$$

The higher the standard deviation of an activity, the more likely it is that the actual time required to complete the activity will differ from the *Te*. PERT also applies a normal probability distribution to estimate the probability of meeting a schedule. The probability is

$$\frac{1}{(2\pi)^{1/2}} \int e^{\frac{Z^2/-2}{}} \ dz$$

in which

$$z = \frac{TS - TE}{S(TE)}$$

where *Ts* represents the scheduled event time *(see Scheduled event time)*, TE the earliest expected event time, and *S(TE)* the accumulated standard deviation.

Figure 3 shows a PERT diagram consisting of seven activities, A, B, C, D, E, F, and G, with three time estimates for each activity before schedule calculation.

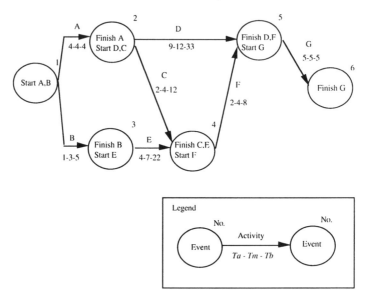

Figure 3 Example of PERT with Three Time Estimates Before Schedule Calculation

From the formula,

$$Te = \frac{Ta + 4 \times Tm + Tb}{6} \qquad \text{and} \qquad S(Te) = \frac{Tb - Ta}{6}$$

Te (A)	=	4	S(A)	=	0	$S2$(A)	=	0
Te (B)	=	3	S(B)	=	0.67	$S2$(B)	=	0.45
Te (C)	=	5	S(C)	=	1.67	$S2$(C)	=	2.79
Te (D)	=	15	S(D)	=	4	$S2$(D)	=	16
Te (E)	=	9	S(E)	=	3	$S2$(E)	=	9
Te (F)	=	4	S(F)	=	1	$S2$(F)	=	1
Te (G)	=	5	S(G)	=	0	$S2$(G)	=	0

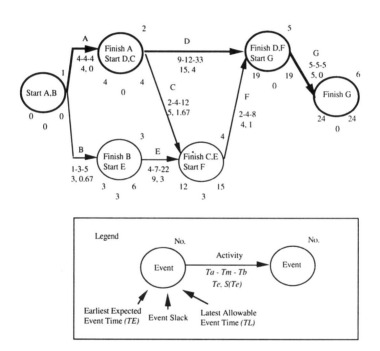

Figure 4 Schedule Calculation from PERT Method

Figure 4 shows a computation schedule for a PERT network. Bold nodes refer to critical events, and bold lines represent critical paths of the network. The earliest expected event time is shown on the bottom left corner, and the latest allowable event time is shown on the bottom right corner of the event. The event slack is shown in the middle under the event. The standard deviation of the path can be obtained from the formula

431

$$S(TE) = [\ \Sigma \ S_i^2 \ (Te) \ ]^{1/2}$$

where i = all activity(s) of the longest path

$$= \ (0 + 16 + 0)^{1/2} \ = \ 4$$

The probability that the project will meet the scheduled event time on day 25 or that the project will be finished in 25 days is

$$Z = \frac{25 - 24}{4} = 0.25$$

From Table 1, at $Z = 0.25$, the probability $= 0.6$. Hence the probability that the project will be finished in 25 days is approximate 60%.

Table 1   Value of Standard Normal Distribution Function

Z	Probability of Meeting Due Date	Z	Probability of Meeting Due Date	Z	Probability of Meeting Due Date
3.0	0.999	0.8	0.788	-1.2	0.115
2.8	0.997	0.6	0.726	-1.4	0.081
2.6	0.995	0.4	0.655	-1.6	0.055
2.4	0.992	0.2	0.579	-1.8	0.036
2.2	0.986	0.0	0.500	-2.0	0.023
2.0	0.977	-0.2	0.421	-2.2	0.014
1.8	0.964	-0.4	0.345	-2.4	0.008
1.6	0.945	-0.6	0.274	-2.6	0.005
1.4	0.919	-0.8	0.212	-2.8	0.003
1.2	0.885	-1.0	0.159	-3.0	0.001
1.0	0.841				

432

PESSIMISTIC TIME ESTIMATE The maximum time required for an activity under adverse conditions.

A pessimistic time estimate *(Tb)* is the maximum estimated time required for an activity to be completed when an unusually bad condition is experienced. There is considered less than a 5% chance of occurrence for repeated activity performance under essentially the same conditions. *Tb* is assumed to be in the worst possible work condition, including an initial failure or delay; however, it does not include unexpected major hazards or acts of God.

Estimates of *Ta (see Optimistic time estimate)*, *Tm (see Most likely time estimate)*, and *Tb* are based on work with enough labor and facilities. Time estimates should be revised when the scope of activity, or labor and facilities assigned to the activity, are changed.

The general *Tb* curve is shown in Figure 1. Table 1 shows *Tb* appearing in the column under Time Estimates, and Figure 2 is a PERT diagram with *Tb* shown for each activity.

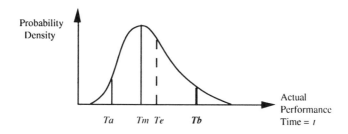

Figure 1 Pessimistic Time Estimate Shown on Distribution of Performance Time

Table 1 Pessimistic Time Estimates

Node	Activity	Time Estimates Ta	Tm	Tb	Te = [Ta+4Tm+Tb]/6	S(Te) = [Tb-Ta]/6	S(Te) x S(Te)
1-2	A	4	4	4	4	0	0
1-3	B	1	3	5	3	0.67	0.44
2-4	C	2	4	12	5	1.67	2.79
2-5	D	9	12	33	15	4	16
3-4	E	4	7	22	9	3	9
4-5	F	2	4	8	4	1	1
5-6	G	5	5	5	5	0	0

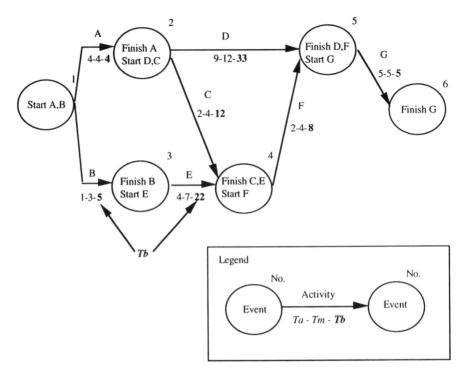

Figure 2 Example of PERT Diagram with T*b* Indicated

PLAN *(see Project plan)*

PLANNED COST The approved planned cost for a work package or summary item. When totaled with the planned costs for all other work packages, this cost results in the total cost estimate committed under a contract for the program or project. It includes direct and indirect costs associated with the project.

Planned cost must to be submitted to authorized bodies for approval. The responsible bodies for project performance should be authorized to spend up to the planned cost without notification or request for any change in appropriation. This planned cost, called the *budget,* should be an accurate and realistic target. The planned cost is prepared using project breakdown structure and work packages as shown in Figures 1 to 3.

434

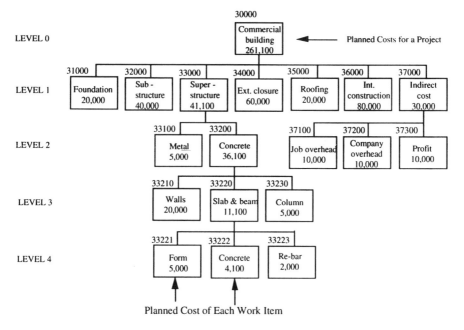

Figure 1 Project Breakdown Structure with Planned Cost for Commercial Building

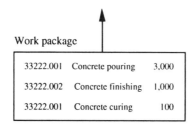

Figure 2 Work Packages Related to Work Item 332222

Cost Estimate

Item Code	Description	Quantity	Material	Labor	Equipment	Subcontr.	Total
33222.001	Concrete pouring	30 C.Y.	1,500	500	1,000	0	3,000
33222.002	Concrete finishing	60 S.F.	0	500	500	0	1,000
33222.003	Concrete curing	60 S.F.	0	100	100	0	100
	Subtotal.	30 C.Y.	1,500	1,100	1,500	0	4,100

Planned Costs

Figure 3 Planned Cost Estimate at Work Package Level

435

PLANNING The establishment of project activities and events, their logical relationships and interrelations to each other, and the sequences in which they are to be accomplished.

Project planning is one of the most important functions determining the success or failure of a project. Prior to the start of a project, a project plan, determining the sequential order in which project tasks or activities will be performed, should be developed. The project planning must satisfy the project objectives, which is within time and budget. It involves the determination of project tasks or activities, their logical interrelationships with each other, and the duration and resources required to complete the project.

Planning starts with the identification of activities based on the project breakdown structure (PBS) *(see Project breakdown structure)* and the relationships among activities; then the duration and resources required for each activity are estimated. The final stage of planning is the schedule computation *(see Network analysis)* resulting in the time, money, material, equipment, and labor schedule that are used as baselines for the project execution to be followed.

Planning requires that project personnel with experience, such as superintendents, project engineers, and project managers, think logically to find the best possible way to execute the project from the start to the completion of a project. During the process of identifying the interrelationship among project activities, careful attention should be paid to the technological and managerial constraints *(see Constraint)* of activities that require workers with special skills, scarce resources, long-lead-time resources, tight environmental control, high work quality, and so on. Project planning should reveal problems in advance and minimize uncertainty so that the project can proceed in an efficient manner.

There are several ways to represent the project plan graphically *(see Project plan)* such as bar chart, network diagram, and line of balance technique *(see Bar chart, Network, and Line of balance technique)*. A network and a bar chart are the most popular methods for planning a project.

PLUG DATE (SCHEDULED DATE) *(see Contract date)*

PRECEDENCE DIAGRAM *(see Precedence diagramming method)*

PRECEDENCE DIAGRAMMING METHOD (PDM) A network planning technique wherein activities are represented by a box with connecting lines showing logical dependencies (relationships) among activities.

The precedence diagramming method is a network scheduling technique that uses a node or box to show the logical sequences of work to be performed from the

436

start to finish of a project. A network diagram is drawn after all activities and their relationships are defined. In a precedence diagram, each line shows the relationship between two activities, and each node or box represents one activity.

Normally, a precedence diagram is developed before the activity data are input into a computer and a network computation *(see Network analysis)* is performed. The critical aspect of precedence diagram development is accurate definition of project activities and their logical constraints. A precedence diagram developed for a particular project will be used as a means of communicating the work sequences among project participants.

The relationships among activities in PDM is defined as dependency *(see Dependency)*. Four types of relationships are start to start, finish to start, finish to finish, and start to finish. They can be with or without time delay. A precedence diagram is shown in Figure 1, where activity 200, Excavation, is the immediately preceding activity to activity 700, Foundation, with a finish-to-start relationship and no delay.

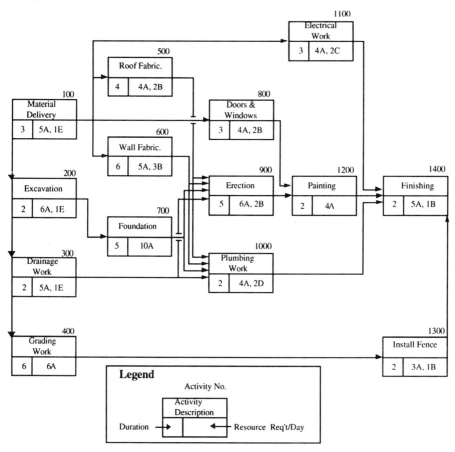

Figure 1 Precedence Diagram

When a precedence diagram cannot be drawn on one page, a connection has to be made from one page to another. Figure 2 shows how to make the page connection.

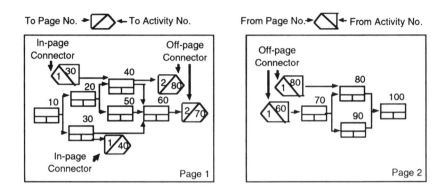

Figure 2 Connection Between Two Network Drawing Pages

PRECEDING ACTIVITY (PREDECESSOR) An activity that has some measurable portion of its duration logically restraining a subsequent activity or activities.

A preceding activity imposes a logical constraint on a particular activity. In an arrow diagram network *(see Arrow diagramming method),* a particular activity cannot start unless its preceding activities are completed, as shown in Figure 1. The activity Painting will start after completion of the activities Doors and Windows and Erection. Unlike an arrow diagram network, a precedence diagram network *(see Precedence diagramming method)* allows a preceding activity to impose either start to start, start to finish, finish to start, or finish to finish *(see Dependency)* on a particular activity as shown in Figures 2 and 3.

A preceding activity can be classified as one of two types: (1) an immediately preceding activity and (2) a non-immediately preceding activity. Any activities that are on the same path as a particular activity and are scheduled to start or to be completed before the start or completion of that activity are considered as its preceding activities. But the ones directly preceding a particular activity are immediately preceding activities.

Developing a network requires all activities to be listed; then a logical constraint between activities is defined to represent the work sequencing. As a result, any activity in the network can be located along with its preceding and succeeding activities *(see Succeeding activity).*

438

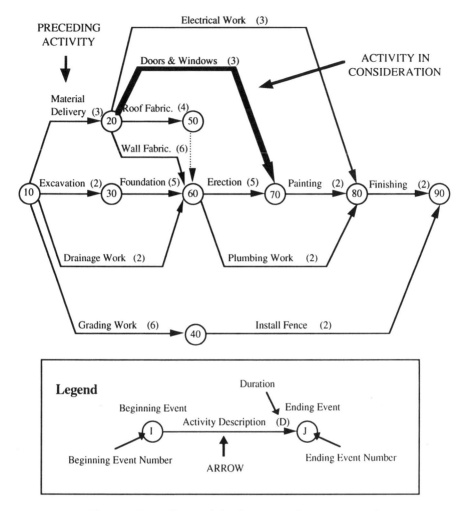

Figure 1 Preceding Activity in Arrow Diagram Network

439

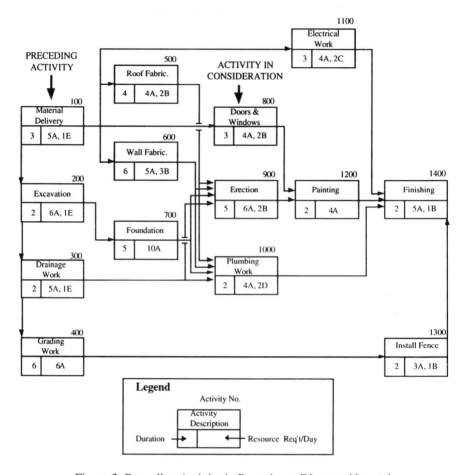

Figure 2 Preceding Activity in Precedence Diagram Network

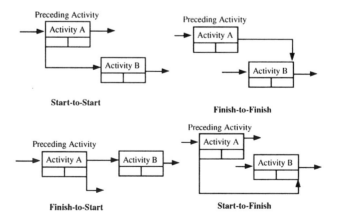

Figure 3 Possible Dependencies in Precedence Diagram Network

Figure 4 illustrates an activity list showing activities and their preceding and succeeding activities. The letters P.L.D.F. and S.L.D.F. indicate the immediately preceding and succeeding activities of a particular activity.

```
-------------------------------------------------------------------------------------------------------------------
CIVIL ENGINEERING DEPARTMENT                    PRIMAVERA PROJECT PLANNER

REPORT DATE 23NOV93  RUN NO.   21           PROJECT SCHEDULE REPORT                   START DATE 13MAR93  FIN DATE  9APR93
             22:32
SCHEDULE REPORT - PREDECESSORS AND SUCCESSORS                                          DATA DATE  17MAR93  PAGE NO.   1
----- -----   ---- ---- - --- ----------   -------------------------------------------  -------- -------- -------- -------- -----
ACTIVITY    ORIG REM                             ACTIVITY DESCRIPTION                    EARLY    EARLY    LATE     LATE   TOTAL
   ID       DUR  DUR   %  CODE                                                           START    FINISH   START    FINISH FLOAT
----- -----   ---- ---- - --- ----------   -------------------------------------------  -------- -------- -------- -------- -----

   100      3    3    0         MATERIAL DELIVERY                                       17MAR93* 19MAR93  17MAR93* 19MAR93    0

                S.L.D.F.,*     200.SS   0.   2.   2,*     300.SS   0.   2.    7
                S.L.D.F.,*     400.SS   0.   6.  10,*     500.FS   0.   4.    2
                S.L.D.F.,*     600.FS   0.   6.   0,*     800.FS   0.   3.    8
                S.L.D.F.,*    1100.FS   0.   3.  10,


                P.L.D.F.,*     100.SS   0.   3.   0,

   200      2    2    0         EXCAVATION                                             17MAR93  18MAR93  19MAR93  22MAR93    2

                S.L.D.F.,*     700.FS   0.   5.   2,


                P.L.D.F.,*     100.SS   0.   3.   0,

   300      2    2    0         DRAINAGE WORK                                          17MAR93  18MAR93  26MAR93  29MAR93    7

                S.L.D.F.,      900.FS   0.   5.   0,     1000.FS   0.   2.    5


                P.L.D.F.,*     100.SS   0.   3.   0,

   400      6    6    0         GRADING WORKE                                          17MAR93  24MAR93  31MAR93   7APR93   10

                S.L.D.F.,*    1300.FS   0.   2.  10,
```

P = Preceding Dependency Type
S = Succeeding Dependency Type
L = Lag
D = Duration
F = Total Float

Figure 4 Activity List Showing Preceding and Succeeding Activities

PRECONSTRUCTION CPM A plan schedule for the conceptual and design phase preceding the awarding of a contract.

A preconstruction CPM is the schedule developed in the early stages of a project using the arrow or precedence diagramming technique. It is used in planning and controlling the project during the conceptual and design phases. Preconstruction CPM activities could be as follows:

441

1. *Predesign activities:* approval of A/E, feasibility study, budget preparation, site selection and investigation, surveying, etc.
2. *Design activities:* conceptual design, preliminary design, detailed design, value engineering, constructibility study, etc.
3. *Long-lead procurement activities*

The development of a preconstruction CPM is a difficult task that requires the participation of an owner, consultant, A/E, supplier and/or potential contractors involved in design-built or fast-track projects to create the best possible plan for the project. Figure 1 shows the span of a preconstruction schedule correlated with a traditional construction schedule (not a fast-track approach).

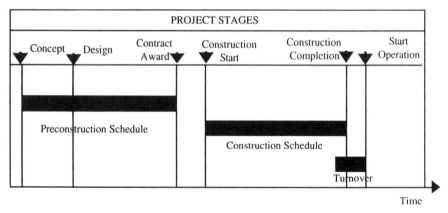

Figure 1 Span of Preconstruction Schedule in
Correlation with Construction Schedule

PRELIMINARY SCHEDULE The schedule that is developed to define all of the contractor's planned operations during an initial stage of the project's execution.

A preliminary schedule is developed to serve as a basis for the early stage of a project. The duration of this schedule depends on the size and complexity of the project, but the maximum time span is not recommended to be more than 3 months. The preliminary network diagram shows the sequence and interdependent relations of all activities that are expected to begin within the time frame specified.

The preliminary schedule should be developed based on the project breakdown structure (PBS) *(see Project breakdown structure)* by using the lowest level, called *work packages (see Work package),* with one work package representing an activity in the network. It should contain the same amount of detail regarding the project as a detailed schedule *(see Detailed schedule).* The preliminary schedule should be submitted for approval by the owner or owner's representative within contractually specified number of days after the notice of contract award.

When the detailed schedule *(see Detailed schedule)* is submitted at a later date, there is an overlap in time between the preliminary and detailed schedules. Therefore, the detailed schedule should include the as-built progress of certain activities in the preliminary construction schedule. Work activities in the preliminary schedule normally include project mobilization, procurement, site work, utilities, excavation, foundation, initial submittals, fabrications, and permits. Figure 1 shows a time span for a preliminary schedule related to other schedules.

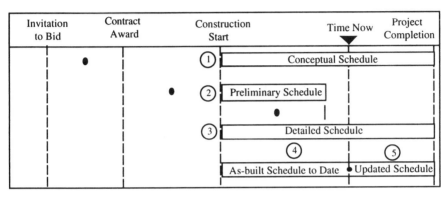

● Submitted Date

Figure 1 Time Span of Project Schedules

PROGRESS TREND A comparison of actual versus projected (scheduled) degrees of completion for a project over a given period of time used to determine whether the overall rate of project performance is increasing, remaining the same, or decreasing.

During the project execution phase, management must regularly evaluate the project status to identify the project's progress trend as increasing, decreasing, or remaining the same. If the progress trend is decreasing or remaining the same over a given period of time, management must attempt to improve the project progress or to indicate the problem areas impeding the project progress. Time is a major factor indicating the progress trend. The project progress is therefore measured at a specified time, such as weekly or monthly.

The most common indicators of progress trend are cost and schedule. The schedule variance (SV) *(see Schedule variance)* defined by the difference between the budgeted cost of work performed (BCWP) and the budgeted cost of work scheduled (BCWS) at any point in time indicates project performance to be behind or ahead of schedule for the work accomplished to date. Figure 1 shows a schedule variance plotted against a monthly time interval. It can be seen that in the first month the project progress is behind schedule for $1000 cost of work. The progress trend in the second and third months shows the improvement from the first month and the current status is $1500 cost of work ahead of schedule. The projected variance for the fourth

update is $3000 cost of work ahead of schedule. The same principle described in using schedule variance (SV) to identify the cost progress trend can also be applied to time variance, as shown in Figure 2 (see *Schedule variance*).

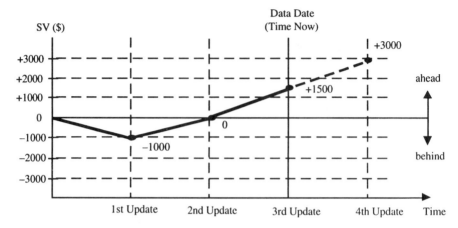

Figure 1 Schedule Variance Trend

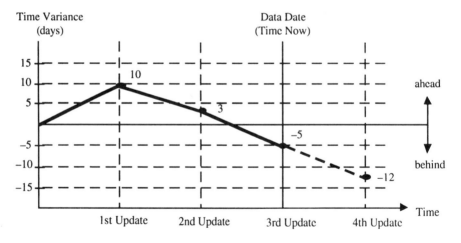

Figure 2 Time Variance Trend

PROJECT The overall work being planned. It must have one start point and one finish point in time. A project plan describing how, when, and by what means the project will be carried out must be created for a particular project.

Some projects are large, complex, and long in duration: for example, a nuclear power plant, a dam, a new plane, an oil refinery, or environmental cleanup. Others are average-sized projects with a duration of 2 to 3 years, such as a multistory residential or commercial building complex, a few miles of highway, a bridge or bridge rehabilitation, international convention preparation, or writing a book with many participants. Small projects are those that require limited resources for a short period of time, such as building a residential house, relocating to a new city, or building a deck for a house.

There are common attributes that make the above-mentioned endeavors "projects":

1. There is a starting and an ending point in time at which it is said that a project is complete. In other words, a project has a well-defined life span.
2. There is a well-defined scope and objective.
3. In most situations a project is nonrepetitive and viewed as a one-time effort.
4. A project cuts across many organization and functional lines in each organization.
5. To be called a project the endeavor must be an interplay of materials, equipment, and various human trades and management skills.

A project can have one or more objectives; achieving them represents the project completion. These objectives often involve research, design, development, manufacturing and construction, or hardware installation. A project involves a team of people committed at least full time primarily to project work. A project imposes a unique set of requirements on the organization.

PROJECT BREAKDOWN STRUCTURE (PBS) A task-oriented family tree of activities that organize, define, and display the work to be accomplished.

A PBS is composed of three components: work items, levels, and work packages *(see Work item, Level,* and *Work package).* The PBS partitions the project into manageable elements of work for which costs, budgets, and schedules can be established. The integration of a project's organization structure with the PBS helps the project manager to assign responsibility for the various technical tasks to specific project personnel.

The PBS also provides an easy-to-follow numbering system to allow for a hierarchical tracking of progress. Formation of the PBS family tree begins by subdividing, or partitioning, the project objective into successively smaller work elements until the lowest level to be reported on or controlled is reached. A PBS developed for a commercial building is shown in Figure 1.

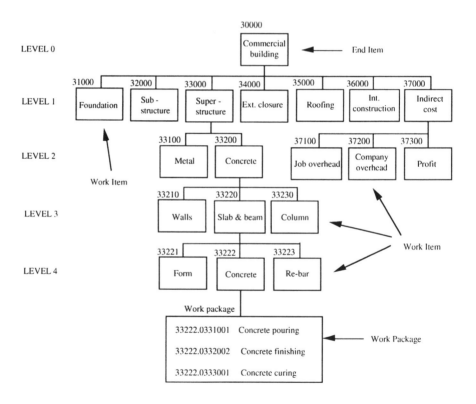

Figure 1 Project Breakdown Structure for Commercial Building

446

PROJECT CONTROL The ability to determine a project status as it relates to the time and schedule selected.

Since a project plan can never be maintained as originally developed, it is necessary to have a control system that will enable managers to take corrective actions. The result of external influences requires continuous planning and control. A control cycle is repetitive over the entire project life, as presented in Figure 1.

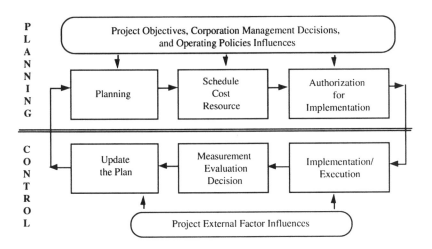

Figure 1 Project Planning and Control Cycle

The need for cyclical approach requires the use of computers, which are of great help in reviewing the operation (activities) periodically and to update or replan. There are two approaches in project control using a CPM technique *(see Critical path method).*

1. *Loose control.* Resources and cost are not considered (assuming unlimited availability).
2. *Close control.* Resources required and available to a project are considered in order to obtain the best possible resource utilization. In this approach the concept of critical path, critical activity, and float is lost once the schedule is established.

A project control level dictates the review frequency, which depends on the project complexity as overall project duration, average activity duration, time units used, percentage project completion, and the number of organization involved in the project. The project control plan and control must be able to provide answers to the following questions: What is the setback for meeting a project schedule, cost, and quality performance? What is the project outlook—improving or not? If not, what are the major factors controlling project planned performance?

PROJECT DURATION The elapsed duration from project start date through project finish date.

Project duration can be defined as the minimum of total time required from the project start date [early start date of start activity(s)] to project completion date [early finish date of ending activity(s)]. Network analysis *(see Network analysis)* determines how long the project duration is from the chain of activities that requires the longest period to perform. In other words, it is the total duration of the critical path of the project provided that there are no total floats along the critical path and that each activity starts as soon as possible.

The purpose of the project duration is to identify how long an anticipated duration a project needs. For example, if project duration is 50 days, the project requires 50 working days to reach the end of the project. However, it is very rare for any project duration to be the same as the project's actual duration unless the activity duration is estimated properly and accurately.

The project duration will be recalculated when change orders are required or available resources are changed. Although the allowance of lost time caused by inclement weather may affect activity duration, overall weather contingency is not included in project duration. Some contractors will plan to add the contingency on the overall project duration for time overrun. Nevertheless, it depends on what has been written in the contract.

Figures 1 and 2 show a project in arrow and precedence diagrams, respectively. For this example, the project duration is 18 days. However, the early finish date and late finish date of the ending activity, Finishing, is 19 because the activity will have to be completed by the beginning of day 19 of the project working calendar, as noted at the bottom of the figures.

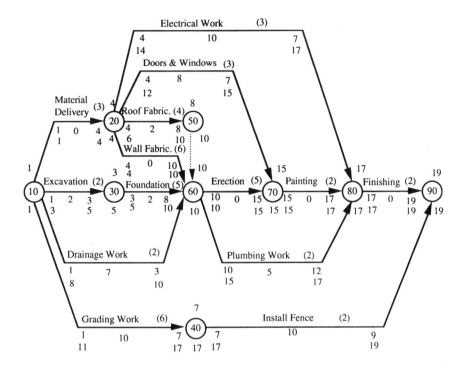

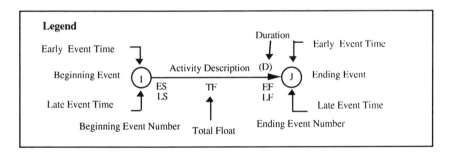

Legend

NOTE: The schedule computation in time units is based on the following assumption:

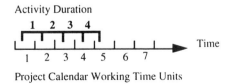

Activity Duration

Project Calendar Working Time Units

Activity with 4 time units starts at working time unit 1 and is finished at working time unit 5.

Figure 1 Project Duration Shown in Arrow Diagram Network

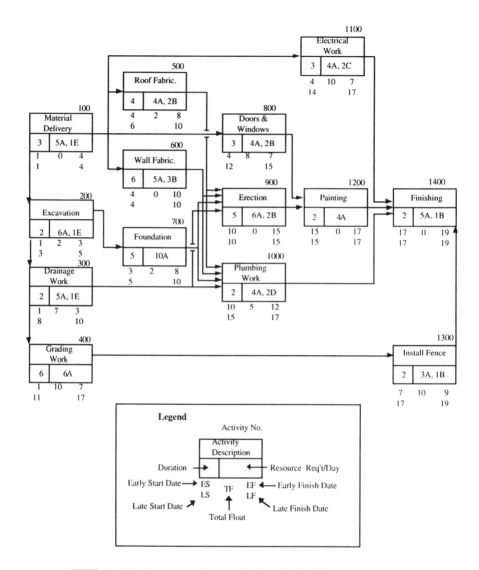

			1100
			Electrical Work
			3	4A, 2C
			4	10	7
			14		17

Legend

Activity No.

Activity Description

Duration → ← Resource Req't/Day

Early Start Date → ES   TF   EF ← Early Finish Date
LS   LF
Late Start Date ↗   Late Finish Date
Total Float

NOTE: The schedule computation in time units is based on the following assumption:

Activity Duration

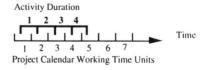

Time

Project Calendar Working Time Units

Activity with 4 time units starts at working time unit 1 and is finished at working time unit 5.

Figure 2 Project Duration Shown in Precedence Diagram Network

PROJECTED TOTAL COST *(see Estimate at completion)*

PROJECT FINISH DATE (1) The current estimate of the calendar date when all project activities will be completed. (2) The completion date specified by contract or other restraint *(see Required completion date).*

The project finish date is the last date that all work on the project needs to be done, and therefore it is equal to the late finish date of the last activity of the project *[see Late finish date (project)].* It can be calculated at any data date or time now *(see Data date).* It is not necessary to remain the same, as the project is under way and may not be the same as the project required completion date *(see Required completion date).* If the project finish date is earlier than the project required completion date, the project will have some positive total floats. On the other hand, if it is later than the project required completion date, negative total floats will appear on the project schedule.

For a project consisting of a single ending activity *(see Ending activity),* the project finish date will be the same as the late finish date of the last activity; however, for a project with multiple ending activities, there may be more than one project finish date since the late finish dates of the ending activities can be different unless the late finish dates of all ending activities or events are assumed to be the greatest late finish date of the ending activities or events. The networks shown in Figures 1 to 4 illustrate different project finish dates for a single ending activity and for multiple ending activities in both the arrow and precedence diagram techniques.

451

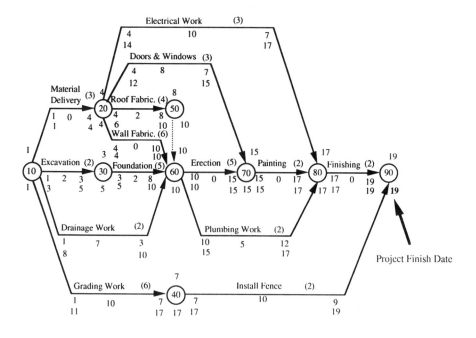

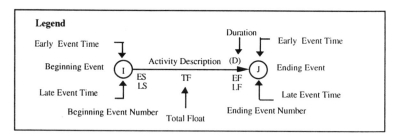

Legend

Early Event Time
Beginning Event
Late Event Time

Duration
Early Event Time
Ending Event
Late Event Time

Activity Description (D)

ES   TF
LS

EF
LF

Beginning Event Number
Total Float
Ending Event Number

NOTE: The schedule computation in time units is based on the following assumption:

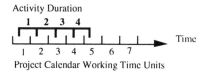

Activity Duration

Time

Project Calendar Working Time Units

Activity with 4 time units starts at working time unit 1 and is finished at working time unit 5.

Figure 1 Project Finish Date Shown on Network with
Single Ending Activity in Arrow Diagram Technique

Note: The late finish date of event 90 is assumed to be equal to its early finish date.

452

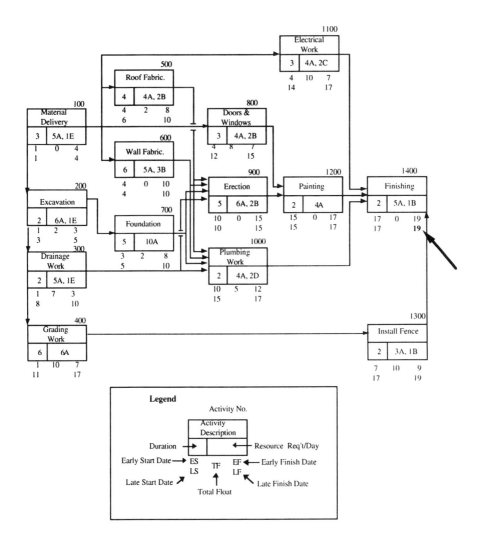

NOTE: The schedule computation in time units is based on the following assumption:

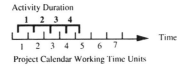

Activity Duration

Project Calendar Working Time Units

Activity with 4 time units starts at working time unit 1 and is finished at working time unit 5.

Figure 2 Project Finish Date Shown on Network with
Single Ending Activity in Precedence Diagram Technique

Note: The late finish date of activity 1400 is assumed to be equal to its early finish
date.

453

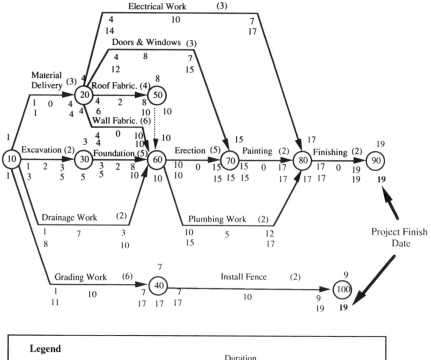

Electrical Work (3)

Doors & Windows (3)

Material Delivery (3)
Roof Fabric. (4)
Wall Fabric. (6)

Excavation (2)
Foundation (5)
Erection (5)
Painting (2)
Finishing (2)

Drainage Work (2)
Plumbing Work (2)

Grading Work (6)
Install Fence (2)

Project Finish Date

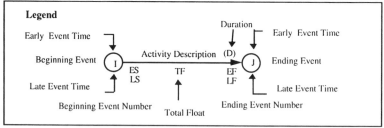

Legend

Duration

Early Event Time
Early Event Time

Beginning Event
Activity Description (D)
Ending Event

I
ES
LS
TF
EF
LF
J

Late Event Time
Late Event Time

Beginning Event Number
Total Float
Ending Event Number

NOTE: The schedule computation in time units is based on the following assumption:

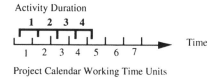

Activity Duration

Time

Project Calendar Working Time Units

Activity with 4 time units starts at working time unit 1 and is finished at working time unit 5.

Figure 3 Project Finish Date Shown on Network with
Multiple Ending Activities in Arrow Diagram Technique

Note: The late finish dates of all ending events (90 and 100) are assumed to be the
same as the greatest late event time among all the ending events.

454

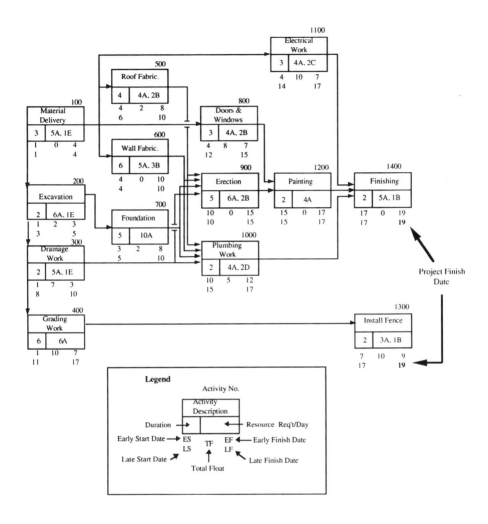

NOTE: The schedule computation in time units is based on the following assumption:

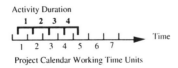

Activity Duration

Project Calendar Working Time Units

Activity with 4 time units starts at working time unit 1 and is finished at working time unit 5.

Figure 4   Project Finish Date Shown on Network with
Multiple Ending Activities in Precedence Diagram Technique

Note: The late finish dates of all ending activities (1400 and 1500) are assumed to be
the same as the greatest late finish date among all the ending activities.

PROJECT PLAN A primary document for project activities. It covers the project from initiation through completion or a future time span of the project.

The project plan is developed using the CPM technique *(see Critical path method).* It is to be remembered that any project is a dynamic entity responding to changing internal and external conditions. The project plan is prepared to meet the project primary and secondary objectives and is largely influenced by corporate executive decisions that initiate, finance, and control a project. From the plan representing the sequences of project activities, a project schedule can be prepared once duration, resources, and cost requirements associated with all operations are estimated and further allocated for project implementation.

The company in charge of project implementation must consider operating policies, which in turn may influence the project plan and schedule. The final stage of a project planning phase is authorization of the plan developed for implementation. The project planning phase—planning, scheduling, and authorization—is shown in Figure 1. The two aspects of the project that are critical for success are planning and control. Practically, comprehensive initial planning is required and flexible control during project implementation is recommended.

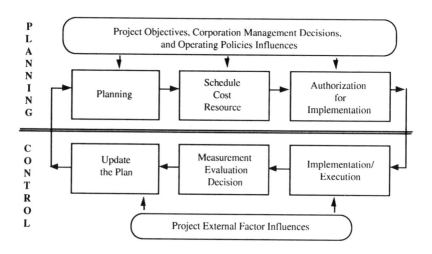

Figure 1 Project Planning and Control Cycle

PROJECT PLANNING *(see Planning)*

PROJECT PROGRESSING *(see Monitoring)*

PROJECT SEGMENT *(see Work item)*

PROJECT START DATE (1) The earliest calendar start date among all activities in the project network. (2) The date specified in the notice to proceed from which contract time periods are calculated.

The project start date is the calendar working date of the start activity(s) *(see Start activity)* of the project. It is normally equal to the early start date of the start activity(s) *[see Early start date (project)]*. It can be changed or modified at any data date or time now *(see Data date)* as long as the time now is earlier than the actual project start date. Once the project is started, the project start date will be recorded and will remain the same until the project is executed. It is important to clarify the project start date to all contractual participants because this date is generally counted as the first day of the project duration. In case of delay claims, the project start date is considered in calculating delays, acceleration, and project time-related expenses. Figures 1 and 2 illustrate the project start date in the arrow and precedence diagram networks, respectively.

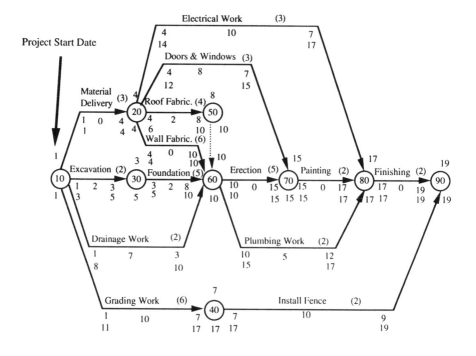

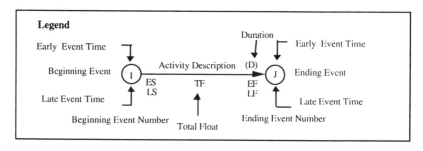

Legend

Early Event Time
Beginning Event
Late Event Time
Duration
Early Event Time
Activity Description (D)
Ending Event
Late Event Time
Beginning Event Number
Ending Event Number
Total Float

ES   TF   EF
LS   LF

NOTE: The schedule computation in time units is based on the following assumption:

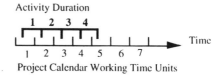

Activity Duration

Time

Project Calendar Working Time Units

Activity with 4 time units starts at working time unit 1 and is finished at working time unit 5.

Figure 1 Project Start Date Shown in Arrow Diagram Network

458

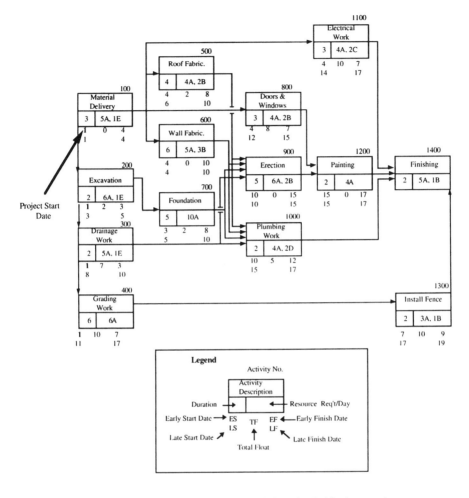

NOTE: The schedule computation in time units is based on the following assumption:

Activity Duration

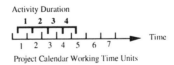

Project Calendar Working Time Units

Activity with 4 time units starts at working time unit 1 and is finished at working time unit 5.

Figure 2 Project Start Date Shown in Precendencce Diagram Network

R

REDUNDANT RELATIONSHIP (REDUNDANT LINK) A direct link between two activities that are already linked through other channels.

Project network development requires the identification of all activities and their interrelationships before the network diagram can be drawn *(see Network)*. Most of the time, in a complex project, a network diagram contains the redundant relationships between two activities. This situation occurs when two connected activities are already connected through their predecessors, as shown in Figures 1 and 2 *(see Preceding activity)*.

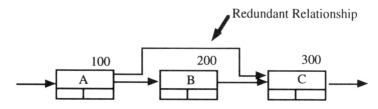

Figure 1 Redundant Relationship in Precedence Diagram

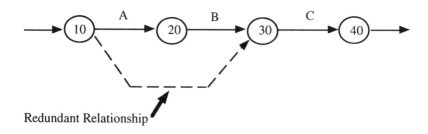

Figure 2 Redundant Relationship in Arrow Diagram

In Figure 1, activity 300 is preceded by activities 200 and 100. In this case, activity 200 is preceded directly by activity 100; therefore, the relationship between activities 100 and 300 is a redundant relationship. In Figure 2, activity C is preceded by activities A and B. In this case, activity B is preceded directly by activity A;

therefore, the relationship between activities A and C is a redundant relationship. Figure 3 shows a redundant relationship in a precedence diagram network as activity 800 directly preceding activity 1400. This relationship already exists through activity 1200.

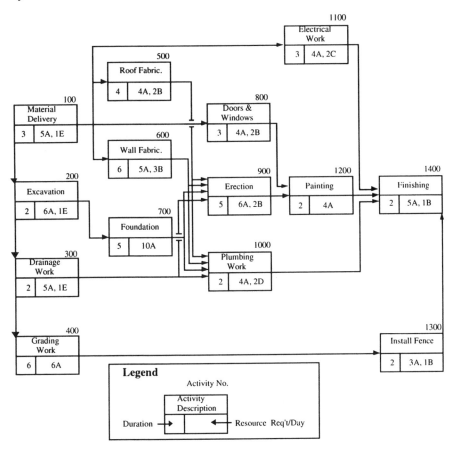

Figure 3 Redundant Relationship in Precedence Diagram Network

Redundant relationships do not result in schedule computation errors *(see Network analysis)* but cause unnecessary computation and confusion and congestion in a network diagram. It is recommended that redundant relationships be removed and only necessary relationships kept. When the sequencing logic is expected to change, it is justifiable to incorporate the redundant relationship in the network.

RELATIONSHIP *(see Dependency)*

REMAINING AVAILABLE RESOURCES The difference between the resources available and the resources required in any given project day (time unit).

By their nature project resources are of two main types: stockable (materials) and nonstockable (labor and construction equipment). Stockable resources not used will increase the inventory for a given period at an additional cost, reflected in the inventory carrying cost. The nonstockable resources allocated to be used in a given project day and not used are considered lost. It will increase the project cost and resource demand for the coming project period(s). During project duration, the following situations regarding resource availability and requirement (need) can occur:

1. *The required resource(s) are satisfied by the available resources.* This means that no resources remain in the availability pool. This is the optimum solution when the available resources are fully utilized.
2. *The daily resource requirement is less than the available resources.* Idle trades and equipment lead to a higher project direct cost. A buffer work zone needs to be created to utilize all capacities.
3. *The daily required resources exceed the availability levels.* Project activities cannot be scheduled and some of them need to be postponed, leading to delays.

Hypotheses 1, 2, and 3 are related to nonstockable resources. To verify project planning and scheduling performance for the key resources, the utilization factor (U), see *Resource utilization factor,* can be computed or projected for the remaining duration from the formula

$$U = \frac{\text{summation of daily project resource allocated}}{\text{summation of daily project resource required}}$$

Figure 1 shows calculation of the utilization factor for a given resource. This example shows that 41% of the available resources allocated to the project have been not utilized (nonstockable resources).

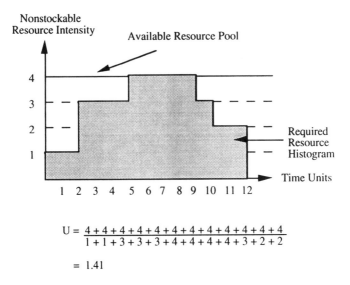

$$U = \frac{4+4+4+4+4+4+4+4+4+4+4+4}{1+1+3+3+3+4+4+4+4+3+2+2}$$

$$= 1.41$$

Figure 1 Calculation of Resource Utilization Factor

REMAINING DURATION The estimated number of time units needed to complete an activity in progress from the data date.

Remaining duration is the current amount of time left for an activity to be completed. Remaining duration can be calculated at any time now or data date to monitor the progress or status of activities.

During the progress of the project, the project scheduler can track the progress of activities whether they are started or completed. The scheduler can compute the remaining duration of the activity already started and not yet completed by subtracting the time units already consumed from the activity original duration.

Since the duration for any activity can be obtained from the following formula, other information (i.e., quantity to complete and % completed to date) can also be used to calculate the remaining duration of activities and certainly will lead to the same results.

$$\text{Activity duration} \quad = \quad \frac{\text{quantity of work}}{\text{sustained production rate/time unit}}$$

If the amount of quantity of work left is known, the remaining duration can be calculated easily by substituting the amount to the quantity of work. Similarly, if the percent completed to date is known, the remaining duration is simply equal to (100 − percent completed) x original duration, provided that the original duration remains the same. If the activity is not started, its remaining duration will be equal to its original duration unless the original duration is reestimated.

Figures 1 and 2 show the remaining duration of each activity in tabular format

for the arrow diagram and precedence diagram methods, respectively. The activities Material Delivery and Drainage Work are completed; therefore, their remaining durations are 0. The activity Excavation is started and for some reason needs 3 more days to be completed, which is more than initially expected. The activity Grading Work is in progress and its remaining duration is 4 days. The remaining durations of the other activities are the same as their original durations.

		ORIG	REM			ACTIVITY DESCRIPTION	EARLY	EARLY	LATE	LATE	TOTAL
PRED	SUCC	DUR	DUR	%	CODE		START	FINISH	START	FINISH	FLOAT
10	20	3	0	100		MATERIAL DELIVERY	17MAR93A	19MAR93A			
10	60	2	0	100		DRAINAGE WORK	17MAR93A	18MAR93A			
10	30	2	3	0		EXCAVATION	17MAR93A	23MAR93		23MAR93	0
10	40	6	4	33		GRADING WORK	17MAR93A	24MAR93		8APR93	11
20	60	6	6	0		WALL FABRICATION	19MAR93	26MAR93	23MAR93	30MAR93	2
20	50	4	4	0		ROOF FABRICATION	19MAR93	24MAR93	25MAR93	30MAR93	4
20	70	3	3	0		DOORS AND WINDOWS	19MAR93	23MAR93	2APR93	6APR93	10
20	80	3	3	0		ELECTRICAL WORK	19MAR93	23MAR93	6APR93	8APR93	12
30	60	5	5	0		FOUNDATION	24MAR93	30MAR93	24MAR93	30MAR93	0
50	60	0	0	0		DUMMY 1	25MAR93	24MAR93	31MAR93	30MAR93	4
40	90	2	2	0		INSTALL FENCE	25MAR93	26MAR93	9APR93	12APR93	11
60	70	5	5	0		ERECTION	31MAR93	6APR93	31MAR93	6APR93	0
60	80	2	2	0		PLUMBING WORK	31MAR93	1APR93	7APR93	8APR93	5
70	80	2	2	0		PAINTING	7APR93	8APR93	7APR93	8APR93	0
80	90	2	2	0		FINISHING	9APR93	12APR93	9APR93	12APR93	0

Figure 1 Remaining Duration Shown
in Tabular Format for Arrow Diagram Method

ACTIVITY	ORIG	REM			ACTIVITY DESCRIPTION	EARLY	EARLY	LATE	LATE	TOTAL
ID	DUR	DUR	%	CODE		START	FINISH	START	FINISH	FLOAT
100	3	0	100		MATERIAL DELIVERY	17MAR93A	19MAR93A			
300	2	0	100		DRAINAGE WORK	17MAR93A	18MAR93A			
200	2	3	0		EXCAVATION	17MAR93A	23MAR93		23MAR93	0
400	6	4	33		GRADING WORK	17MAR93A	24MAR93		8APR93	11
600	6	6	0		WALL FABRICATION	19MAR93	26MAR93	23MAR93	30MAR93	2
500	4	4	0		ROOF FABRICATION	19MAR93	24MAR93	25MAR93	30MAR93	4
800	3	3	0		DOORS & WINDOWS	19MAR93	23MAR93	2APR93	6APR93	10
1100	3	3	0		ELECTRICAL WORK	19MAR93	23MAR93	6APR93	8APR93	12
700	5	5	0		FOUNDATION	24MAR93	30MAR93	24MAR93	30MAR93	0
1300	2	2	0		INSTALL FENCE	25MAR93	26MAR93	9APR93	12APR93	11
900	5	5	0		ERECTION	31MAR93	6APR93	31MAR93	6APR93	0
1000	2	2	0		PLUMBING WORK	31MAR93	1APR93	7APR93	8APR93	5
1200	2	2	0		PAINTING	7APR93	8APR93	7APR93	8APR93	0
1400	2	2	0		FINISHING	9APR93	12APR93	9APR93	12APR93	0

Figure 2 Remaining Duration Shown
in Tabular Format for Precedence Diagram Method

REMAINING TOTAL FLOAT The difference between the early finish time and late finish time of an activity at the data date (time now) when the project is in progress.

The remaining float is the current amount of total float left for an activity. It is calculated when the project is updated. The scheduler will track the progress of activities, assess activity remaining duration *(see Remaining duration),* and reschedule the project. In each update, some activities' total float may be used, others may not be. The number of total float time units for each activity after rescheduling is called the remaining total float.

The remaining total float concept is not applied to completed activities. For activities not started, if the logic of network and estimate duration are not changed, the remaining total float will be the same as the original total float. With the same conditions, if some total floats of in-progress activity(s) are consumed, the remaining float will be equal to the original total float less the consumed total float. On the other hand, if the total float is not consumed and no jobs are accelerated, the remaining float will be the same as the original total float.

Figures 1 and 2 show the original total float for each activity in tabular format for the arrow diagram and precedence diagram methods, respectively. Figures 3 and 4 show the remaining float of each activity in tabular format for the arrow diagram and precedence diagram methods, respectively, compare them with Figures 1 and 2. As mentioned earlier, the remaining floats do not appear for the completed activities as shown in Figures 3 and 4. The remaining floats of some activities are different from the original total floats whereas those of others are the same, depending on the activity status at the data date.

CIVIL ENGINEERING DEPARTMENT   PRIMAVERA PROJECT PLANNER

REPORT DATE 23NOV93   RUN NO.   6   PROJECT SCHEDULE REPORT   START DATE 17MAR93   FIN DATE 9APR93
                17:06
CLASSIC SCHEDULE REPORT - SORT BY ES, TF   DATA DATE 17MAR93   PAGE NO.   1

PRED	SUCC	ORIG DUR	REM DUR	%	CODE	ACTIVITY DESCRIPTION	EARLY START	EARLY FINISH	LATE START	LATE FINISH	TOTAL FLOAT
10	20	3	3	0		MATERIAL DELIVERY	17MAR93*	19MAR93	17MAR93*	19MAR93	0
10	30	2	2	0		EXCAVATION	17MAR93	18MAR93	19MAR93	22MAR93	2
10	60	2	2	0		DRAINAGE WORK	17MAR93	18MAR93	26MAR93	29MAR93	7
10	40	6	6	0		GRADING WORK	17MAR93	24MAR93	31MAR93	7APR93	10
30	60	5	5	0		FOUNDATION	19MAR93	25MAR93	23MAR93	29MAR93	2
20	60	6	6	0		WALL FABRICATION	22MAR93	29MAR93	22MAR93	29MAR93	0
20	50	4	4	0		ROOF FABRICATION	22MAR93	25MAR93	24MAR93	29MAR93	2
20	70	3	3	0		DOORS AND WINDOWS	22MAR93	24MAR93	1APR93	5APR93	8
20	80	3	3	0		ELECTRICAL WORK	22MAR93	24MAR93	5APR93	7APR93	10
40	90	2	2	0		INSTALL FENCE	25MAR93	26MAR93	8APR93	9APR93	10
50	60	0	0	0		DUMMY 1	26MAR93	25MAR93	30APR93	29MAR93	2
60	70	5	5	0		ERECTION	30MAR93	5APR93	30MAR93	5APR93	0
60	80	2	2	0		PLUMBING WORK	30MAR93	31MAR93	6APR93	7APR93	5
70	80	2	2	0		PAINTING	6APR93	7APR93	6APR93	7APR93	0
80	90	2	2	0		FINISHING	8APR93	9APR93	8APR93	9APR93	0

Figure 1 Total Float Shown in Tabular Format for Arrow Diagram Method

CIVIL ENGINEERING DEPARTMENT   PRIMAVERA PROJECT PLANNER

REPORT DATE 23NOV93 RUN NO.   12   PROJECT SCHEDULE REPORT   START DATE 13MAR93 FIN DATE 9APR93
         17:02
CLASSIC SCHEDULE REPORT - SORT BY ES, TF   DATA DATE 17MAR93 PAGE NO.   1

ACTIVITY ID	ORIG DUR	REM DUR	%	CODE	ACTIVITY DESCRIPTION	EARLY START	EARLY FINISH	LATE START	LATE FINISH	TOTAL FLOAT
100	3	3	0		MATERIAL DELIVERY	17MAR93*	19MAR93	17MAR93*	19MAR93	0
200	2	2	0		EXCAVATION	17MAR93	18MAR93	19MAR93	22MAR93	2
300	2	2	0		DRAINAGE WORK	17MAR93	18MAR93	26MAR93	29MAR93	7
400	6	6	0		GRADING WORK	17MAR93	24MAR93	31MAR93	7APR93	10
700	5	5	0		FOUNDATION	19MAR93	25MAR93	23MAR93	29MAR93	2
600	6	6	0		WALL FABRICATION	22MAR93	29MAR93	22MAR93	29MAR93	0
500	4	4	0		ROOF FABRICATION	22MAR93	25MAR93	24MAR93	29MAR93	2
800	3	3	0		DOORS & WINDOWS	22MAR93	24MAR93	1APR93	5APR93	8
1100	3	3	0		ELECTRICAL WORK	22MAR93	24MAR93	5APR93	7APR93	10
1300	2	2	0		INSTALL FENCE	25MAR93	26MAR93	8APR93	9APR93	10
900	5	5	0		ERECTION	30MAR93	5APR93	30MAR93	5APR93	0
1000	2	2	0		PLUMBING WORK	30MAR93	31MAR93	6APR93	7APR93	5
1200	2	2	0		PAINTING	6APR93	7APR93	6APR93	7APR93	0
1400	2	2	0		FINISHING	8APR93	9APR93	8APR93	9APR93	0

Figure 2 Total Float Shown in Tabular Format for Precedence Diagram Method

CIVIL ENGINEERING DEPARTMENT   PRIMAVERA PROJECT PLANNER

REPORT DATE 23NOV93 RUN NO.   9   PROJECT SCHEDULE REPORT   START DATE 17MAR93 FIN DATE 12APR93
         17:16
CLASSIC SCHEDULE REPORT - SORT BY ES, TF   DATA DATE 19MAR93 PAGE NO.   1

PRED	SUCC	ORIG DUR	REM DUR	%	CODE	ACTIVITY DESCRIPTION	EARLY START	EARLY FINISH	LATE START	LATE FINISH	TOTAL FLOAT
10	20	3	0	100		MATERIAL DELIVERY	17MAR93A	19MAR93A			
10	60	2	0	100		DRAINAGE WORK	17MAR93A	18MAR93A			
10	30	2	3	0		EXCAVATION	17MAR93A	23MAR93		23MAR93	0
10	40	6	4	33		GRADING WORK	17MAR93A	24MAR93		8APR93	11
20	60	6	6	0		WALL FABRICATION	19MAR93	26MAR93	23MAR93	30MAR93	2
20	50	4	4	0		ROOF FABRICATION	19MAR93	24MAR93	25MAR93	30MAR93	4
20	70	3	3	0		DOORS AND WINDOWS	19MAR93	23MAR93	2APR93	6APR93	10
20	80	3	3	0		ELECTRICAL WORK	19MAR93	23MAR93	6APR93	8APR93	12
30	60	5	5	0		FOUNDATION	24MAR93	30MAR93	24MAR93	30MAR93	0
50	60	0	0	0		DUMMY 1	25MAR93	24MAR93	31MAR93	30MAR93	4
40	90	2	2	0		INSTALL FENCE	25MAR93	26MAR93	9APR93	12APR93	11
60	70	5	5	0		ERECTION	31MAR93	6APR93	31MAR93	6APR93	0
60	80	2	2	0		PLUMBING WORK	31MAR93	1APR93	7APR93	8APR93	5
70	80	2	2	0		PAINTING	7APR93	8APR93	7APR93	8APR93	0
80	90	2	2	0		FINISHING	9APR93	12APR93	9APR93	12APR93	0

Figure 3 Remaining Float Shown in Tabular Format for Arrow Diagram Method

----
CIVIL ENGINEERING DEPARTMENT   PRIMAVERA PROJECT PLANNER

REPORT DATE 23NOV93 RUN NO. 11   PROJECT SCHEDULE REPORT   START DATE 13MAR93 FIN DATE 12APR93
        17:21
CLASSIC SCHEDULE REPORT - SORT BY ES, TF   DATA DATE 19MAR93 PAGE NO.   1

ACTIVITY ID	ORIG DUR	REM DUR	%	CODE	ACTIVITY DESCRIPTION	EARLY START	EARLY FINISH	LATE START	LATE FINISH	TOTAL FLOAT
100	3	0	100		MATERIAL DELIVERY	17MAR93A	19MAR93A			
300	2	0	100		DRAINAGE WORK	17MAR93A	18MAR93A			
200	2	3	0		EXCAVATION	17MAR93A	23MAR93		23MAR93	0
400	6	4	33		GRADING WORK	17MAR93A	24MAR93		8APR93	11
600	6	6	0		WALL FABRICATION	19MAR93	26MAR93	23MAR93	30MAR93	2
500	4	4	0		ROOF FABRICATION	19MAR93	24MAR93	25MAR93	30MAR93	4
800	3	3	0		DOORS & WINDOWS	19MAR93	23MAR93	2APR93	6APR93	10
1100	3	3	0		ELECTRICAL WORK	19MAR93	23MAR93	6APR93	8APR93	12
700	5	5	0		FOUNDATION	24MAR93	30MAR93	24MAR93	30MAR93	0
1300	2	2	0		INSTALL FENCE	25MAR93	26MAR93	9APR93	12APR93	11
900	5	5	0		ERECTION	31MAR93	6APR93	31MAR93	6APR93	0
1000	2	2	0		PLUMBING WORK	31MAR93	1APR93	7APR93	8APR93	5
1200	2	2	0		PAINTING	7APR93	8APR93	7APR93	8APR93	0
1400	2	2	0		FINISHING	9APR93	12APR93	9APR93	12APR93	0

Figure 4 Remaining Float Shown
in Tabular Format for Precedence Diagram Method

REQUEST FOR INFORMATION (RFI) A form sent by a contractor requesting action or clarification by the owner or his agents (engineer, architect) for items on the contract documents which are unclear or incorrect. An RFI can have an effect on the project schedule. Some RFI forms have a section for the contractor to identify the latest response date that would not affect schedule performance, based on the most recent CPM schedule update.

REQUIRED COMPLETION DATE A date assigned for completion of an activity or accomplishment of an event for purposes of meeting a specified schedule requirement.

The required completion date or scheduled completion date is the date of completion assigned to a specific activity, event, or project. The required completion date is similar to the contract date *(see Contract date)*. The difference is that the contract date consists of constraints for both start and completion of activities or events, but the required completion date imposes only the completion date of the activities or events.

Most of the time, project participants, such as the owner, contractors, or subcontractors, require a specific completion date for an activity or event to utilize their resources efficiently. Generally, the required completion date of activities or events is more focused than the required start dates of the activities or events. Note that if many completion dates on the network are required, the schedule calculation may result in more critical paths, which provides less flexibility for the project.

If the required completion date is assigned to activities for the precedence diagram technique, it will affect only completion of the activity. On the other hand, if this date is imposed on an event (or nodes) in the arrow diagram technique, it will affect all preceding activities of the event.

It is necessary to point out that the required scheduled date may or may not be the same as the early finish date of the project. If they are the same, no floats will be available to management prior to project submission. If it is later than the early finish date of the project, some positive float will be available. If it is earlier than the early finish date of the project, negative floats will appear in the schedule. The networks in Figures 1 to 6 show different required scheduled dates imposed on the last activity of a project with three different situations for the required completion date.

468

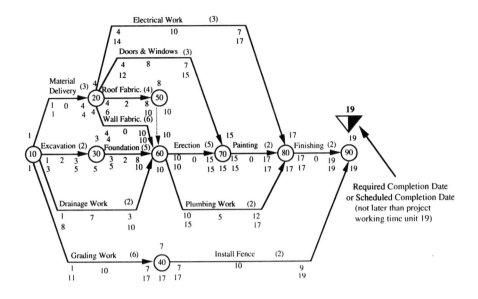

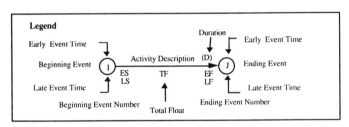

Legend

Early Event Time

Beginning Event

Late Event Time

Beginning Event Number

Duration

Activity Description (D)

Early Event Time

Ending Event

Late Event Time

Ending Event Number

Total Float

ES TF EF
LS LF

NOTE: The schedule computation in time units is based on the following assumption:

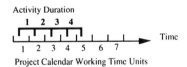

Activity Duration

1 2 3 4

Time

1 2 3 4 5 6 7

Project Calendar Working Time Units

Activity with 4 time units starts at working time unit 1 and is finished at working time unit 5.

Figure 1 Schedule Computation of Project Whose Required Completion Date
Is the Same as Early Finish Date in Arrow Diagram

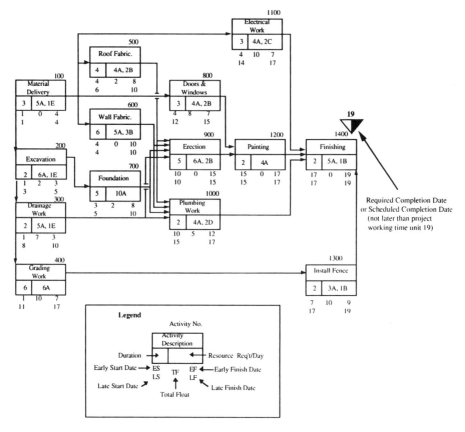

NOTE: The schedule computation in time units is based on the following assumption:

Activity with 4 time units starts at working time unit 1 and is finished at working time unit 5.

Figure 2 Schedule Computation of Project Whose Required Completion Date
Is the Same as Early Finish Date in Precedence Diagram

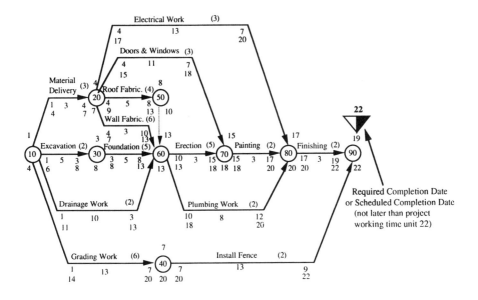

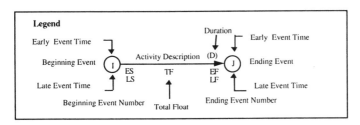

Legend

Early Event Time — Beginning Event — Late Event Time — Beginning Event Number — Activity Description (D) — Total Float — Duration — Early Event Time — Ending Event — Late Event Time — Ending Event Number — ES LS — TF — EF LF

NOTE: The schedule computation in time units is based on the following assumption:

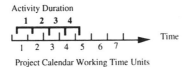

Activity Duration
Time
Project Calendar Working Time Units

Activity with 4 time units starts at working time unit 1 and is finished at working time unit 5.

Figure 3 Schedule Computation of Project Whose Required Completion Date Is Later Than Early Finish Date in Arrow Diagram

Note: A 3-time-unit reserve will be reflected as total float in the schedule calculation.

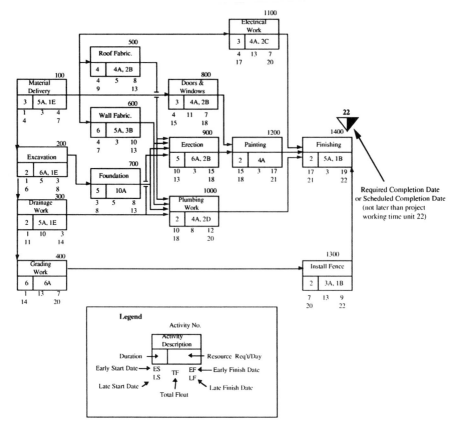

NOTE: The schedule computation in time units is based on the following assumption:

Activity with 4 time units starts at working time unit 1 and is finished at working time unit 5.

Figure 4 Schedule Computation of Project Whose Required Completion Date
Is Later Than Early Finish Date in Precedence Diagram

Note: A 3-time-unit reserve will be reflected as total float in the schedule calculation.

472

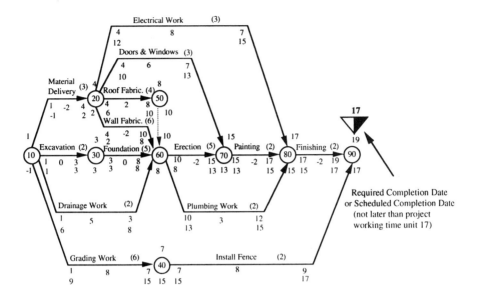

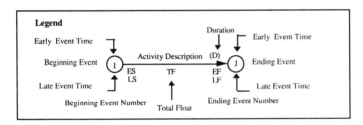

Legend

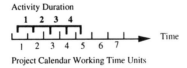

NOTE: The schedule computation in time units is based on the following assumption:

Activity Duration

Activity with 4 time units starts at working time unit 1 and is finished at working time unit 5.

Figure 5 Schedule Computation of Project Whose Required Completion Date
Is Earlier Than Early Finish Date in Arrow Diagram

Note: −2 time units of total float will be reflected in the schedule calculation.

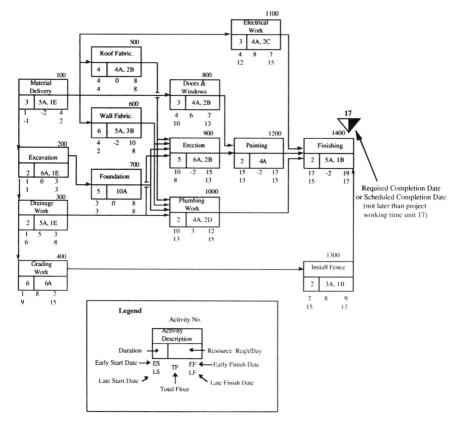

NOTE: The schedule computation in time units is based on the following assumption:

Activity with 4 time units starts at working time unit 1 and is finished at working time unit 5.

Figure 6 Schedule Computation of Project Whose Required Completion Date Is Later Than Early Finish Date in Precedence Diagram

Note: −2 time units of total float will be reflected in the schedule calculation.

RESCHEDULE The process of changing the logic, duration, and/or dates of an existing schedule in response to imposed external conditions.

A project schedule is normally prepared by a general contractor or project manager to show the intention to fulfill the contractual agreement that a project be finished within a specified time frame or date. Due to imposed external conditions, the original or as-planned schedule may be changed before or after the start of project execution. These imposed conditions can include, among others:

1. Unavailability of required resources, such as materials, equipment, or specialized crews
2. Adverse severe weather conditions
3. Improper logical constraints among activities
4. Change in the scope of work

As soon as management discovers that the project schedule may not be carried out as originally planned, rescheduling is triggered. Typically, the rescheduling process involves change in the activity logical constraints, activity duration, and imposed dates or the elimination or addition of project activities. After changes are made in the as-planned project schedule, the revised project schedule needs to be recalculated to determine the new start and completion dates of a project and its activities.

Figures 1 and 2 show the as-planned and revised precedence logic diagram prior to the project start. It is assumed that the following conditions cause the project to be rescheduled:

1. Activity 500, Roof Fabrication, cannot be started until activity 600, Wall Fabrication, is started.
2. Activity 400, Grading Work, cannot be started until activity 300, Drainage Work, is completed.
3. Activity 1400, Finishing, will take 7 days to complete instead of 2 days.

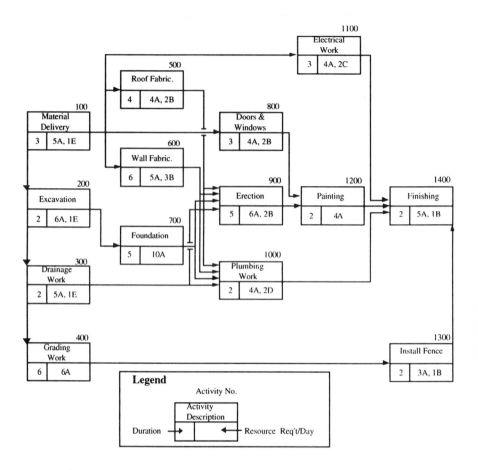

Figure 1 As-planned Precedence Diagram Network

476

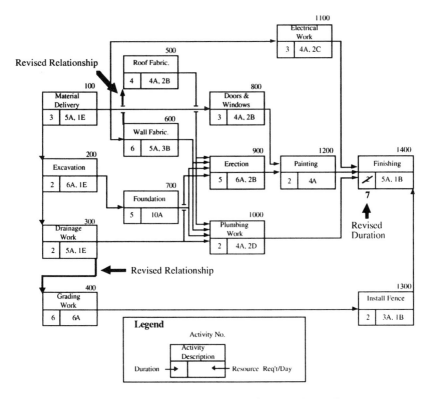

Figure 2 Revised Precedence Diagram Network

Figures 3 and 4 show the original and revised schedules in tabular format.

```
-----------------------------------------------------------------------------------------------------------
CIVIL ENGINEERING DEPARTMENT              PRIMAVERA PROJECT PLANNER

REPORT DATE 23NOV93  RUN NO.   22         PROJECT SCHEDULE REPORT              START DATE 13MAR93  FIN DATE  9APR93
          22:35
CLASSIC SCHEDULE REPORT - SORT BY ES, TF                                      DATA DATE  17MAR93  PAGE NO.   1
```

ACTIVITY ID	ORIG DUR	REM DUR	%	CODE	ACTIVITY DESCRIPTION	EARLY START	EARLY FINISH	LATE START	LATE FINISH	TOTAL FLOAT
100	3	3	0		MATERIAL DELIVERY	17MAR93*	19MAR93	17MAR93*	19MAR93	0
200	2	2	0		EXCAVATION	17MAR93	18MAR93	19MAR93	22MAR93	2
300	2	2	0		DRAINAGE WORK	17MAR93	18MAR93	26MAR93	29MAR93	7
400	6	6	0		GRADING WORK	17MAR93	24MAR93	23MAR93	31MAR93	10
700	5	5	0		FOUNDATION	19MAR93	25MAR93	23MAR93	29MAR93	2
600	6	6	0		WALL FABRICATION	22MAR93	29MAR93	22MAR93	29MAR93	0
500	4	4	0		ROOF FABRICATION	22MAR93	25MAR93	24MAR93	29MAR93	2
800	3	3	0		DOORS & WINDOWS	22MAR93	24MAR93	1APR93	5APR93	8
1100	3	3	0		ELECTRICAL WORK	22MAR93	24MAR93	5APR93	7APR93	10
1300	2	2	0		INSTALL FENCE	25MAR93	26MAR93	8APR93	9APR93	10
900	5	5	0		ERECTION	30MAR93	5APR93	30MAR93	5APR93	0
1000	2	2	0		PLUMBING WORK	30MAR93	31MAR93	6APR93	7APR93	5
1200	2	2	0		PAINTING	6APR93	7APR93	6APR93	7APR93	0
1400	2	2	0		FINISHING	8APR93	9APR93	8APR93	9APR93	0

Figure 3 As-planned Schedule Report in Tabular Format

REPORT DATE 24NOV93 RUN NO.   6   PROJECT SCHEDULE REPORT   START DATE 13MAR93 FIN DATE 16APR93
           1:15
CLASSIC SCHEDULE REPORT - SORT BY ES, TF   DATA DATE 17MAR93 PAGE NO.   1

ACTIVITY ID	ORIG DUR	REM DUR	%	CODE	ACTIVITY DESCRIPTION	EARLY START	EARLY FINISH	LATE START	LATE FINISH	TOTAL FLOAT
100	3	3	0		MATERIAL DELIVERY	17MAR93*	19MAR93	17MAR93*	19MAR93	0
200	2	2	0		EXCAVATION	17MAR93	18MAR93	19MAR93	22MAR93	2
300	2	2	0		DRAINAGE WORK	17MAR93	18MAR93	26MAR93	29MAR93	7
700	5	5	0		FOUNDATION	19MAR93	25MAR93	23MAR93	29MAR93	2
400	6	6	0		GRADING WORK	19MAR93	26MAR93	7APR93	14APR93	13
600	6	6	0		WALL FABRICATION	22MAR93	29MAR93	22MAR93	29MAR93	0
500	4	4	0		ROOF FABRICATION	22MAR93	25MAR93	24MAR93	29MAR93	2
800	3	3	0		DOORS & WINDOWS	22MAR93	24MAR93	1APR93	5APR93	8
1100	3	3	0		ELECTRICAL WORK	22MAR93	24MAR93	5APR93	7APR93	10
1300	2	2	0		INSTALL FENCE	29MAR93	30MAR93	15APR93	16APR93	13
900	5	5	0		ERECTION	30MAR93	5APR93	30MAR93	5APR93	0
1000	2	2	0		PLUMBING WORK	30MAR93	31MAR93	6APR93	7APR93	5
1200	2	2	0		PAINTING	6APR93	7APR93	6APR93	7APR93	0
1400	7	7	0		FINISHING	8APR93	16APR93	8APR93	16APR93	0

Figure 4 Revised Schedule Report in Tabular Format

RESOURCE Any consumable, except time, required to accomplish an activity, such as labor, materials, equipment, machinery, tools, and shop space.

Resource requirements are estimated at the activity level based on project/activity documentation and achieved productivity or production rates under the stated environmental and managerial conditions or similar conditions. More than one resource/activity needed can be estimated. Based on the CPM software selected, the maximum number of different resources/activities for the entire project are specified, and this limitation must be considered by the user. These resources may be constant over an activity's duration or may be variable in intensity. Having estimated the resources needed for the activities considered, and based on the CPM diagram already developed, the project time analyzed will produce a schedule with no resource constraints. Project resources all have certain characteristics that differentiate them in specific categories for further analysis:

1. *Stockability.* The available resources for each project day cannot exceed the requirements because unused resources cannot be stored until the next day without loss in quality or excess quality. Figure 1 shows the daily loss of nonstockable resources at the project level.

2. *Required/availability resource intensity*

a. *Constant intensity (uniform requirement).* This is a resource that over the activity or project duration is required in a uniform manner. In project planning situations are encountered when the same resource is uniformly required for an activity but is not required for similar activities.

b. *Nonuniform intensity (nonuniform requirement).* A nonuniform required resource is one that over a given activity duration is required at a variable rate. To simplify the analysis, it is recommended those activities be split to provide a uniform rate for each portion of activity under consideration. Sometimes, the resource is not required in a uniform pattern over the entire unit of time (e.g., day, shift, or hour).

3. *Resource complexity.* A category of work, such as reinforce concrete in place, may need to be broken down into components such as concrete, reinforcing bars, formwork, labor for reinforcing installation, and so on.

It is recommended that resources be considered at the macro level, for simplification of resource analysis. If this approach does not lead to desired managerial information, micro resources analysis is the alternative.

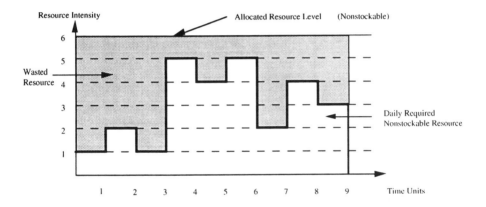

Figure 1   Daily Resource Loss of Nonstockable Resources

RESOURCE AGGREGATION *(see Aggregation)*

RESOURCE AGGREGATION PROCESS   The step-by-step procedures for obtaining resources histograms for a given project with or without computer usage.

The process of calculating required resources to implement a plan is a step-by-step operation:

1. A project CPM network using the arrow or precedence diagram technique is developed (Figure 1).

479

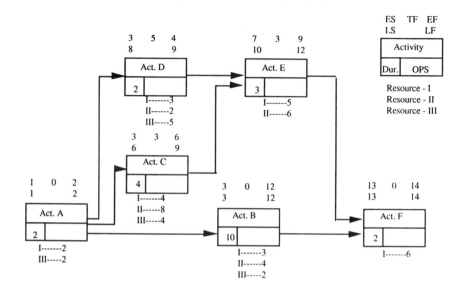

Figure 1 Sample Precedence Network Calculation Indicating
Required Resources for Each Activity at Constant Daily Rate

2. The required resources/activity are limited; sometimes more than one re-
   source/activity is specified.
3. The CPM network is computed considering all contract dates.
4. A bar chart in ascending order of early or late activity start is developed
   (Figure 2).

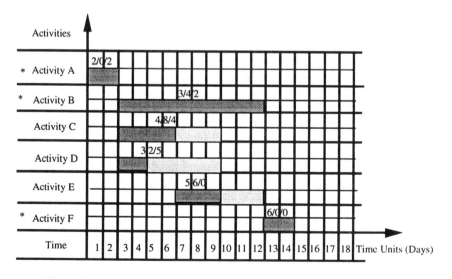

Figure 2 Bar Chart in Ascending Order of Daily Start and Total Float

5. For each specified resource for each project time unit, the total daily required resource is calculated (Figure 3).

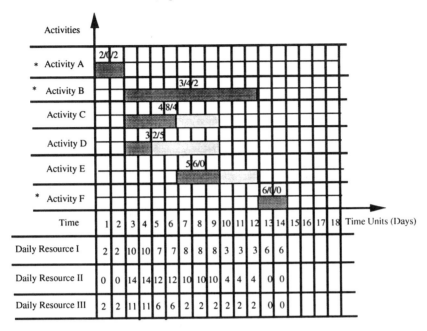

Figure 3 Bar Chart (Early Start) with Project Daily Requirement for Resources I, II, and III

6. For each resource, based on requirements per project time unit, histograms *(see Resource histogram)* are developed (Figures 4 to 6). Resource histograms can be developed for early or late start schedules.

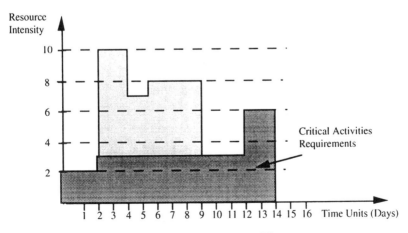

Figure 4 Required Resource I – Histogram

481

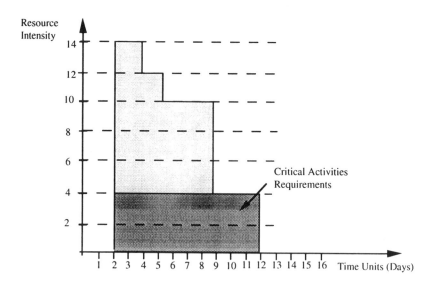

Figure 5 Required Resource II – Histogram

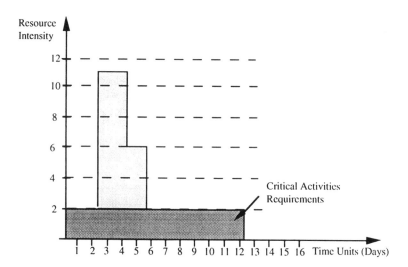

Figure 6 Required Resource III – Histogram

7. For each resource analyzed cumulative curves for early start and late start schedules are developed on the same plot (Figures 7 and 8).

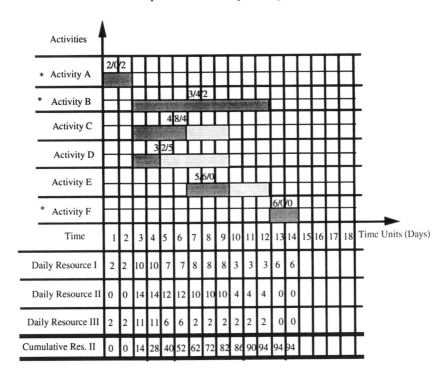

Figure 7 Bar Chart (Early Start) Indicating
Cumulative Requirement for Resource II

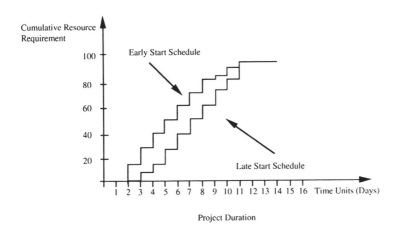

Figure 8 Cumulative Resource II Requirements
Based on Early and Late Start Schedules

RESOURCE ALLOCATION PROCESS The scheduling of activities in a network with the knowledge of certain constraints imposed on resource availability.

Many of the constraints hedging in a plan will have origins in the limitation of total resources or in the proportion of total resources that can be committed to a project. Resource allocation procedures or programs allocate available resources to project activities in an attempt to obtain the shortest project schedule (completion date) consistent with fixed resource limits. For project managers, one of the most challenging tasks is allocating resources that are available only in a fixed amount for a given period.

The first step of the process is to identify project required resources with limited availability and the periods of availability:

1. *Available resources with constant intensity for long period of time or project duration* (e.g., a tower crane for lifting materials for a multistory building).

2. *Available resources for shorter or cyclic periods.* Available resources can be stockable (if they are not used today, they can be used in the future) or nonstockable (if they are not used when allocated, they will incur costs to the project in addition to a productivity loss). The most challenging situation for project managers is to allocate nonstockable resources to project activities from the project resource pool.

To reach the best use of resources, the utilization factor *(see Resource utilization factor)* of the allocated resources is desired. The additional cost directly associated with utilization of allocated resources fall in one of the following categories:

1. Cost of hiring or requiring additional resources
2. Cost of idle labor or equipment
3. Cost of delay to other activities not having enough resources to schedule the work

The quality of allocation decision can be measured through a resource utilization factor. A utilization factor of 1 for the project is considered best. A simple procedure is presented for understanding the allocation procedures that lead to a more realistic schedule.

1. The resource availability limits are known for the entire project duration (not probable in a real-life situation).
2. Starting with the first day of the project, consider all activities that can be scheduled and select the one with the highest priority. The most common priority criterion is the earliest start and minimum total float.
3. If an activity meets the allocation criterion and there are available resources in the pool, schedule and assume that the activity will not be interrupted until completion. A more complex allocation procedure considers the possibility of interrupting an activity after its start and rescheduling the remaining portion at a later date.

Various priority rules for scheduling resources in a project on a given day are incorporated in various programs available. A few of them are listed below.

1. Schedule activities from the sorted list on the basis of first come, first served.
2. Schedule activities requiring more than one resource first.
3. Schedule activities with the longest duration first.
4. Schedule activities with the shortest duration first.
5. Schedule activities with the least amount of free float first.
6. Schedule activities with the least amount of total float first.
7. Schedule activities to maximize the allocated resource utilization factor.
8. Employ any combination of the criteria listed above.

A simple heuristic resource allocation program for only one nonstockable resource using the early start and total float rule considering no splitting activities is presented in Figure 1. This procedure allocates one available resource type to complete activities serially in time. Starting from the first day of the project, schedule all activities that satisfy resource availability; then, if necessary, recompute the network and consider the second day of the project, and so on.

The highest priority on any given day is given to all in-progress critical or noncritical activities that have already been scheduled. In this serial procedure, rescheduling of noncritical activities in-progress to create resources reserved for scheduling critical activities is not considered (parallel procedure).

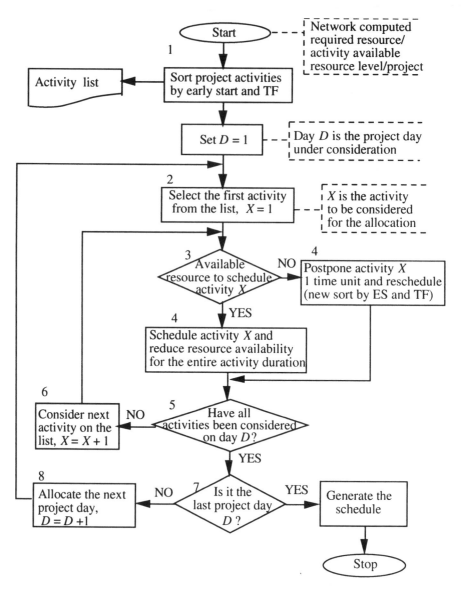

Figure 1 Heuristic Program for Allocating One Resource to Project Activities

486

RESOURCE AVAILABILITY DATE The calendar date when a resource quantity becomes available to be allocated to project activity(s).

Ordinary CPM techniques do not take explicit account of resource requirements and limitations on available resources to be allocated. Most calculations consider only constraints imposed on the start of the activity under consideration — as soon as its direct predecessors are completed, the activity is scheduled to start. When available resources are limited, activity start time and/or completion time are constrained not only by the precedence relationships but also by resource availability.

For the scheduler it is important to know the level and date of resource availability if the resource requirements/activity are known. Resources available at a constant intensity during a long project duration are shown in Figure 1 (e.g., three tower cranes available from the fourth month for the entire project duration). Available resources with a constant level but for a shorter project duration are shown in Figure 2 (e.g., two tower cranes available starting in month 4, four tower cranes available starting in month 8, and one tower crane available starting in month 12).

Most CPM-oriented computer programs accept as input multiple levels of availability for a limited or unlimited number of available resources. In this case for each resource under consideration the user must specify the start of availability date and the length of availability. Referring to Figure 2, consider the resource "tower crane" the user (scheduler) will specify at the project level. For the resource tower crane,

From the project start of	month 4	2 units	for 4 months
From the project start of	month 8	4 units	for 4 months
From the project start of	month 12	1 unit	for 2 months

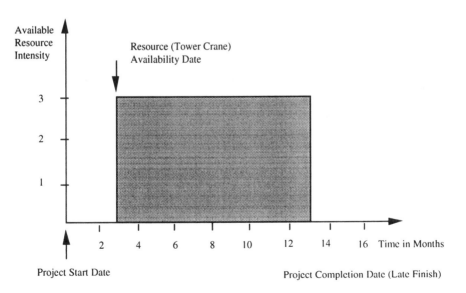

Figure 1 Constant-level Resource Availability

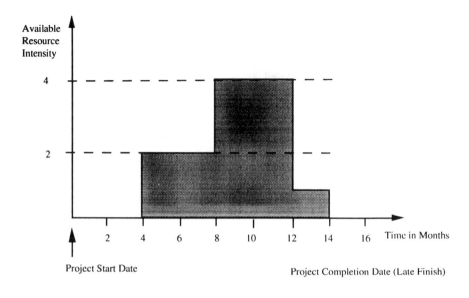

Figure 2 Variable Resource Availability

RESOURCE AVAILABILITY POOL The amount of resources available for any given allocation period.

Ordinary CPM techniques do not take explicit account of activity resource requirements or limitations on their availability to be allocated to various project activities on any given day.

While the assumption of unlimited resources may be justified in some situations, most project managers are facing the problem of relatively fixed trade availability, a specific amount of construction equipment or specialized tools, and sometimes limitations on materials delivery or availability at the planned rate.

When resources are limited, the scheduler will face a new set of scheduling constraints. It is of prime importance to determine the level of available resources from a common project resource pool on a daily basis. The resource availability pool plays a major role in balancing project resource requirements and the project resources availability level(s). The following steps are recommended in deciding the resource(s) available pool.

1. Starting with the CPM early start plan, determine the project early finish and resource demand level for each project day.
2. Determine the resource available pool at two rates: normal price (e.g., 8 hours/day or 40 hours/week) and at least one level of premium price (e.g., 10 hours/day or 50 hours/week — cost of overtime).

488

3. Compare project resource demand *(see Resource histogram)* with the available resource pool at the normal price (minimum) and find out if the project can be completed by the project early finish date. (Assume that the project early finish date is equal to or less than the project contractual due date.)
4. If available resources at normal price levels cannot satisfy project requirements, determine the causes of delay and the cost associated with it, such as penalties, lost bonuses, and project-duration associated expenses (job and general overhead).
5. Calculate the additional cost of using available premium-price resources (e.g., overtime) and the outcome of the project completion date.
6. Compare the additional cost of using the premium-price available pool with the savings of finishing the project early and decide on the most economical solution.

The resource availability pool certainly has a major impact on project duration. Figure 1 shows the typical situation in CPM scheduling technique for a single resource when plenty of resources are available.

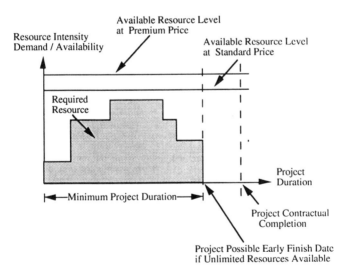

Figure 1 Resource Available Level Greater Than
Resource Required to Complete Project by Early Finish Date

Another project scenario is that when resource(s) available at the standard price are less than the required resource(s) for the early start schedule. In this situation, two possible solutions emerge.

1. Use resources at the standard level. The project will be delayed beyond the early finish, but the project will be completed by the contractual due date (Figure 2).

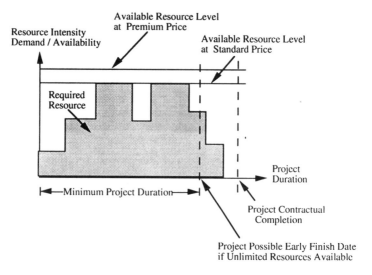

Figure 2 Standard Price Resources Availability Approach

2. Use premium-price resources at the additional cost. The project can be completed by the due date (Figure 3).

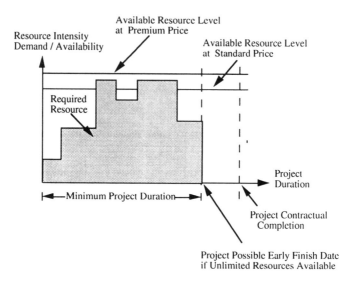

Figure 3 Premium-price Resource Availability Approach

In this case, some allocation procedures are needed to obtain a schedule with requirements less than or equal to the available resources. The analysis problem will be more complex if more than one resource is considered during the allocation procedure.

Most of the time, resource availability at the standard or premium price over activity duration is not constant over the projected project duration due to variations in crew size (absenteeism), idle or broken construction equipment, or disruption in the flow of materials. When available resource(s) pool fluctuates in time, it increases the complexity of developing a realistic project plan.

RESOURCE CALENDAR A graphical representation showing the availability of a particular resource throughout the duration of the project.

RESOURCE CODE The code assigned to a particular labor skill, material, or equipment type. The code is used to identify a given resource type at an activity or project level. Any resource code used during activity data input should be matched.

Various CPM software limits the resource codes to a few characters. Generally, only two characters will be used: AA, A1, 1A are valid characters. The planner should check the CPM software reference manual before assigning codes to project resources. Figure 1 shows a computerized printout of resource files, indicating resource description, code, and associated calendar.

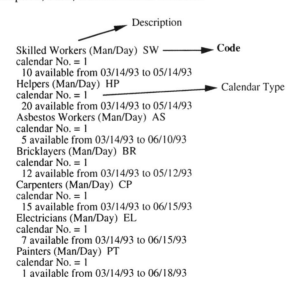

Skilled Workers (Man/Day) SW ⟶ **Code**
calendar No. = 1
  10 available from 03/14/93 to 05/14/93
Helpers (Man/Day) HP
calendar No. = 1 ⟶ Calendar Type
  20 available from 03/14/93 to 05/14/93
Asbestos Workers (Man/Day) AS
calendar No. = 1
  5 available from 03/14/93 to 06/10/93
Bricklayers (Man/Day) BR
calendar No. = 1
  12 available from 03/14/93 to 05/12/93
Carpenters (Man/Day) CP
calendar No. = 1
  15 available from 03/14/93 to 06/15/93
Electricians (Man/Day) EL
calendar No. = 1
  7 available from 03/14/93 to 06/15/93
Painters (Man/Day) PT
calendar No. = 1
  1 available from 03/14/93 to 06/18/93

Figure 1 Resource File

Any computer-generated resource report will show the resource code and resource description for easy identification. An example of labor skill resources for construction projects indicating the code, description, and most common unit of measure is shown in Table 1. It is recommended that resource coding and description be standardized for the company, to facilitate communication between managers at various organizational levels, between projects, between areas of operation, and so on.

Table 1 Resource Codes, Trade Description, and Resources' Unit of Measure

Code	Trade Description	Unit
SW	Skilled Workers	Man/Day
HP	Helpers	Man/Day
AS	Asbestos Workers	Man/Day
BH	Bricklayer Helpers	Man/Day
BL	Boilermakers	Man/Day
BR	Bricklayers	Man/Day
CF	Cement Finishers	Man/Day
CP	Carpenters	Man/Day
EL	Electricians	Man/Day
EO	Equipment Operators	Man/Day
EV	Elevator Constructors	Man/Day
GL	Glaziers	Man/Day
LA	Lathers	Man/Day
ML	Millwrights	Man/Day
MT	Mosaic & Terrazzo Workers	Man/Day
MS	Marble Setters	Man/Day
PB	Plumbers	Man/Day
PD	Pile Drivers	Man/Day
PH	Paper Hangers	Man/Day
PL	Plasterers	Man/Day
PT	Painters	Man/Day
RB	Rodmen (Reinforcing)	Man/Day
RF	Roofers	Man/Day
SF	Steamfitters or Pipefitters	Man/Day
SM	Stone Masons	Man/Day
SP	Sprinkler Installers	Man/Day
SS	Structural Steel Makers	Man/Day
SW	Sheet Metal Workers	Man/Day
TL	Tile Layers (Floor)	Man/Day
WS	Welders, Structural Steel	Man/Day

RESOURCE DESCRIPTION A concise description, including the unit of measure for a given resource code.

For computerized resource analysis, a resource file must be created. For each resource type that will be estimated or allocated to the project, a resource code *(see Resource code),* resource description, and weekly calendar type (number) associated with resource availabilities are generally required.

Based on selected CPM software, a limited number of characters may be used for the description. Furthermore, any resource description related to generating a computerized report will be printed for clarification. In the resource description, a unit of measure must be included during the requirements estimate and allocation process. The units of measure for requirements (aggregation) and allocation (limited resource) must be identical.

A printout indicating resources considered for an example project is shown in Figure 1. An example of coding and description for the most common trades for construction projects is shown in Table 1.

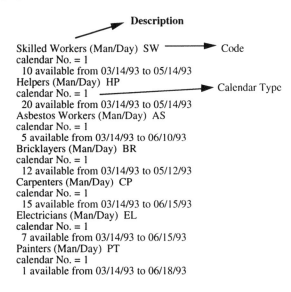

Figure 1 Resource File

493

Table 1 Trade Description

Code	Trade Description	Unit
SW	Skilled Workers	Man/Day
HP	Helpers	Man/Day
AS	Asbestos Workers	Man/Day
BH	Bricklayer Helpers	Man/Day
BL	Boilermakers	Man/Day
BR	Bricklayers	Man/Day
CF	Cement Finishers	Man/Day
CP	Carpenters	Man/Day
EL	Electricians	Man/Day
EO	Equipment Operators	Man/Day
EV	Elevator Constructors	Man/Day
GL	Glaziers	Man/Day
LA	Lathers	Man/Day
ML	Millwrights	Man/Day
MT	Mosaic & Terrazzo Workers	Man/Day
MS	Marble Setters	Man/Day
PB	Plumbers	Man/Day
PD	Pile Drivers	Man/Day
PH	Paper Hangers	Man/Day
PL	Plasterers	Man/Day
PT	Painters	Man/Day
RB	Rodmen (Reinforcing)	Man/Day
RF	Roofers	Man/Day
SF	Steamfitters or Pipefitters	Man/Day
SM	Stone Masons	Man/Day
SP	Sprinkler Installers	Man/Day
SS	Structural Steel Makers	Man/Day
SW	Sheet Metal Workers	Man/Day
TL	Tile Layers (Floor)	Man/Day
WS	Welders, Structural Steel	Man/Day

RESOURCE HISTOGRAM A graphic display of the amount of resource required and/or available as a function of time on a graph. Individual summary, incremental, and cumulative resource curve levels can be shown.

Resource histograms are developed at the project or multiproject level for any of the following situations:

1. *Resource requirements only as needed during each project working time unit or cumulative from the project start.* Figure 1 shows daily requirements for resource AA assuming that all activities will start at the earliest dates. Figure 2 shows the daily requirements for the resource AA late start schedule.

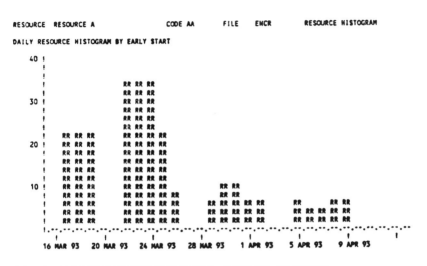

Figure 1 Daily Requirement for Resource AA at Early Start Schedule

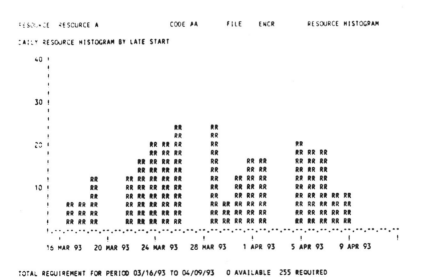

Figure 2 Daily Requirements for Resource AA Late Start Schedule

495

2. *Resource requirement histograms for project resources after leveling (see Leveling) procedures.* Figure 3 shows the histogram for resource AA after leveling.

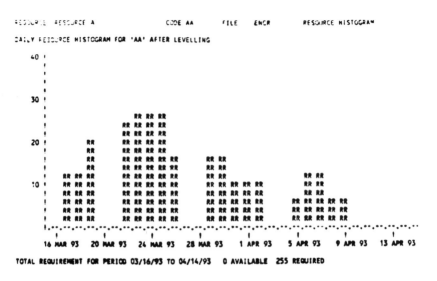

Figure 3 Histogram for Resource AA after Leveling

3. *Resource histogram considering the same available time and required resources (see Allocation).* See Figure 4.

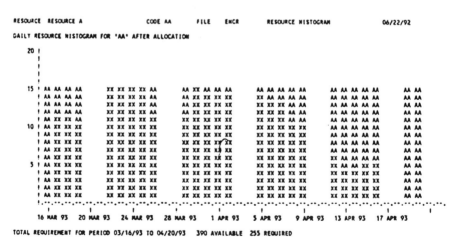

Figure 4 Resource Histogram for Resource AA for Schedule
After Limited Resource Consideration

A precedence diagram, including the resources estimated for each activity at a constant intensity for the entire activity duration, is shown in Figure 5. It was the base for developing the histograms in Figures 1 to 4.

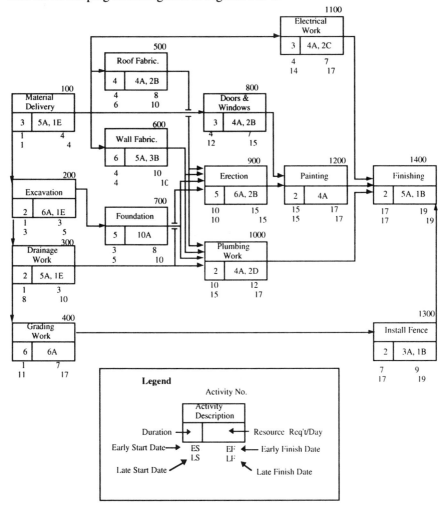

NOTE: The schedule computation in time units is based on the following assumption:

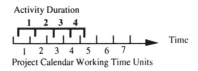

Activity Duration

Activity with 4 time units starts at working time unit 1 and is finished at working time unit 5.

Figure 5 Precedence Diagram with Resources/Activity

RESOURCE LEVELING *(see Leveling)*

RESOURCE-LIMITED SCHEDULING *(see Allocation)*

RESOURCE PLOTS *(see Resource histogram)*

RESOURCE PRIORITY The process of assigning or listing the required project resources in the order of scarcity, cost/unit, or other managerial criteria.

Some required project resources have priority over others and the hierarchy of resources by priority will play a major role in improving an original (initial) project plan using leveling procedures. For example, nonstockable resources (skilled and unskilled labor, construction equipment, fresh concrete, and hot asphalt mixture) have priority over stockable resources (most materials). This aspect of activities scheduling would not be important if the project could be scheduled to use each resource allocated *(see Resource allocation process)* to its fullest extent. In most cases, a schedule with the optimum utilization *(see Resource utilization factor)* of each resource is nearly impossible, and for this reason the project manager must choose the project resources to be optimized and the ones (ready available) that must fall where they may.

When the scope of planning is to generate a schedule that will lead to more uniform resource requirements over the entire project duration, the project manager must develop a list of project resources in order of priority, with the highest priority generally placed at the top of the list (Table 1). The resource priority list should not contain more than 10 resources for very large and complex networks because the effectiveness of leveling procedures *(see Leveling)* will be reflected in only three or four resources on the top of the priority list.

Table 1 Project Resources Priority List for Industrial Construction Project

Resource Priority	Resource Description	Unit of Measure
1	Welders	Day
2	200-Ton Hydraulic Crane	Day
3	Brick Mason	Day
4	Structural Steel Erectors	Day
5	Concrete	C.Y.

RESOURCE REQUIREMENT (ACTIVITY) The amount of the resource(s) required (estimated) for each time unit during activity duration. The required resource amount can be constant or variable over the activity's duration.

1. Required at constant or uniform intensity (Figure 1).

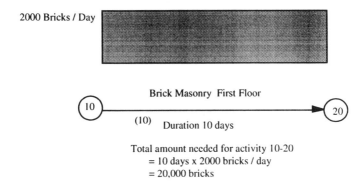

Figure 1 Uniform Resource Requirement

2. Nonuniform requirements (Figure 2), that is, requirements that vary over the activity's duration.

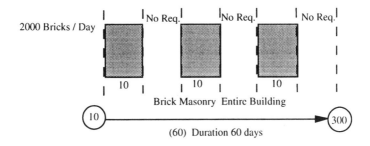

Figure 2 Nonuniform Requirements over Activity 10–300 Duration, but Uniform for Short Periods

Note: There are periods with no requirement. Resource bricks are required at a rate of 2000 bricks/day during three periods of 10 days each for the activity Brick Masonry for Entire Building (10–300).

3. Variable requirements over activity's duration (Figure 3).

Resource Intensity

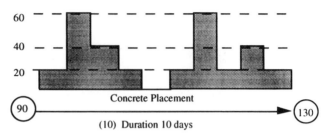

Figure 3 Concrete Placement in Structural Frame
(Columns, Girders/Beams, and Slabs)

Note: Reason for fluctuation: small quantities in columns, more in beams and slabs.

4. Breakdown of a category of work (considered as a resource) into its components, sometimes called micro or simple resources.
   a. Simple resources (micro) are those that cannot be broken down further. In construction the followings are typical: bricks, concrete blocks, concrete, trades, and construction equipment. Considering simple resources (micro) when estimating requirements/activity is very useful to the purchasing department and for inventory, hiring, and acquisition policy and practice.
   b. Macro or complex resources are those categories of work that contain an assembly of simple resources. In construction, "reinforced concrete" is a complex resource because it represents the following simple resources: formwork in place, carpenter and carpenter-helper hours, reinforcing bars, road labor hours, concrete, concrete mason hours, concrete finisher hours, and concrete pumps or other placement equipment.

A more detailed breakdown is possible. For example, the simple resource "concrete" component of the complex resource "reinforced concrete" can be broken down into:

500

- Cement (tons/C.Y.)
- Course aggregate (C.Y.)
- Fine aggregate (C.Y.)
- Additives (lb or gallons/C.Y.)
- Water (gallon/C.Y.)

Also, a macro resource may be required at a constant level during the activity's duration. Generally, its components are nonuniform.

The planner/scheduler should give consideration to the level of complexity during the estimating process because, for the long range, overall project planning may be adequate, but for short-interval planning, oversimplification will lead to erroneous results and erroneous decisions.

RESOURCE UTILIZATION FACTOR An indicator of the utilization of allocated *(see Allocation)* nonstockable resources during project duration.

$$U = \frac{\Sigma \text{ daily project resource allocated}}{\Sigma \text{ daily project resource required}}$$

if the daily resource requirement is less than or equal to resource availability.

During the perfect (ideal) schedule, allocated resources should equal resource requirements, leading to a resource utilization factor of 1. A utilization factor of greater than 1 means that more nonstockable resources (labor or production capacity) have been allocated than are needed.

For the schedule shown in Figure 1 with a project duration from March 17, 1993 to April 16, 1993 and allocated resource AA at a constant rate of 15 units for each working day between March 16, 1993 and April 20, 1993, the utilization factor for resource AA is

$$U(\text{AA}) = \frac{390 \text{ allocated units}}{255 \text{ required units}} = 1.73$$

The daily resource histogram for resource AA after allocation which was the base for calculating the utilization factor of nonstockable resource AA is shown in Figure 2.

CIVIL ENGINEERING DEPARTMENT   PRIMAVERA PROJECT PLANNER

REPORT DATE 24NOV93 RUN NO.   41   PROJECT SCHEDULE   START DATE 13MAR93 FIN DATE 09APR93
          06:26
CLASSIC SCHEDULE REPORT (RESOURCE)   DATA DATE 17MAR93 PAGE NO.   1

ACTIVITY ID	ORIG DUR	REM DUR	CAL ID	%	CODE	ACTIVITY DESCRIPTION	EARLY START	EARLY FINISH	LATE START	LATE FINISH	TOTAL FLOAT	RESOURC	QUNTY
100	3	3	1	0		MATERIAL DELIVERY	17MAR93	19MAR93	17MAR93	19MAR93	0	AA	15
												BB	3
200	2	2	1	0		EXCAVATION	17MAR93	18MAR93	19MAR93	22MAR93	2	AA	12
												BB	2
300	2	2	1	0		DRAINAGE WORK	17MAR93	18MAR93	26MAR93	29MAR93	7	AA	10
												BB	2
400	6	6	1	0		GRADING WORK	17MAR93	24MAR93	31MAR93	07APR93	10	AA	36
700	5	5	1	0		FOUNDATION	19MAR93	25MAR93	23MAR93	29MAR93	2	AA	50
600	6	6	1	0		WALL FABRICATION	22MAR93	29MAR93	22MAR93	29MAR93	0	AA	30
												BB	18
500	4	4	1	0		ROOF FABRICATION	22MAR93	25MAR93	24MAR93	29MAR93	2	AA	16
												BB	8
800	3	3	1	0		DOORS & WINDOWS	22MAR93	24MAR93	01APR93	05APR93	8	AA	12
												BB	6
1100	3	3	1	0		ELECTRICAL WORK	22MAR93	24MAR93	05APR93	07APR93	10	AA	12
												CC	6
1300	2	2	1	0		INSTALL FENCE	25MAR93	26MAR93	08APR93	09APR93	10	AA	6
												BB	2
900	5	5	1	0		ERECTION	30MAR93	05APR93	30MAR93	05APR93	0	AA	30
												BB	10
1000	2	2	1	0		PLUMBING WORK	30MAR93	31MAR93	06APR93	07APR93	5	AA	8
												DD	4
1200	2	2	1	0		PAINTING	06APR93	07APR93	06APR93	07APR93	0	AA	8
1400	2	2	1	0		FINISHING	08APR93	09APR93	08APR93	09APR93	0	AA	10
												BB	2

Figure 1 Allocated Resource Schedule

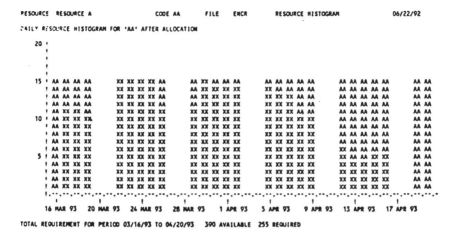

Figure 2 Daily Resource Histogram of Resource AA

RESTRAINT *(see Constraint)*

REVISION (REVIEW) *(see Monitoring)*

S

SCHEDULE The plan for completion of a project based on a logical arrangement of activities.

A schedule comprises of several activities or tasks representing a project plan in a logical order. A project schedule is normally prepared by a general contractor to show the intention of fulfilling the contractual agreement that a project will be finished within a specified time frame or date. A schedule is used as a means to communicate a project plan *(see Project plan)* to various project participants, to control a project, and to provide management with project information for decision making. Schedule development requires a project network *(see Network)* to be constructed showing technological and managerial constraints *(see Dependency)* among activities as shown in Figures 1 (arrow diagram network) and 2 (precedence diagram network). After the network is developed, the schedule computation is performed *(see Network analysis)* to identify the time required for each activity to be started and finished and overall project duration and interim completion dates.

The most common methods for schedule representation are a tabular format report and a bar chart format report as shown in Figures 3 and 4 *(see Bar chart)*. The critical path method (CPM) and program evaluation and review technique (PERT) are the most widely used methods for project schedule development.

Before the start of project execution, the detailed schedule *(see Detailed schedule)* must be developed to represent the planned sequences of operations. During project execution, the schedule must be updated regularly to reflect changes and their impact on overall project picture *(see Updated schedule)*.

504

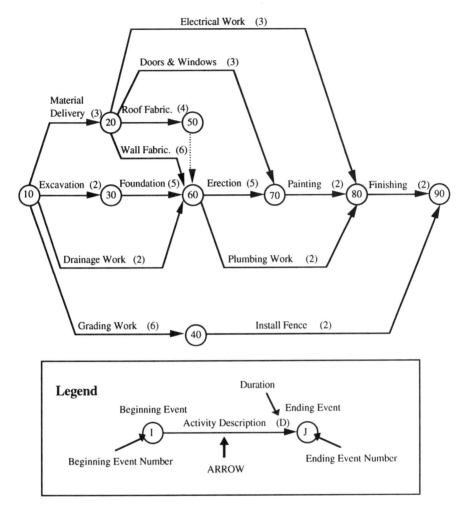

Figure 1 Arrow Diagram Network

505

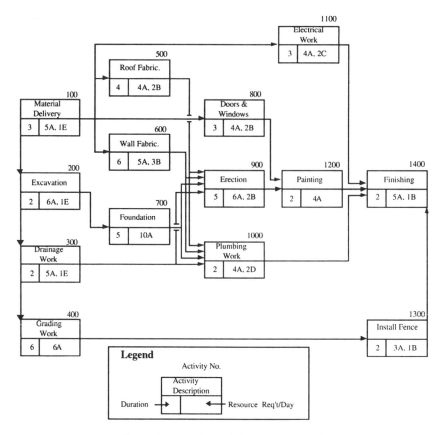

Figure 2 Precedence Diagram Network

CIVIL ENGINEERING DEPARTMENT   PRIMAVERA PROJECT PLANNER

REPORT DATE 23NOV93   RUN NO.   22   PROJECT SCHEDULE REPORT   START DATE 13MAR93   FIN DATE   9APR93
              22:35
CLASSIC SCHEDULE REPORT - SORT BY ES, TF   DATA DATE   17MAR93   PAGE NO.   1

ACTIVITY ID	ORIG DUR	REM DUR	% 	CODE	ACTIVITY DESCRIPTION	EARLY START	EARLY FINISH	LATE START	LATE FINISH	TOTAL FLOAT
100	3	3	0		MATERIAL DELIVERY	17MAR93*	19MAR93	17MAR93*	19MAR93	0
200	2	2	0		EXCAVATION	17MAR93	18MAR93	19MAR93	22MAR93	2
300	2	2	0		DRAINAGE WORK	17MAR93	18MAR93	26MAR93	29MAR93	7
400	6	6	0		GRADING WORKS	17MAR93	24MAR93	31MAR93	7APR93	10
700	5	5	0		FOUNDATION	19MAR93	25MAR93	23MAR93	29MAR93	2
600	6	6	0		WALL FABRICATION	22MAR93	29MAR93	22MAR93	29MAR93	0
500	4	4	0		ROOF FABRICATION	22MAR93	25MAR93	24MAR93	29MAR93	2
800	3	3	0		DOORS & WINDOWS	22MAR93	24MAR93	1APR93	5APR93	8
1100	3	3	0		ELECTRICAL WORK	22MAR93	24MAR93	5APR93	7APR93	10
1300	2	2	0		INSTALL FENCE	25MAR93	26MAR93	8APR93	9APR93	10
900	5	5	0		ERECTION	30MAR93	5APR93	30MAR93	5APR93	0
1000	2	2	0		PLUMBING WORK	30MAR93	31MAR93	6APR93	7APR93	5
1200	2	2	0		PAINTING	6APR93	7APR93	6APR93	7APR93	0
1400	2	2	0		FINISHING	8APR93	9APR93	8APR93	9APR93	0

Figure 3 Tabular Schedule Report (Precedence Diagram)

506

```
CIVIL ENGINEERING DEPARTMENT                  PRIMAVERA PROJECT PLANNER


REPORT DATE 23NOV93  RUN NO.   23           PROJECT SCHEDULE REPORT                    START DATE 13MAR93  FIN DATE  9APR93
           22:38
BAR CHART BY ES, EF, TF                                                               DATA DATE 17MAR93   PAGE NO.   1


                                                                                                         DAILY-TIME PER.   1
----------------------------------------------------------------------------------------------------------------------------
 ............ACTIVITY DESCRIPTION.............       15    22    29    05    12    19    26    03    10    17
ACTIVITY ID  OD  RD  PCT   CODES    FLOAT   SCHEDULE  MAR   MAR   MAR   APR   APR   APR   APR   MAY   MAY   MAY
----------- ---- ---- --- ----------- -----  --------  93    93    93    93    93    93    93    93    93    93
                                                      ----------------------------------------------------------
EXCAVATION                                   EARLY    . EE  .     .     .     .     .     .     .     .     .
        200   2    2    0               2             . *   .     .     .     .     .     .     .     .     .

DRAINAGE WORK                                EARLY    . EE  .     .     .     .     .     .     .     .     .
        300   2    2    0               7             . *   .     .     .     .     .     .     .     .     .

MATERIAL DELIVERY                            EARLY    . EEE .     .     .     .     .     .     .     .     .
        100   3    3    0               0             . *   .     .     .     .     .     .     .     .     .

GRADING WORK                                 EARLY    . EEE..EEE  .     .     .     .     .     .     .     .
        400   6    6    0              10             . *   .     .     .     .     .     .     .     .     .

FOUNDATION                                   EARLY    . * E..EEEE .     .     .     .     .     .     .     .
        700   5    5    0               2             . *   .     .     .     .     .     .     .     .     .

DOORS & WINDOWS                              EARLY    . *   EEE   .     .     .     .     .     .     .     .
        800   3    3    0               8             . *   .     .     .     .     .     .     .     .     .

ELECTRICAL WORK                              EARLY    . *   EEE   .     .     .     .     .     .     .     .
       1100   3    3    0              10             . *   .     .     .     .     .     .     .     .     .

ROOF FABRICATION                             EARLY    . *   EEEE  .     .     .     .     .     .     .     .
        500   4    4    0               2             . *   .     .     .     .     .     .     .     .     .

WALL FABRICATION                             EARLY    . *   EEEEE..E     .     .     .     .     .     .     .
        600   6    6    0               0             . *   .     .     .     .     .     .     .     .     .

INSTALL FENCE                                EARLY    . *   . EE  .     .     .     .     .     .     .     .
       1300   2    2    0              10             . *   .     .     .     .     .     .     .     .     .

PLUMBING WORK                                EARLY    . *   .     .EE   .     .     .     .     .     .     .
       1000   2    2    0               5             . *   .     .     .     .     .     .     .     .     .

ERECTION                                     EARLY    . *   .     .EEEE..E    .     .     .     .     .     .
        900   5    5    0               0             . *   .     .     .     .     .     .     .     .     .

PAINTING                                     EARLY    . *   .     .     .EE   .     .     .     .     .     .
       1200   2    2    0               0             . *   .     .     .     .     .     .     .     .     .

FINISHING                                    EARLY    . *   .     .     . EE  .     .     .     .     .     .
       1400   2    2    0               0             . *   .     .     .     .     .     .     .     .     .
```

Figure 4 Bar Chart Schedule Report (Precedence Diagram)

SCHEDULED EVENT TIME In PERT, an arbitrary schedule time that can be imposed on any event but is equally used only at a certain milestone or the last event, to find the probability of meeting the imposed date.

The scheduled event time *(TS)* is a time assigned to any particular event for the purpose of achieving the status of work in PERT. The scheduled event time can be used as a milestone for any event in a PERT diagram; however, it becomes widely used only on the last event of the network.

Since PERT applies nondeterministic system, it involves uncertainty and the probability of completing work. The scheduled event time is therefore used to calculate the probability or chance of meeting the designated scheduled event time.

To compute the probability of meeting the *TS*, the activity expected duration *(Te) (see Expected duration)* and its standard deviation *[S(Te)]* must be precalculated. The event completion time is considered to have a normal probability distribution with mean value of *TE* and standard deviation of the mean of *TE* as shown in Figure 1.

Probability

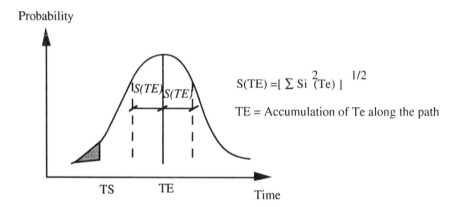

$$S(TE) = [\, \Sigma \, Si^2(Te)\,]^{1/2}$$

TE = Accumulation of Te along the path

Figure 1 Normal Probability Distribution of Event Completion Time
with Mean *TE* and Standard Deviation *S(TE)*

$$\text{probability of meeting } TS, \; P[TS] \quad = \quad \frac{\text{shaded area}}{\text{total area under curve}} \quad \%$$

508

The calculation can be simplified by using the formula

$$Z = \frac{TS - TE}{S(TE)}$$

where TS represents the scheduled event time and TE represents the earliest expected event time.

$$S(TE) = [\ \Sigma\ Si^2\ (TE)\ ]^{1/2}$$

where i = all activity(s) of the longest path(s).

If there is more than one longest path, the Z value will be calculated based on the largest $S(TE)$, the highest standard deviation. After calculating the Z values, the probability of meeting TS for the specific event can be obtained by matching the Z value to the probability from Table 1.

Table 1 Value of the Standard Normal Distribution Function

Z	Probability of Meeting Due Date	Z	Probaility of Meeting Due Date	Z	Probaility of Meeting Due Date
3.0	0.999	0.8	0.788	-1.2	0.115
2.8	0.997	0.6	0.726	-1.4	0.081
2.6	0.995	0.4	0.655	-1.6	0.055
2.4	0.992	0.2	0.579	-1.8	0.036
2.2	0.986	0.0	0.500	-2.0	0.023
2.0	0.977	-0.2	0.421	-2.2	0.014
1.8	0.964	-0.4	0.345	-2.4	0.008
1.6	0.945	-0.6	0.274	-2.6	0.005
1.4	0.919	-0.8	0.212	-2.8	0.003
1.2	0.885	-1.0	0.159	-3.0	0.001
1.0	0.841				

Figure 2 shows an example of a PERT network consisting of six events with schedule calculation, and Table 2 shows the probability of meeting TS on day 11 for event 4 and on day 25 for event 6. Table 2 indicates that based on the given network and associated time information, the probability of achieving event 4 at $TS = 11$ is 38%, whereas the probability of achieving event 6 at $TS = 25$ is 60%. Table 3 shows various TS values imposed on events for comparing different probabilities of meeting TS.

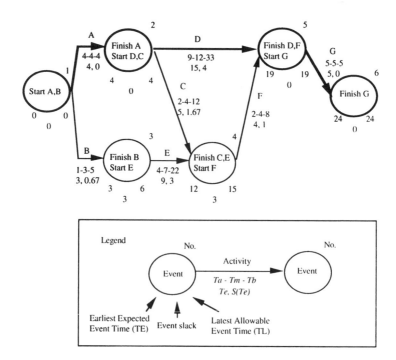

Figure 2 Schedule Calculation of PERT Network

Table 2 Calculation for Probability of Meeting *TS* from Sample PERT Network

Event	TE	S(TE)	TS	Z	Probability of Meeting TS
1	0	0	N/A	N/A	N/A
2	4	0	N/A	N/A	N/A
3	3	Sqrt.[(0.67)(0.67)] = 0.67	N/A	N/A	N/A
4	12	Sqrt.[(0.67)(0.67) +(3)(3)] = 3.07	11	-0.33	0.38
5	19	Sqrt. [(0)(0)+(4)(4)] = 4	N/A	N/A	N/A
6	24	Sqrt.[(0)(0)+(4)(4)+(0)(0)] = 4	25	0.25	0.60

Table 3 Comparison of Different Probabilities of Meeting *TS* for Event 6, the Last Network Event

Event	TE	S(TE)	TS	Z	Probability of Meeting TS
6	24	Sqrt.[(0)(0)+(4)(4)+(0)(0)] = 4	22	-0.50	0.31
			23	-0.25	0.40
			24	0.00	0.50
			25	0.25	0.60
			26	0.50	0.62

Note: For the completion of a project at the computed early finish time, one standard deviation beyond the early finish date is recommended as a minimal safety margin, giving the project manager the probability of completing the project within the agreed-upon duration at 84.1%.

SCHEDULE PERFORMANCE INDEX (SPI) This index provides the schedule efficiency of the work accomplished. The SPI is defined as the ratio of the cumulative budgeted cost of work performed over the cumulative budgeted cost of work scheduled.

The schedule performance index (SPI) is one of the C/SCSC elements and represents a performance measurement of whether the work accomplished to date is behind schedule or ahead of schedule based on the earned value *(see Cost and schedule control system criteria).* While variances such as the SV provide significant information about the progress of the project to date in absolute terms, an index such as the SPI provides additional insight into the status of the project in relative terms. The SPI is calculated using the formula:

$$SPI = BCWP/BCWS$$

(See Budgeted cost of work performed and Budgeted cost of work scheduled.) Interpretation of the SPI is as follows:

VARIANCE	Less Than 1.00	Equal to 1.00	More Than 1.00
SPI	Behind Schedule	On Schedule	Ahead of Schedule

For example, an SPI of 0.9 (90%) indicates that $90 worth of work has been accomplished to date for $100 worth of work scheduled.

To develop the SPI in a project, the following steps are generally taken:

Step 1: Use the original PBS of a project.

Step 2: Use the original cost estimation of the project based on the PBS.

Step 3: Use the latest updated CPM schedule by the data date *(see Data date).*

Step 4: Calculate the BCWP of the activity level at the point of the data date (time now).

Step 5: Calculate the BCWS of the activity level at the point of the data date (time now).

Step 6: Calculate the SPI by dividing the BCWP by the BCWS at the project level. The results of steps 4 to 6 are shown in the monthly cost performance report.

The preceding six steps are explained using a residential house as an example.

1. *Step 1:* PBS

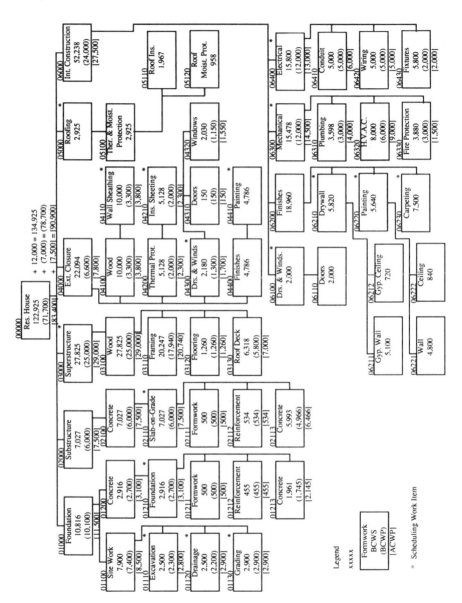

2. *Step 2:* Cost Estimation

DESCRIPTION	U/M	U/P	Q'TY	TOTAL
01 General Requirements				
Subtotal	L/S	12,000	1	12,000
02 Site Work				
01110 Excavation	C.Y.	1.89	1,323	2,500
01120 Drainage	L.F.	3.11	804	2,500
01130 Grading	C.Y.	1.06	2,736	2,900
Subtotal				7,900
03 Concrete				
01211 Foundation Form	SFCA	2.89	173	500
01212 Foundation Re-bar	Tons	910	0.5	455
01213 Foundation Concrete	C.Y.	103.2	19	1,961
02111 Slab-on-Grade Form	L.F.	1.38	362	500
02112 Slab-on-Grade Re-bar	Tons	890	0.6	534
02113 Slab-on-Grade Concrete	C.Y.	90.8	66	5,993
Subtotal				9,943
06 Wood				
03110 Superstructure Framing	M.B.F.	1,191	17	20,247
03120 Superstructure Flooring	S.F.	0.63	2,000	1,260
03130 Roof Decking	S.F.	5.01	1,261	6,318
04110 Wall Sheathing	S.F.	1.17	8,547	10,000
Subtotal				37,825
07 Thermal & Moisture Protection				
04210 Wall Insulating Sheeting	S.F.	0.6	8,547	5,128
05110 Roof Insulating	S.F.	1.56	1,261	1,967
05120 Roof Moisture Protection	S.F.	0.76	1,261	958
Subtotal				8,053
08 Doors & Windows				
04310 Ext. Doors	Ea.	150	1	150
04320 Ext. Windows	Ea.	203	10	2,030
06110 Int. Doors	Ea.	200	10	2,000
Subtotal				4,180
09 Finishes				
04410 Ext. Painting	S.F.	0.56	8,547	4,786
06211 Int. Gyp. Wallboard	S.F.	0.34	15,000	5,100
06212 Int. Gyp. Ceiling Board	S.F.	0.36	2,000	720
06311 Int. Wall Painting	S.F.	0.32	15,000	4,800
06312 Int. Ceiling Painting	S.F.	0.42	2,000	840
06230 Carpeting	S.Y.	5	1,500	7,500
Subtotal				23,746
15 Mechanical				
06310 Plumbing	L.F.	3.98	904	3,598
06320 H.V.A.C.	L/S	8,000	1	8,000
06330 Fire Protection	Ea.	776	5	3,880
Subtotal				15,478
16 Electrical				
06410 Electrical Conduit	L/S	5,000	1	5,000
06420 Electrical Wiring	L/S	5,000	1	5,000
06430 Electrical Fixtures	L/S	5,800	1	5,800
Subtotal				15,800
GRAND TOTAL				134,925

3. *Step 3:* Updated CPM Network Schedule

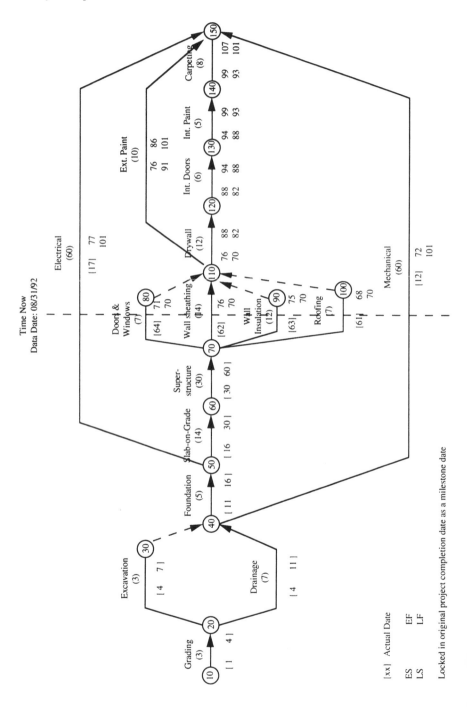

514

4. *Steps 4 to 6:* Monthly Cost Performance Report

MONTHLY COST PERFORMANCE REPORT: CPM ITEM

PROJECT: RESIDENTIAL HOUSE   REPORT DATE: 08/31/92

ITEM		CURRENT PERIOD									CUMULATIVE TO DATE									AT COMPLETION		
		BUDGET COST		ACWP	VARIANCE			INDEX		PC	BUDGET COST		ACWP	VARIANCE			INDEX		PC	BAC	EAC	ACV
PBS	CPM	BCWS	BCWP		SV	CV	AV	SPI	CPI		BCWS	BCWP		SV	CV	AV	SPI	CPI				
01130	10 - 20										2,900	2,900	2,900	0	0	0	1.00	1.00	1.00	2,900	2,900	
01110	20 - 30										2,500	2,300	2,800	-200	-500	-300	0.92	0.82	0.92	2,500	2,800	-300
01120	20 - 40										2,500	2,200	2,800	-300	-600	-300	0.88	0.79	0.88	2,500	2,800	-300
01210	40 - 50										2,916	2,700	3,100	-216	-400	-184	0.93	0.87	0.93	2,916	3,100	-184
06300	40 - 150	5,160	5,000	5,200	-160	-200	-40	0.97	0.96	0.32	13,416	12,000	14,500	-1,416	-2,500	-1,084	0.89	0.83	0.78	15,478	16,562	-1,084
02110	50 - 60										7,027	6,000	7,500	-1,027	-1,500	-473	0.85	0.80	0.85	7,027	7,500	-473
03000	60 - 70	14,903	13,900	15,000	-1,003	-1,100	-97	0.93	0.93	0.50	27,825	25,000	29,000	-2,825	-4,000	-1,175	0.90	0.86	0.90	27,825	29,000	-1,175
06400	50 - 150	5,260	5,100	5,300	-160	-200	-40	0.97	0.96	0.32	12,361	12,000	13,000	-361	-1,000	-639	0.97	0.92	0.76	15,800	16,439	-639
04300	70 - 80	1,555	1,300	1,700	-255	-400	-145	0.84	0.76	0.60	1,555	1,300	1,700	-255	-400	-145	0.84	0.76	0.60	2,180	2,325	-145
04210	70 - 90	2,135	2,000	2,300	-135	-300	-165	0.94	0.87	0.39	2,135	2,000	2,300	-135	-300	-165	0.94	0.87	0.39	5,128	5,293	-165
04110	70 - 110	3,570	3,300	3,800	-270	-500	-230	0.92	0.87	0.33	3,570	3,300	3,800	-270	-500	-230	0.92	0.87	0.33	10,000	10,230	-230
05000	70 - 100																			2,925	2,925	0
06210	110 - 120																			5,820	5,820	0
04410	110 - 150																			4,786	4,786	0
06100	120 - 130																			2,000	2,000	0
06220	130 - 140																			5,640	5,640	0
06230	140 - 150																			7,500	7,500	0
SUBTOTAL		32,583	30,600	33,300	-1,983	-2,700	-717	0.94	0.92	0.25	78,705	71,700	83,400	-7,005	-11,700	-4,695	0.91	0.86	0.58	122,925	127,620	-4,695
GEN. REQ.		2,400	2,400	2,500		-100	-100	1.00	0.96	0.20	7,200	7,000	7,500	-200	-500	-300	0.97	0.93	0.58	12,000	12,300	-300
TOTAL		34,983	33,000	35,800	-1,983	-2,800	-817	0.94	0.92	0.24	85,905	78,700	90,900	-7,205	-12,200	-4,995	0.92	0.87	0.58	134,925	139,920	-4,995

Note: The project at the data date has 0.94 in the current period and 0.92 cumulatively as the scheduling performance index (SPI). It shows that the project is behind schedule because the SPI is less than 1. The same comment is valid for all work items except for item 01130, which is on schedule (SPI=0).

515

SCHEDULE VARIANCE (SV) The difference between the project budgeted cost of work performed (BCWP) and the project budgeted cost of work schedule (BCWS) at any point in time or data date. Schedule variance measures the project schedule progress in contract cost unit.

Schedule variance (SV) is one of the C/SCSC elements and represents a performance measure of whether the work accomplished to date is behind schedule or ahead of schedule *(see Cost and schedule control system criteria)*.

At a given time now, the SV is obtained using the formula

$$SV = BCWP \text{ minus } BCWS$$

(See Budgeted cost of work performed and Budgeted cost of work schedule.) Interpretation of the SV is as follows:

VARIANCE	-	0	+
SV	Behind Schedule	On Schedule	Ahead of Schedule

As shown in the definition, the SV measures the scheduled progress of the project in equivalent dollars, not in time units. So the SV cannot positively indicate whether or not any specific task or milestone has been accomplished because the actual sequence and timing of various activities may be different from those planned.

To develop the SV in a project, the following steps are generally taken:

Step 1: Use the original PBS of a project.
Step 2: Use the original cost estimation of the project based on the PBS.
Step 3: Use the latest updated CPM schedule by the data date (time now) *(see Data date)*.
Step 4: Calculate the BCWP of the activity and project level on the data date (time now).
Step 5: Calculate the BCWS of the activity and project level on the data date (time now).
Step 6: Calculate the SV by subtracting BCWS from BCWP at the project level. The results of the steps 4 to 6 are shown in the monthly cost performance report.
Step 7: Draw the SV on the graph using the time-phased and cumulative values by the data date (time now).

The preceding seven steps are explained using a residential house as an example.

516

1. *Step 1:* PBS

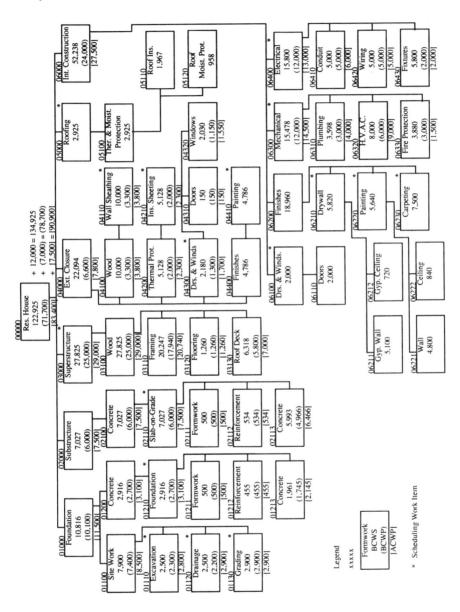

517

2. *Step 2:* Cost Estimation

DESCRIPTION	U/M	U/P	QTY	TOTAL
01 General Requirements				
Subtotal	L/S	12,000	1	12,000
02 Site Work				
01110 Excavation	C.Y.	1.89	1,323	2,500
01120 Drainage	L.F.	3.11	804	2,500
01130 Grading	C.Y.	1.06	2,736	2,900
Subtotal				7,900
03 Concrete				
01211 Foundation Form	SFCA	2.89	173	500
01212 Foundation Re-bar	Tons	910	0.5	455
01213 Foundation Concrete	C.Y.	103.2	19	1,961
02111 Slab-on-Grade Form	L.F.	1.38	362	500
02112 Slab-on-Grade Re-bar	Tons	890	0.6	534
02113 Slab-on-Grade Concrete	C.Y.	90.8	66	5,993
Subtotal				9,943
06 Wood				
03110 Superstructure Framing	M.B.F.	1,191	17	20,247
03120 Superstructure Flooring	S.F.	0.63	2,000	1,260
03130 Roof Decking	S.F.	5.01	1,261	6,318
04110 Wall Sheathing	S.F.	1.17	8,547	10,000
Subtotal				37,825
07 Thermal & Moisture Protection				
04210 Wall Insulating Sheeting	S.F.	0.6	8,547	5,128
05110 Roof Insulating	S.F.	1.56	1,261	1,967
05120 Roof Moisture Protection	S.F.	0.76	1,261	958
Subtotal				8,053
08 Doors & Windows				
04310 Ext. Doors	Ea.	150	1	150
04320 Ext. Windows	Ea.	203	10	2,030
06110 Int. Doors	Ea.	200	10	2,000
Subtotal				4,180
09 Finishes				
04410 Ext. Painting	S.F.	0.56	8,547	4,786
06211 Int. Gyp. Wallboard	S.F.	0.34	15,000	5,100
06212 Int. Gyp. Ceiling Board	S.F.	0.36	2,000	720
06311 Int. Wall Painting	S.F.	0.32	15,000	4,800
06312 Int. Ceiling Painting	S.F.	0.42	2,000	840
06230 Carpeting	S.Y.	5	1,500	7,500
Subtotal				23,746
15 Mechanical				
06310 Plumbing	L.F.	3.98	904	3,598
06320 H.V.A.C.	L/S	8,000	1	8,000
06330 Fire Protection	Ea.	776	5	3,880
Subtotal				15,478
16 Electrical				
06410 Electrical Conduit	L/S	5,000	1	5,000
06420 Electrical Wiring	L/S	5,000	1	5,000
06430 Electrical Fixtures	L/S	5,800	1	5,800
Subtotal				15,800
GRAND TOTAL				134,925

3. *Step 3:* Updated CPM Network Schedule

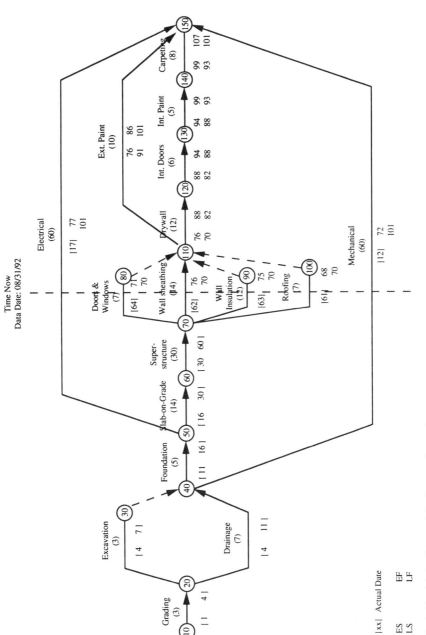

519

4. *Steps 4 to 6:* Monthly Cost Performance Report

MONTHLY COST PERFORMANCE REPORT: CPM ITEM

PROJECT : RESIDENTIAL HOUSE

REPORT DATE: 08/31/92

ITEM PBS	CPM	Current Period BCWS	BCWP	ACWP	SV	CV	AV	SPI	CPI	PC	Cumulative to Date BCWS	BCWP	ACWP	SV	CV	AV	SPI	CPI	PC	At Completion BAC	EAC	ACV
01130	10 - 20										2.900	2.900	2.900	0	0	0	1.00	1.00	1.00	2.900	2.900	
01110	20 - 30										2.500	2.300	2.800	-200	-500	-300	0.92	0.82	0.92	2.500	2.800	-300
01120	20 - 40										2.500	2.200	2.800	-300	-600	-300	0.88	0.79	0.88	2.500	2.800	-300
01210	40 - 50										2.916	2.700	3.100	-216	-400	-184	0.93	0.87	0.93	2.916	3.100	-184
06300	40 - 150	5.160	5.000	5.200	-160	-200	-40	0.97	0.96	0.32	13.416	12.000	14.500	-1.416	-2.500	-1.084	0.89	0.83	0.78	15.478	16.562	-1.084
02110	50 - 60										7.027	6.000	7.500	-1.027	-1.500	-473	0.85	0.80	0.85	7.027	7.500	-473
03000	60 - 70	14.903	13.900	15.000	-1.003	-1.100	-97	0.93	0.93	0.50	27.825	25.000	29.000	-2.825	-4.000	-1.175	0.90	0.86	0.90	27.825	29.000	-1.175
06400	50 - 150	5.260	5.100	5.300	-160	-200	-40	0.97	0.96	0.32	12.361	12.000	13.000	-361	-1.000	-639	0.97	0.92	0.76	15.800	16.439	-639
04300	70 - 80	1.555	1.300	1.700	-255	-400	-145	0.84	0.76	0.60	1.555	1.300	1.700	-255	-400	-145	0.84	0.76	0.60	2.180	2.325	-145
04210	70 - 90	2.135	2.000	2.300	-135	-300	-165	0.94	0.87	0.39	2.135	2.000	2.300	-135	-300	-165	0.94	0.87	0.39	5.128	5.293	-165
04110	70 - 110	3.570	3.300	3.800	-270	-500	-230	0.92	0.87	0.33	3.570	3.300	3.800	-270	-500	-230	0.92	0.87	0.33	10.000	10.230	-230
05000	70 - 100																			2.925	2.925	0
06210	110-120																			5.820	5.820	0
04410	110-150																			4.786	4.786	0
06100	120-130																			2.000	2.000	0
06220	130-140																			5.640	5.640	0
06230	140-150																			7.500	7.500	0
SUBTOTAL		32.583	30.600	33.300	-1.983	-2.700	-717	0.94	0.92	0.25	78.705	71.700	83.400	-7.005	-11.700	-4.695	0.91	0.86	0.58	122.925	127.620	-4.695
GEN. REQ.		2.400	2.400	2.500		-100	-100	1.00	0.96	0.20	7.200	7.000	7.500	-200	-500	-300	0.97	0.93	0.58	12.000	12.300	-300
TOTAL		34.983	33.000	35.800	-1.983	-2.800	-817	0.94	0.92	0.24	85.905	78.700	90.900	-7.205	-12.200	-4.995	0.92	0.87	0.58	134.925	139.920	-4.995

Note: The project at the data date has −1,983 in the current period and −7,205 cumulatively as the schedule variance (SV). It shows that the project is behind schedule because the SV is less than 0.

The same comment is valid for all work items except for item 01130, which is on schedule (SV=0).

5. *Step 7:* SV

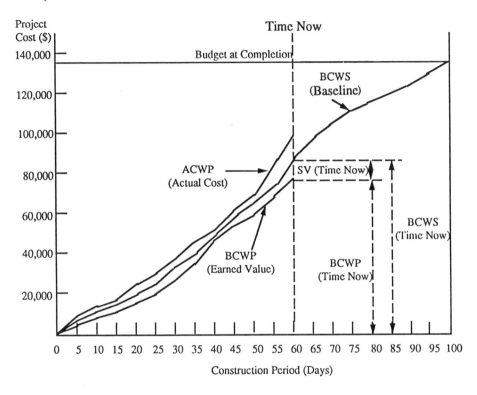

SCHEDULING The process of assigning the schedule start and finish calendar dates to all or a group of activities that belong to a project.

Project scheduling is to identify the start and finish times in calendar dates for each activity comprised in a project. During the planning stage, after all activities and their logical relationships are defined, the scheduling process will determine when each activity is scheduled to start and be completed. Project scheduling can be performed by different scheduling techniques. Four common scheduling techniques are (1) the critical path method (CPM) *(see Critical path method)* using an arrow or precedence diagramming method, (2) a bar chart or Gantt chart *(see Bar chart)*, (3) the program evaluation and review technique (PERT) *(see PERT)*, and (4) the line of balance technique (LOB) *(see Line of balance technique)*.

The most popular scheduling techniques are CPM and bar chart. The network analysis *(see Network analysis)* is employed in CPM to perform scheduling to define the scheduled dates of a project and its activities. Unlike CPM, bar chart scheduling does not require the establishment of interrelationships among activities. The activity duration along with the scheduled start and completion dates of each activity will

be assigned based on management experience. Therefore, the scheduling process for a bar chart technique does not require mathematical computation.

SCHEDULING SPECIFICATIONS *(see Specifications)*

SCHEDULING VARIANCE (TIME) The difference in time units between planned start and finish activity or project dates and actual or revised start and finish activity dates.

During the project planning stage, an as-planned project schedule *(see As-planned schedule)* is developed to identify the planned start and completion of a project and its activities. The as-planned project schedule, however, is seldom followed exactly from the project start to completion. After the project is started, the as-planned schedule is regularly updated or revised so that the actual and forecasted project schedule can be compared against the planned schedule.

The indicator of the difference in time units between planned and actual or forecasted schedules for a project and all activities is called *scheduling variance*. The scheduling variance will identify the project schedule status as being behind or ahead of schedule. Typically, a schedule variance can be identified for a project and each activity. A project schedule variance is the difference between the planned and updated or forecasted project completion dates. For the scheduling variance of each activity, it is also the difference between the planned and updated or forecasted activity completion dates. Figure 1 shows the scheduling variance in tabular format. The project scheduling variance is one day behind schedule, and for activity 200, the scheduling variance is 3 days behind schedule.

```
---------------------------------------------------------------------------------------------------------
CIVIL ENGINEERING DEPARTMENT                      PRIMAVERA PROJECT PLANNER

REPORT DATE 24NOV93   RUN NO.    18              PROJECT SCHEDULE              START DATE 13MAR93  FIN DATE 12APR93
                      00:57
CLASSIC SCHEDULE REPORT                                                       DATA DATE  19MAR93  PAGE NO.   1
---------------------------------------------------------------------------------------------------------
ACTIVITY  TARG CUR  CAL          ACTIVITY                          EARLY     EARLY     TARGET    TARGET    VAR.
  ID      DUR  DUR  ID  %  CODE  DESCRIPTION                       START     FINISH    START     FINISH    FINIS
--------  ---- ---  --- --- ---- --------------------------------  --------  --------  --------  --------  -----
     100   3   2    1 100        MATERIAL DELIVERY                 17MAR93A  19MAR93A  17MAR93   19MAR93     0
     300   2   2    1 100        DRAINAGE WORK                     17MAR93A  18MAR93A  17MAR93   18MAR93     0
     200   2   2    1   0        EXCAVATION                        17MAR93A  23MAR93   17MAR93   18MAR93    -3
     400   6   2    1  33        GRADING WORK                      17MAR93A  24MAR93   17MAR93   24MAR93     0
     500   4   0    1   0        ROOF FABRICATION                  19MAR93    29MAR93  22MAR93   25MAR93    -2
     600   6   0    1   0        WALL FABRICATION                  19MAR93    26MAR93  22MAR93   29MAR93     1
     800   3   0    1   0        DOORS & WINDOWS                   19MAR93    23MAR93  22MAR93   24MAR93     1
    1100   3   0    1   0        ELECTRICAL WORK                   19MAR93    23MAR93  22MAR93   24MAR93     1
     700   5   0    1   0        FOUNDATION                        24MAR93    30MAR93  19MAR93   25MAR93    -3
    1300   2   0    1   0        INSTALL FENCE                     25MAR93    26MAR93  25MAR93   26MAR93     0
     900   5   0    1   0        ERECTION                          31MAR93    06APR93  30MAR93   05APR93    -1
    1000   2   0    1   0        PLUMBING WORK                     31MAR93    01APR93  30MAR93   31MAR93    -1
    1200   2   0    1   0        PAINTING                          07APR93    08APR93  06APR93   07APR93    -1
    1400   2   0    1   0        FINISHING                         09APR93    12APR93  08APR93   09APR93    -1
```

Figure 1 Planned vs. Updated Schedule in Tabular Format

522

SCOPE CHANGE *(see Change order)*

SCOPE (PROJECT SCHEDULING) Defines the work to be done and is documented by the contract parameters or a project to which the company is committed. A system boundary indicating activities that define the project.

From the planning and scheduling point of view, the scope defined in the CPM network content can be developed for construction activities only or for incorporating design, submittals, procurement, and startup operations. The planning and scheduling scope is defined through specifications *(see Planning and Specifications)*. If they are omitted from contract documents or are in a summary format, ambiguity will exist and the entire purpose of planning and scheduling will be lost. The project scheduling scope is used to inform the project participating parties about the work to be done, and most important to record progress and to update the schedule for the coming period.

SHORT-INTERVAL PLAN (SHORT-INTERVAL SCHEDULE) A 2 to 3-week span schedule used to detail work activities from the master schedule in an area for analysis or to plan work assigned at the crew level.

A short-interval plan is used to divide the updated master plan into smaller workable segments. Short-interval planning involves dividing the project into tasks that take 1 to 2 weeks to complete, then meeting at the end of every 1 or 2 weeks to review progress and plan the next interval. It gives management a step-by-step strategy for achieving objectives in a timely and orderly fashion. It also gives all project personnel specific goals that they can accomplish in 5 to 10 working days. This approach is a practical method for planning and monitoring job progress and detecting and correcting problems before they develop into crises.

Development of a short-interval plan requires a project manager, job superintendents, field foremen, subcontractors, purchasing agents, and so on, to meet and discuss the plan. A sample of a short-interval planning form is shown in Figure 1. The process for developing a short-interval plan is:

1. Set measurable objectives for the next short-interval plan.
2. Review the status of work scheduled for completion during the last interval.
3. Adjust projected cost to complete each activity in the job cost control system.
4. Modify the job schedules to reflect the impact of the previous interval.
5. Plan the work for the next short interval.
6. Make specific personnel assignments.
7. Schedule subcontractors and related trades.
8. Plan fabrication requirements.

9. Satisfy tool and equipment requirements.
10. Assign one person to make sure that suppliers will meet all delivery dates for materials and equipment needed during the upcoming intervals, allowing adequate lead times.
11. Establish material and equipment handling procedures on the job site.
12. Discuss the interval that will follow the upcoming one.
13. Review the short-interval planning process itself and revise it, as necessary, to make it more efficient and responsive to job needs.
14. Identify the date of the next general contractor's job conference. If appropriate, plan to hold a short meeting prior to the conference to help the project manager prepare.

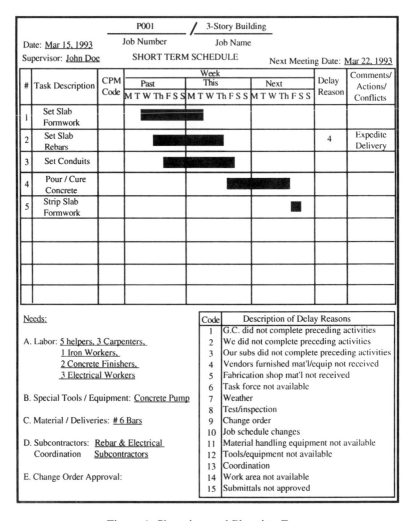

P001 / 3-Story Building									
Date: Mar 15, 1993		Job Number		Job Name					
Supervisor: John Doe		SHORT TERM SCHEDULE				Next Meeting Date: Mar 22, 1993			

#	Task Description	CPM Code	Week Past	Week This	Week Next	Delay Reason	Comments/ Actions/ Conflicts
			M T W Th F S S	M T W Th F S S	M T W Th F S S		
1	Set Slab Formwork		▬▬▬▬				
2	Set Slab Rebars		▬▬▬	▬▬		4	Expedite Delivery
3	Set Conduits		▬▬	▬▬			
4	Pour / Cure Concrete			▬▬▬			
5	Strip Slab Formwork				▪		

Needs:	Code	Description of Delay Reasons
	1	G.C. did not complete preceding activities
A. Labor: 5 helpers, 3 Carpenters,	2	We did not complete preceding activities
1 Iron Workers,	3	Our subs did not complete preceding activities
2 Concrete Finishers,	4	Vendors furnished mat'l/equip not received
3 Electrical Workers	5	Fabrication shop mat'l not received
	6	Task force not available
B. Special Tools / Equipment: Concrete Pump	7	Weather
	8	Test/inspection
C. Material / Deliveries: # 6 Bars	9	Change order
	10	Job schedule changes
D. Subcontractors: Rebar & Electrical	11	Material handling equipment not available
Coordination Subcontractors	12	Tools/equipment not available
	13	Coordination
E. Change Order Approval:	14	Work area not available
	15	Submittals not approved

Figure 1 Short-interval Planning Form

SLACK *(see Activity total float and Free float)*

SLACK PATHS The sequence of activities and events that do not lie on the critical path(s), but have the same total slack (total float).

A slack path is a noncritical path of a network consisting of noncritical activity(s) with the same amount of total float *(see Activity total float)* along the path. In each network it is common to have more than one slack path, and in each slack path it is possible to have more than one noncritical activity.

Figures 1 and 2 illustrate a simple network consisting of several slack paths in arrow and precedence diagrams, respectively. Each slack path can be determined in four different ways:

1. The path lying between two critical activities (e.g., the path of the activity Doors and Windows with 8 days of total float).
2. The path between beginning event and ended event or starting activities and ended activity (e.g., the path of the activities Grading Work and Install Fence with 10 days of total float).
3. The path lying between beginning event and critical activity (e.g., the path of the activities Excavation and Foundation with 2 days of total float).
4. The path lying between critical activity and ended event.

The purpose of identifying the slack paths is for the project manager to differentiate from critical path(s) so that the project manager can give priority to the critical activities. However, the slack path should not be totally ignored because the slack path(s) with or less than a small certain amount of total float, determined by the project manager, is considered as a nearly critical path *(see Nearly critical path)* and has the chance to become a critical path during project progress.

525

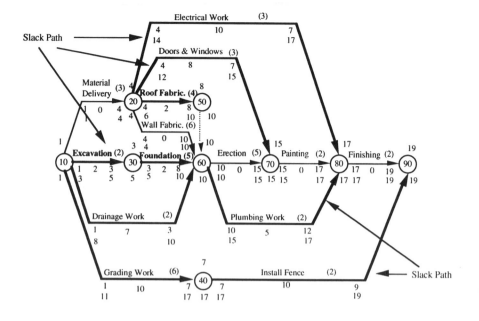

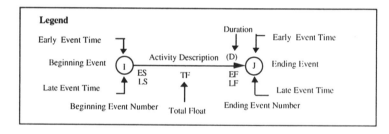

Legend

Early Event Time

Beginning Event

Late Event Time

Activity Description (D)

ES
LS

TF

EF
LF

Duration

Early Event Time

Ending Event

Late Event Time

Beginning Event Number   Total Float   Ending Event Number

NOTE: The schedule computation in time units is based on the following assumption:

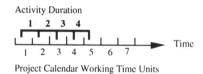

Activity Duration

Project Calendar Working Time Units

Activity with 4 time units starts at working time unit 1 and is finished at working time unit 5.

Figure 1 Slack Paths Shown in Arrow Diagram Network

526

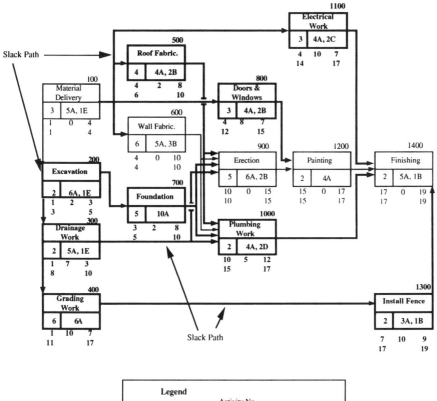

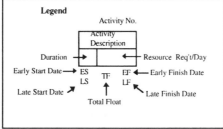

Slack Path

Slack Path

NOTE: The schedule computation in time units is based on the following assumption:

Activity Duration

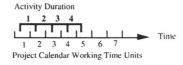

Time

Project Calendar Working Time Units

Activity with 4 time units starts at working time unit 1 and is finished at working time unit 5.

Figure 2 Slack Paths Shown in Precedence Diagram Network

527

SPECIFICATIONS (PLANNING AND SCHEDULING SPECIFICATIONS)
The contractual requirements regarding project schedule and control to be performed by a contractor or consultant.

Planning and scheduling specifications are the contract documents identifying what level of schedule control is desired and to what level of detail that schedule must be developed. The specifications are used to ensure that the owner will be provided with the project schedule that will be utilized to control the work. It is recommended that the specifications focus on developing, updating, and reporting the schedule for planning, coordinating, and performing the work, including the work of subcontractors, suppliers, and vendors. The specifications should be divided into three sections as described below.

Section 1 General Organization and Responsibilities

1.1 Description, Reference, and Standard
1.2 Scheduling Responsibility
1.3 Minimum Qualifications for Planning and Scheduling Staff
1.4 CPM Training Requirements for Contractor, Subcontractor, and/or Owner
1.5 Preliminary Network Submission Deadline
1.6 Detailed Network Submission Deadline
1.7 Review and Approval Process
1.8 Cost of Planning, Scheduling, and Monitoring
1.9 Progress Payments for Planning, Scheduling, and Monitoring
1.10 Subcontractor Input
1.11 Contractor's Scheduling Plan
1.12 Planning, Scheduling, and Monitoring Audits
1.13 Confidentiality and Schedule Ownership
1.14 Computer Access and Security

Section 2 Scope and Products

2.1 Network Analysis Technique
2.2 CPM Computer Software (or Equivalent) to Be Used
2.3 Activity Related Information
    2.3.1 Activity Description
    2.3.2 Activity Duration (Time Units)
    2.3.3 Activity Coding System (PBS, CSI, or Owner Provided)

STANDARD DEVIATION A measure of the expected variation of an activity or project's duration in PERT.

PERT, the program evaluation review technique *(see PERT),* is a scheduling computation technique using expected value *(Te)* as an estimated duration of activity performance time. The expected value *Te* is obtained from three time estimates from the equation

$$Te = \frac{Ta + 4 \times Tm + Tb}{6}$$

where *Ta represents optimistic time (see Optimistic time estimate), Tm most Likely time (see Most likely time estimate), and Tb pessimistic time (see Pessimistic time estimate).*

The PERT computation is based on the judgment of the three time estimates given by experts. Based on the formula, the standard deviation must be considered because the higher the standard deviation of an activity, the more likely the actual time to perform the activity will differ from *Te.* An example is:

Activity X: *Ta* =10, *Tm* = 30, *Tb* = 20.
Hence, *Te* for activity X is $\frac{10 + 4(20) + 30}{6}$ = 20

Activity Y: *Ta* =18, *Tm* = 20, *Tb* = 22.
Hence, *Te* for activity Y is $\frac{18 + 4(20) + 22}{6}$ = 20

The standard deviation of an activity is calculated with the following equation:

$$S(Te) = \frac{Tb - Ta}{6}$$

In this case,

standard deviation of activity X = $\frac{(30-20)}{6}$ = 1.67

standard deviation of activity Y = $\frac{(22-18)}{6}$ = 0.67

530

It is clearly seen that the estimate time of activities X and Y are the same; however, activity X will have a higher chance of reaching an actual duration different from its expected duration than activity Y because the standard deviation of activity X = 1.67, which is greater than the standard deviation of activity Y. Figure 1 shows the standard deviation for all activities.

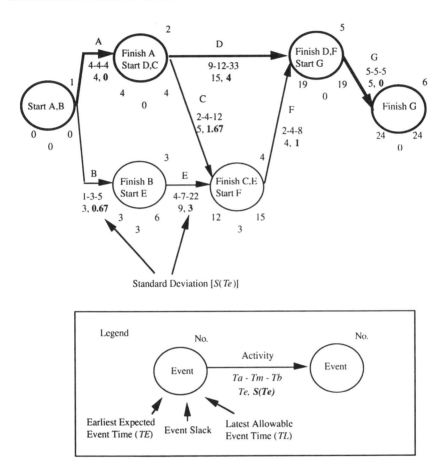

Figure 1 Schedule Calculation from PERT Network Showing Standard Deviation

STANDARD NETWORK DIAGRAM *(see Library network)*

STANDARD PROPOSAL SCHEDULE A preestablished conceptual network on file.

A standard proposal schedule is a condensed schedule that has been developed to be used repetitively for submission with a bid with or without the owner's request *(see Conceptual schedule).* When a general contractor performs the same type of work repeatedly, the standard proposal schedule should be developed so that it can be retrieved for future use. For a large and complex project, the proposal schedule is normally requested when the bid is submitted. The standard proposal schedule is simplified by using the higher level of project breakdown structure (PBS) *(see Project breakdown structure)* to develop a condensed schedule in order to reduce the number of activities in the schedule, as shown in Figure 1. A standard proposal schedule can be kept in a computer in the form of a fragnet or library network *(see Fragnet and Library network).* For a particular project, this schedule will be retrieved and customarily modified to meet the project objectives.

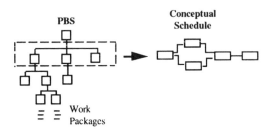

Figure 1 Correlation Between PBS and Proposal Schedule

The standard proposal schedule shows major activities and may or may not contain their duration. Once the bid is awarded, the proposal schedule can be expanded to show each major activity in more detail at a lower level of PBS, as shown in Figure 2. The standard proposal schedule can be represented by a network diagram or bar chart *(see Network* and *Bar chart)* unless specified in the contract.

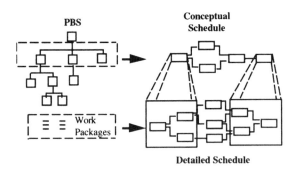

Figure 2 Correlation Between PBS and Proposal and Detailed Schedule

532

START ACTIVITY(S) An initiating activity in the CPM network which has no preceding activity(s).

In CPM network scheduling, there can be more than one start activity in the network. In the case of multiple start activities, all start activities will possess the same early start date unless there are contract start dates *(see Contract date)*. Figures 1 and 2 show arrow and precedence diagram networks with four start activities.

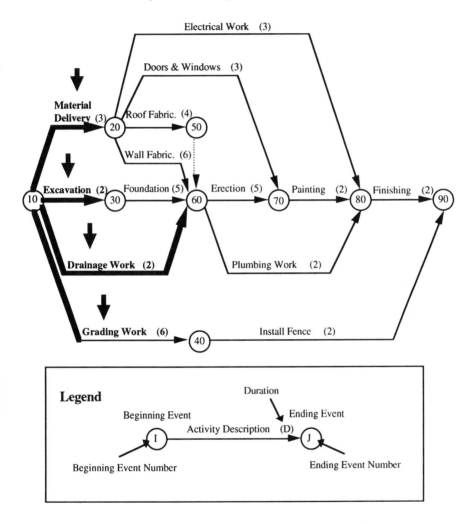

Figure 1 Arrow Diagram Network with Four Start Activities

533

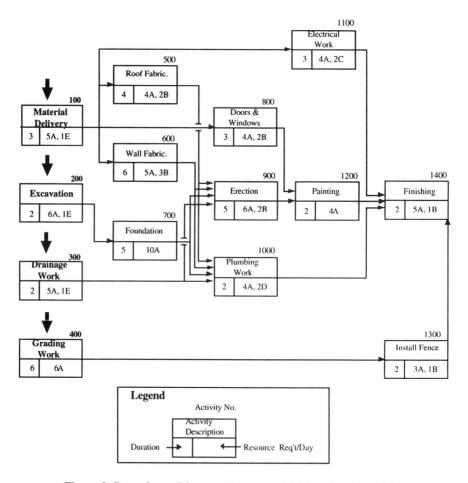

Figure 2 Precedence Diagram Network with Four Start Activities

STATUS The condition of an activity or project at a specified point in time (data date). The activity or project status is obtained through the updating project schedule by indicating progress at regular intervals.

An activity or project progress status is assessed by gathering and validating schedule data periodically. The schedule data are then analyzed and recomputed for schedule updating *(see Updated schedule).* After a project has been started, all activities should be monitored and updated regularly at a specified lapsed time so that

534

the status can be indicated. Typical activity statuses are completed, in-progress, and unstarted activities.

By assessing the status of activities or a project, management can identify which activities are behind or ahead of an as-planned schedule. The problems impeding the project progress can then be indicated, and corrective actions can be taken to accelerate the project schedule in the event of schedule delay. Figure 1 shows the planned work sequences in a precedence diagram network. Figure 2 indicates the status of all activities at a specific data date in a precedence diagram network. Activities 100 and 300 are completed activities, activities 200 and 400 are in-progress activities, and the remaining activities are unstarted activities. Figures 3 and 4 show the status of all activities in tabular and bar chart format.

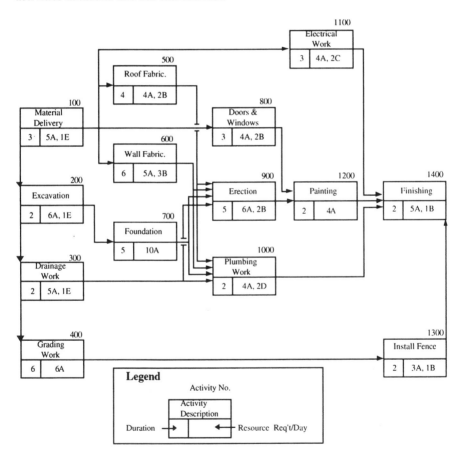

Figure 1 As-planned Precedence Diagram Network

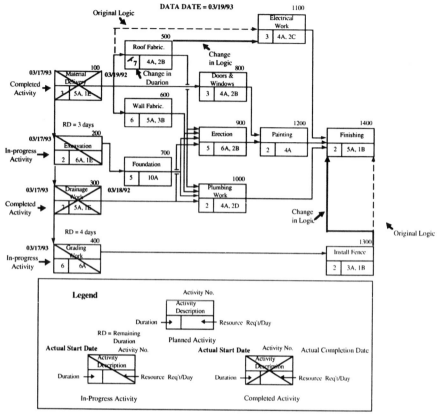

Figure 2 Updated Precedence Diagram Network

CIVIL ENGINEERING DEPARTMENT   PRIMAVERA PROJECT PLANNER

REPORT DATE 24NOV93   RUN NO.   12   PROJECT SCHEDULE REPORT   START DATE 13MAR93   FIN DATE 12APR93
        0:03

CLASSIC SCHEDULE REPORT - SORT BY ES, TF   DATA DATE 19MAR93   PAGE NO.   1

ACTIVITY ID	ORIG DUR	REM DUR	%	CODE	ACTIVITY DESCRIPTION	EARLY START	EARLY FINISH	LATE START	LATE FINISH	TOTAL FLOAT
100	3	0	100		MATERIAL DELIVERY	17MAR93A	19MAR93A			
300	2	0	100		DRAINAGE WORK	17MAR93A	18MAR93A			
200	2	3	0		EXCAVATION	17MAR93A	23MAR93		23MAR93	0
400	6	4	33		GRADING WORK	17MAR93A	24MAR93		8APR93	11
600	6	6	0		WALL FABRICATION	19MAR93	26MAR93	23MAR93	30MAR93	2
500	4	4	0		ROOF FABRICATION	19MAR93	24MAR93	25MAR93	30MAR93	4
800	3	3	0		DOORS & WINDOWS	19MAR93	23MAR93	2APR93	6APR93	10
1100	3	3	0		ELECTRICAL WORK	19MAR93	23MAR93	6APR93	8APR93	12
700	5	5	0		FOUNDATION	24MAR93	30MAR93	24MAR93	30MAR93	0
1300	2	2	0		INSTALL FENCE	25MAR93	26MAR93	9APR93	12APR93	11
900	5	5	0		ERECTION	31MAR93	6APR93	31MAR93	6APR93	0
1000	2	2	0		PLUMBING WORK	31MAR93	1APR93	7APR93	8APR93	5
1200	2	2	0		PAINTING	7APR93	8APR93	7APR93	8APR93	0
1400	2	2	0		FINISHING	9APR93	12APR93	9APR93	12APR93	0

Figure 3 Updated Tabular Report

```
--------------------------------------------------------------------------------------------------------
CIVIL ENGINEERING DEPARTMENT              PRIMAVERA PROJECT PLANNER

REPORT DATE 23NOV93  RUN NO.   10         PROJECT SCHEDULE REPORT                    START DATE 13MAR93  FIN DATE 12APR93
             22:51
BAR CHART BY ES, EF, TF                                                             DATA DATE 19MAR93  PAGE NO.    1

                                                                                              DAILY-TIME PER.    1
--------------------------------------------------------------------------------------------------------
...........ACTIVITY DESCRIPTION.............     15    22    29    05    12    19    26    03    10    17
ACTIVITY ID  OD   RD  PCT   CODES   FLOAT   SCHEDULE   MAR   MAR   MAR   APR   APR   APR   APR   MAY   MAY   MAY
-----------  ---- ---- --- ------------- -----   --------   93    93    93    93    93    93    93    93    93    93
                                                        --------------------------------------------------------
DRAINAGE WORK                                   CURRENT   . AA*  .     .     .     .     .     .     .     .     .
         300   2   0 100                        PLANNED   . EE*  .     .     .     .     .     .     .     .     .

MATERIAL DELIVERY                               CURRENT   . AA*  .     .     .     .     .     .     .     .     .
         100   3   0 100                        PLANNED   . EEE  .     .     .     .     .     .     .     .     .

EXCAVATION                                      CURRENT   . AAE..EE    .     .     .     .     .     .     .     .
         200   2   3   0              0         PLANNED   . EE*  .     .     .     .     .     .     .     .     .

GRADING WORK                                    CURRENT   . AAE..EEE   .     .     .     .     .     .     .     .
         400   6   4  33             11         PLANNED   . EEE..EEE   .     .     .     .     .     .     .     .

DOORS & WINDOWS                                 CURRENT   .   E..EE    .     .     .     .     .     .     .     .
         800   3   3   0             10         PLANNED   .  *  EEE    .     .     .     .     .     .     .     .

ELECTRICAL WORK                                 CURRENT   .   E..EE    .     .     .     .     .     .     .     .
        1100   3   3   0             12         PLANNED   .  *  EEE    .     .     .     .     .     .     .     .

ROOF FABRICATION                                CURRENT   .   E..EEE   .     .     .     .     .     .     .     .
         500   4   4   0              4         PLANNED   .  *  EEEE   .     .     .     .     .     .     .     .

WALL FABRICATION                                CURRENT   .   E..EEEEE .     .     .     .     .     .     .     .
         600   6   6   0              2         PLANNED   .  *  EEEEE..E      .     .     .     .     .     .     .

FOUNDATION                                      CURRENT   .  *  .  EEE..EE    .     .     .     .     .     .     .
         700   5   5   0              0         PLANNED   . E..EEEE    .     .     .     .     .     .     .     .

INSTALL FENCE                                   CURRENT   .  *  .  EE   .     .     .     .     .     .     .     .
        1300   2   2   0             11         PLANNED   .  *  .  EE   .     .     .     .     .     .     .     .

PLUMBING WORK                                   CURRENT   .  *  .     . EE   .     .     .     .     .     .     .
        1000   2   2   0              5         PLANNED   .  *  .     .EE    .     .     .     .     .     .     .

ERECTION                                        CURRENT   .  *  .     . EEE..EE    .     .     .     .     .     .
         900   5   5   0              0         PLANNED   .  *  .     .EEEE..E     .     .     .     .     .     .

PAINTING                                        CURRENT   .  *  .     .     . EE   .     .     .     .     .     .
        1200   2   2   0              0         PLANNED   .  *  .     .     .EE    .     .     .     .     .     .

FINISHING                                       CURRENT   .  *  .     .     . E..E  .     .     .     .     .     .
        1400   2   2   0              0         PLANNED   .  *  .     .     . EE   .     .     .     .     .     .
```

Figure 4 Updated Bar Chart Report

537

Comparison of as-planned and updated schedule status reveals the difference between the planned and actual progress for each activity and project. Figures 5 and 6 shows this comparison in tabular and bar chart format, indicating the project completion date of 1 day behind a planned completion date.

```
-------------------------------------------------------------------------------------------------------
CIVIL ENGINEERING DEPARTMENT                    PRIMAVERA PROJECT PLANNER

REPORT DATE 24NOV93  RUN NO.   18                 PROJECT SCHEDULE              START DATE 13MAR93  FIN DATE 12APR93
            00:57
CLASSIC SCHEDULE REPORT                                                        DATA DATE  19MAR93  PAGE NO.   1
-------------------------------------------------------------------------------------------------------
ACTIVITY   TARG CUR  CAL            ACTIVITY                         EARLY     EARLY     TARGET    TARGET    VAR.
   ID      DUR  DUR  ID  %  CODE    DESCRIPTION                      START     FINISH    START     FINISH    FINIS
---------- ---- ---- --- ---        ----------------------------     --------  --------  --------  --------  -----
      100    3    2   1  100        MATERIAL DELIVERY                17MAR93A  19MAR93A  17MAR93   19MAR93     0
      300    2    2   1  100        DRAINAGE WORK                    17MAR93A  18MAR93A  17MAR93   18MAR93     0
      200    2    2   1   0         EXCAVATION                       17MAR93A  23MAR93   17MAR93   18MAR93    -3
      400    6    2   1  33         GRADING WORK                     17MAR93A  24MAR93   17MAR93   24MAR93     0
      500    4    0   1   0         ROOF FABRICATION                 19MAR93   29MAR93   22MAR93   25MAR93    -2
      600    6    0   1   0         WALL FABRICATION                 19MAR93   26MAR93   22MAR93   29MAR93     1
      800    3    0   1   0         DOORS & WINDOWS                  19MAR93   23MAR93   22MAR93   24MAR93     1
     1100    3    0   1   0         ELECTRICAL WORK                  19MAR93   23MAR93   22MAR93   24MAR93     1
      700    5    0   1   0         FOUNDATION                       24MAR93   30MAR93   19MAR93   25MAR93    -3
     1300    2    0   1   0         INSTALL FENCE                    25MAR93   26MAR93   25MAR93   26MAR93     0
      900    5    0   1   0         ERECTION                         31MAR93   06APR93   30MAR93   05APR93    -1
     1000    2    0   1   0         PLUMBING WORK                    31MAR93   01APR93   30MAR93   31MAR93    -1
     1200    2    0   1   0         PAINTING                         07APR93   08APR93   06APR93   07APR93    -1
     1400    2    0   1   0         FINISHING                        09APR93   12APR93   08APR93   09APR93    -1
```

Figure 5 As-planned vs. Updated Schedule in Tabular Format

```
CIVIL ENGINEERING DEPARTMENT              PRIMAVERA PROJECT PLANNER


REPORT DATE 24NOV93  RUN NO.   16         PROJECT SCHEDULE REPORT                    START DATE 13MAR93  FIN DATE 12APR93
            0:39
BAR CHART BY ES, EF, TF                                                              DATA DATE 19MAR93   PAGE NO.    1


                                                                                                  DAILY-TIME PER.    1

-----------------------------------------------------------------------------------------------------------------
..............ACTIVITY DESCRIPTION..............    15     22     29     05     12     19     26     03     10     17
ACTIVITY ID  OD  RD  PCT   CODES   FLOAT   SCHEDULE  MAR    MAR    MAR    APR    APR    APR    APR    MAY    MAY    MAY
-----------  ---- ---- ---  ------------  -----     --------   93     93     93     93     93     93     93     93     93     93
-----------------------------------------------------------------------------------------------------------------

DRAINAGE WORK                              CURRENT  . AA*  .       .      .      .      .      .      .      .      .
       300   2   0 100                     PLANNED  . EE*  .       .      .      .      .      .      .      .      .

MATERIAL DELIVERY                          CURRENT  . AA*  .       .      .      .      .      .      .      .      .
       100   3   0 100                     PLANNED  . EEE  .       .      .      .      .      .      .      .      .

EXCAVATION                                 CURRENT  . AAE..EE      .      .      .      .      .      .      .      .
       200   2   3   0             0       PLANNED  . EE*  .       .      .      .      .      .      .      .      .

GRADING WORK                               CURRENT  . AAE..EEE     .      .      .      .      .      .      .      .
       400   6   4  33            11       PLANNED  . EEE..EEE     .      .      .      .      .      .      .      .

DOORS & WINDOWS                            CURRENT  .  E..EE       .      .      .      .      .      .      .      .
       800   3   3   0            10       PLANNED  .  *  EEE      .      .      .      .      .      .      .      .

ELECTRICAL WORK                            CURRENT  .  E..EE       .      .      .      .      .      .      .      .
      1100   3   3   0            12       PLANNED  .  *  EEE      .      .      .      .      .      .      .      .

WALL FABRICATION                           CURRENT  .  E..EEEEE    .      .      .      .      .      .      .      .
       600   6   6   0             2       PLANNED  .  * EEEEE..E  .      .      .      .      .      .      .      .

ROOF FABRICATION                           CURRENT  .  E..EEEE..E  .      .      .      .      .      .      .      .
       500   4   7   0             1       PLANNED  .  *  EEEE  .      .      .      .      .      .      .      .

FOUNDATION                                 CURRENT  .  *  . EEE..EE      .      .      .      .      .      .      .
       700   5   5   0             0       PLANNED  .  E..EEEE     .      .      .      .      .      .      .      .

INSTALL FENCE                              CURRENT  .  *  . EE  .      .      .      .      .      .      .      .
      1300   2   2   0            11       PLANNED  .  *  . EE  .      .      .      .      .      .      .      .

PLUMBING WORK                              CURRENT  .  *  .     . EE  .      .      .      .      .      .      .
      1000   2   2   0             5       PLANNED  .  *  .     .EE  .      .      .      .      .      .      .

ERECTION                                   CURRENT  .  *  .     . EEE..EE     .      .      .      .      .      .
       900   5   5   0             0       PLANNED  .  *  .     .EEEE..E      .      .      .      .      .      .

PAINTING                                   CURRENT  .  *  .     .      . EE  .      .      .      .      .      .
      1200   2   2   0             0       PLANNED  .  *  .     .      .EE  .      .      .      .      .      .

FINISHING                                  CURRENT  .  *  .     .      . E..E      .      .      .      .      .
      1400   2   2   0             0       PLANNED  .  *  .     .      . EE  .      .      .      .      .      .
```

Figure 6 As-planned vs. Updated Schedule in Bar Chart Format

STATUSING *(see Monitoring)*

STATUS LINE (TIME-NOW LINE) A vertical line on a time-scaled schedule (network or bar chart) indicating a reporting or data date for an updated schedule.

A status line indicates the date that activity data are recorded or gathered for the

539

updating process *(see Data date)*. All activities with actual start and completion dates before the status line are considered as completed activities. For an activity with an actual start date and without a completion date before the status line, it is referred as an in-progress activity whose remaining duration must be estimated at the given data date. The remaining activities that have not been started at the data date are identified as unstarted activities.

Figure 1 shows an updated schedule *(see Updated schedule)* in a bar chart format. The status line is on March 19, 1993, showing activities 100 and 300 as completed activities and activities 200 and 400 as in-progress activities.

```
-------------------------------------------------------------------------------------------------------------------
CIVIL ENGINEERING DEPARTMENT                PRIMAVERA PROJECT PLANNER

REPORT DATE 23NOV93  RUN NO.   10           PROJECT SCHEDULE REPORT                START DATE 13MAR93  FIN DATE 12APR93
             22:51
BAR CHART BY ES, EF, TF                                                            DATA DATE 19MAR93   PAGE NO.    1

                                                                                              DAILY-TIME PER.    1
-------------------------------------------------------------------------------------------------------------------
.............ACTIVITY DESCRIPTION.............     15    22    29    05    12    19    26    03    10    17
ACTIVITY ID  OD   RD  PCT   CODES    FLOAT    SCHEDULE     MAR   MAR   MAR   APR   APR   APR   APR   MAY   MAY   MAY
-----------  ---- --- ---  ----------- -----   --------    93    93    93    93    93    93    93    93    93    93
                                                         -------------------------------------------------------
DRAINAGE WORK                                CURRENT   . AA*  .     .     .     .     .     .     .     .     .
        300    2    0 100                    PLANNED   . EE*  .     .     .     .     .     .     .     .     .

MATERIAL DELIVERY                            CURRENT   . AA*  .     .     .     .     .     .     .     .     .
        100    3    0 100                    PLANNED   . EEE  .     .     .     .     .     .     .     .     .

EXCAVATION                                   CURRENT   . AAE..EE   .     .     .     .     .     .     .     .
        200    2    3   0                0   PLANNED   . EE*  .     .     .     .     .     .     .     .     .

GRADING WORK                                 CURRENT   . AAE..EEE  .     .     .     .     .     .     .     .
        400    6    4  33               11   PLANNED   . EEE..EEE  .     .     .     .     .     .     .     .

DOORS & WINDOWS                              CURRENT   . E..EE  .     .     .     .     .     .     .     .
        800    3    3   0               10   PLANNED   .  *  EEE  .     .     .     .     .     .     .     .

ELECTRICAL WORK                              CURRENT   . E..EE  .     .     .     .     .     .     .     .
       1100    3    3   0               12   PLANNED   .  *  EEE  .     .     .     .     .     .     .     .

ROOF FABRICATION                             CURRENT   . E..EEE .     .     .     .     .     .     .     .
        500    4    4   0                4   PLANNED   .  *  EEEE .     .     .     .     .     .     .     .

WALL FABRICATION                             CURRENT   . E..EEEEE .    .     .     .     .     .     .     .
        600    6    6   0                2   PLANNED   .  *  EEEEE..E   .     .     .     .     .     .     .

FOUNDATION                                   CURRENT   .  *  . EEE..EE   .     .     .     .     .     .     .
        700    5    5   0                0   PLANNED   . E..EEEE .     .     .     .     .     .     .     .

INSTALL FENCE                                CURRENT   .  *  . EE  .     .     .     .     .     .     .     .
       1300    2    2   0               11   PLANNED   .  *  . EE  .     .     .     .     .     .     .     .

PLUMBING WORK                                CURRENT   .  *  .     . EE  .     .     .     .     .     .     .
       1000    2    2   0                5   PLANNED   .  *  .     .EE  .     .     .     .     .     .     .

ERECTION                                     CURRENT   .  *  .     . EEE..EE   .     .     .     .     .     .
        900    5    5   0                0   PLANNED   .  *  .     .EEEE..E   .     .     .     .     .     .

PAINTING                                     CURRENT   .  *  .     .     . EE  .     .     .     .     .     .
       1200    2    2   0                0   PLANNED   .  *  .     .     .EE  .     .     .     .     .     .

FINISHING                                    CURRENT   .  *  .     .     . E..E .     .     .     .     .     .
       1400    2    2   0                0   PLANNED   .  *  .     .     . EE  .     .     .     .     .     .
```

Status Line = March 19, 1993

Figure 1 Updated Schedule in Bar Chart Format

SUBNETWORK *(see Fragnet)*

SUBPROJECT A portion of a large and complex master project that can be managed as a separate project.

A large and complex project is normally comprised of many small sections of work that are individually independent. Each small section of work, called a *subproject,* can be represented as a separate group of activities, but when combined together, they form a master project. A subproject has its own starting and ending points in time and they are normally tied into a master project.

Breaking a large and complex project down into several small subprojects enables management to plan and control a project effectively because it can be difficult to manage a project in one continuous process. Completion of all subprojects represents the project completion. According to the nature of a project, subprojects will vary from project to project. For example, a typical building construction can have subprojects according to its major components, such as site work, substructure work, superstructure work, mechanical work, electrical work, and finishing work.

SUCCEEDING ACTIVITY (SUCCESSOR) An activity that has some measurable portion of its duration logically restrained by a preceded activity or activities.

A particular activity in consideration imposes a logical constraint on its succeeding activity. In an arrow diagram network *(see Arrow diagramming method),* a succeeding activity cannot start unless its preceding activities *(see Preceding activity)* are completed, as shown in Figure 1. The activity Painting is the immediately succeeding activity of the activity Doors and Windows. Unlike an arrow diagram network, a precedence diagram network *(see Precedence diagramming method)* allows a preceding activity to impose either start to start, start to finish, finish to start, or finish to finish *(see Dependency)* on its succeeding activity, as shown in Figures 2 and 3.

A succeeding activity can be classified into two types: (1) an immediately succeeding activity, and (2) a non-immediately succeeding activity. Any activities that are on the same path as a particular activity and are scheduled to start or be completed after the start or completion of that activity are considered as its succeeding activities. But the ones directly following a particular activity are immediately succeeding activities.

Developing a network requires all activities to be listed; then a logical constraint between activities is defined to represent the work sequencing. As a result, any activity in the network can be located together with its preceding and succeeding activities. Figure 4 illustrates an activity list showing activities and their preceding and succeeding activities. The letters S.L.D.F. indicate the succeeding activity of a particular activity.

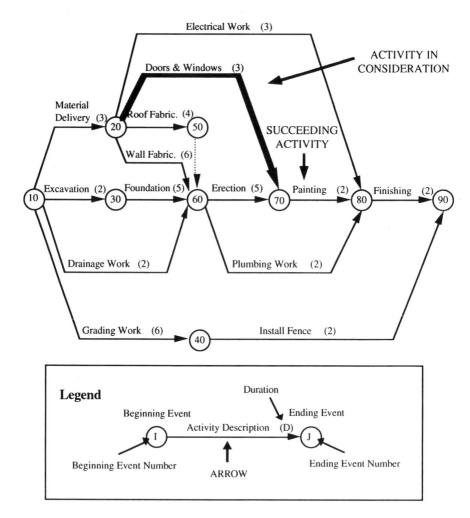

ACTIVITY IN
CONSIDERATION

SUCCEEDING
ACTIVITY

Electrical Work   (3)

Doors & Windows   (3)

Material
Delivery   (3)   Roof Fabric. (4)

20   50

Wall Fabric. (6)

Excavation (2)   Foundation (5)   Erection (5)   Painting   (2)   Finishing   (2)

10   30   60   70   80   90

Drainage Work   (2)   Plumbing Work   (2)

Grading Work   (6)   Install Fence   (2)

40

Legend

Duration

Beginning Event   Ending Event

Activity Description   (D)

I   J

Beginning Event Number   ARROW   Ending Event Number

Figure 1 Succeeding Activity in Arrow Diagram Network

542

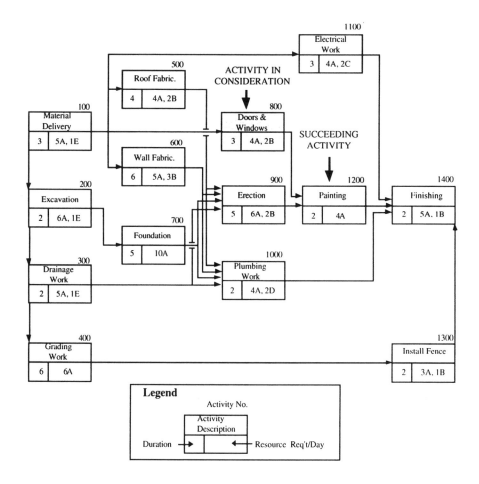

Figure 2 Succeeding Activity in Precedence Diagram Network

543

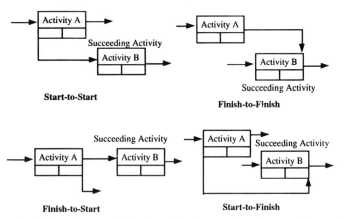

Figure 3 Various Types of Relationships Imposed on Succeeding Activity

```
---------------------------------------------------------------------------------------------------------
CIVIL ENGINEERING DEPARTMENT               PRIMAVERA PROJECT PLANNER

REPORT DATE 23NOV93  RUN NO.   21          PROJECT SCHEDULE REPORT                    START DATE 13MAR93  FIN DATE  9APR93
             22:32
SCHEDULE REPORT - PREDECESSORS AND SUCCESSORS                                         DATA DATE  17MAR93  PAGE NO.     1


----- -----  ---- ---- - --- ----------  -------------------------------------  ------- -------- -------- -------- -----
ACTIVITY  ORIG REM                                ACTIVITY DESCRIPTION              EARLY    EARLY    LATE     LATE    TOTAL
   ID     DUR  DUR   %   CODE                                                       START   FINISH    START   FINISH  FLOAT
----- -----  ---- ---- - --- ----------  -------------------------------------  ------- -------- -------- -------- -----

     100    3    3   0          MATERIAL DELIVERY                                17MAR93* 19MAR93  17MAR93* 19MAR93     0

              S.L.D.F.,*        200.SS   0.   2.    2,*      300.SS   0.   2.    7
              S.L.D.F.,*        400.SS   0.   6.   10,*      500.FS   0.   4.    2
              S.L.D.F.,*        600.FS   0.   6.    0,*      800.FS   0.   3.    8
              S.L.D.F.,*       1100.FS   0.   3.   10,


              P.L.D.F.,*        100.SS   0.   3.    0,


     200    2    2   0          EXCAVATION                                       17MAR93  18MAR93  19MAR93  22MAR93     2

              S.L.D.F.,*        700.FS   0.   5.    2,


              P.L.D.F.,*        100.SS   0.   3.    0,


     300    2    2   0          DRAINAGE WORK                                    17MAR93  18MAR93  26MAR93  29MAR93     7

              S.L.D.F.,         900.FS   0.   5.    0,     1000.FS   0.   2.    5


              P.L.D.F.,*        100.SS   0.   3.    0,


     400    6    6   0          GRADING WORKE                                    17MAR93  24MAR93  31MAR93   7APR93    10

              S.L.D.F.,*       1300.FS   0.   2.   10,
```

P = Preceding Dependency Type   D = Duration
S = Succeeding Dependency Type   F = Total Float
L = Lag

Figure 4 Activity List Showing Preceding and Succeeding Activities

SUCCEEDING EVENT *(see Ending event)*

SUMMARY NETWORK A summarization of the CPM network or a group of activities at a summary level that are used to communicate the plan and program to top management.

A summary network is a network diagram representing the planned sequences of operation at a summary level of detail. For a large and complex project, more than one network can be developed corresponding to different management levels based on the PBS *(see Project breakdown structure),* as shown in Figure 1. A summary network contains project information at a summary level. Each activity in a summary network can be broken down into several activities for a detailed network *(see Detailed schedule).* One subsystem or major function may be represented by one activity in a summary network. A summary network can be developed in an arrow or precedence diagramming technique, but it is not recommended that networks at various levels be developed by using different techniques.

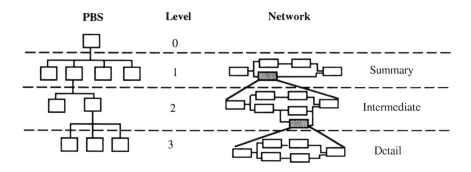

Figure 1 Hierarchy of Network

A summary network is intended to be used as a means of communicating project information to a high management level such as a project manager or owner. At this level, information regarding project plan and progress needs to be concise enough for executive management to have an understanding of overall project views in a brief time period. It is recommended that a summary network is developed as soon as the contract is awarded so that it can be used as a basis for the construction of a detailed network.

SUMMARY NUMBER A number that identifies a work item in the project breakdown structure (PBS).

The summary number is one component of the account code structure *(see Account code structure).* In the structure, a number is assigned to each work item and called the summary number. The summary number is used both for summarizing the cost information of a work package to its work item and for summarizing the cost of information at any level of work items to a summary item. For example, the following account code structure for a commercial building shows use of a summary number.

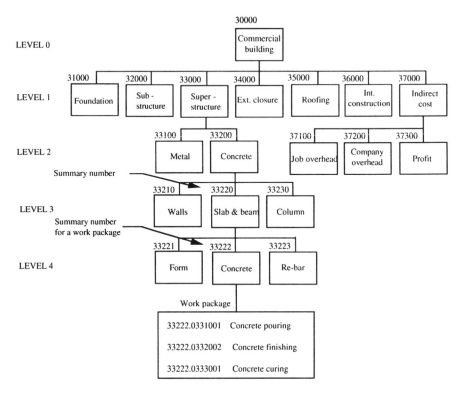

Figure 1 Account Code Structure and Summary Number for Commercial Building

When referring to the "concrete" work package in the account code structure, the charge numbers 33222.0331001, 33222.0332002, and 33222.0333001 are reduced to the summary number, 33222. Also, the subordinate work items 33221, 33222, and 33223 are reduced to the summary number 33220.

T

TARGET DATE The date that an activity or project is scheduled to start or be completed. This date can be imposed or requested by a client or project manager.

A target date specifies the date on which significant or critical activities are scheduled to start or be completed. The start and completion target dates can be imposed by an owner on various systems or subsystems considered as major activities. The most common target date is the final completion date of a project, but the target date for any intermediate activities can be specified by an owner or project manager when they feel that those activities are critical for project completion. The use of a target date allows an owner to execute control over the project schedule so that major activities can be closely monitored.

A target date can be assigned as no later than, no earlier than, or exactly on the specified date. Normally, project scheduling and planning software packages allow the selection of different target date types to be assigned to a specific activity. However, careful consideration must be taken for the selection of target date types because different types of target dates can have an impact on a critical path and project completion date.

Figure 1 shows a precedence diagram network with the target date assigned to a project and activities as follows:

1. Project start date = no earlier than on March 17, 1993
2. Completion of the activity Erection = no later than on March 25, 1993
3. Project completion date = no later than on April 12, 1993

The schedule report in tabular format for a precedence diagram network is shown in Figure 2.

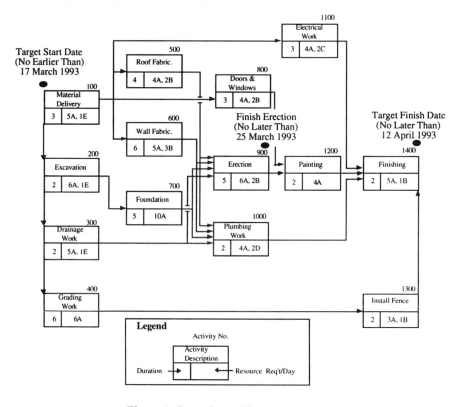

Figure 1 Precedence Diagram Network

548

CIVIL ENGINEERING DEPARTMENT   PRIMAVERA PROJECT PLANNER

REPORT DATE 24NOV93 RUN NO.   2   PROJECT SCHEDULE REPORT   START DATE 13MAR93 FIN DATE 9APR93
          0:51
CLASSIC SCHEDULE REPORT - SORT BY ES, TF   DATA DATE 17MAR93 PAGE NO.   1

ACTIVITY ID	ORIG DUR	REM DUR	%	CODE	ACTIVITY DESCRIPTION	EARLY START	EARLY FINISH	LATE START	LATE FINISH	TOTAL FLOAT
100	3	3	0		MATERIAL DELIVERY	17MAR93	19MAR93	17MAR93	19MAR93	0
200	2	2	0		EXCAVATION	17MAR93	18MAR93	19MAR93	22MAR93	2
300	2	2	0		DRAINAGE WORK	17MAR93	18MAR93	26MAR93	29MAR93	7
400	6	6	0		GRADING WORK	17MAR93	24MAR93	31MAR93	7APR93	10
700	5	5	0		FOUNDATION	19MAR93	25MAR93	23MAR93	29MAR93	2
600	6	6	0		WALL FABRICATION	22MAR93	29MAR93	22MAR93	29MAR93	0
1100	3	3	0		ELECTRICAL WORK	22MAR93	24MAR93	23MAR93	25MAR93*	1
500	4	4	0		ROOF FABRICATION	22MAR93	25MAR93	24MAR93	29MAR93	2
800	3	3	0		DOORS & WINDOWS	22MAR93	24MAR93	1APR93	5APR93	8
1300	2	2	0		INSTALL FENCE	25MAR93	26MAR93	8APR93	9APR93	10
900	5	5	0		ERECTION	30MAR93	5APR93	30MAR93	5APR93	0
1000	2	2	0		PLUMBING WORK	30MAR93	31MAR93	6APR93	7APR93	5
1200	2	2	0		PAINTING	6APR93	7APR93	6APR93	7APR93	0
1400	2	2	0		FINISHING	8APR93	9APR93	8APR93	9APR93*	0

Figure 2 Schedule Report in Tabular Format

TASK A constituent of an activity — a portion of an activity that has a duration of less than the total duration of the activity. A CPM schedule activity could be: "Connect Drops from Piping Mains to Pumps" 4 day duration. The constituent tasks could be: "Weld Downcommer to First Flange," 0.75 day; "Install Check Valve," 0.25 day; "Install Butterfly Valve," 0.25 day; "Install Flex at Pump," 0.5 day; and the like. CPM schedules are developed by activity, not generally by task. Tasks can be tracked by systems other than a CPM schedule, such as labor-hour reporting. The terms *task* and *activity* are not synonymous.

TIED ACTIVITY An activity that must start within a specified time or immediately after its immediate predecessor(s) is completed.

In a situation when two activities are tied together by a logical constraint, a tied activity is the one that must start immediately after its preceding activity *(see Preceding activity)* is completed. Consideration must be taken when two activities are tied together to identify if those two activities are tied closely or loosely.

A closely tied activity is the one that must start immediately upon the completion of its immediate predecessor. This occurs when a long continuous task is broken down into two or more smaller tasks. Figures 1 and 2 show a series of closely tied activities for concrete work. In this sample, concrete pouring must start immediately after concrete is mixed; then, as soon as concrete is poured, concrete finishing must start. Otherwise, the concrete can be hardened, which will make it difficult to pour or finish as required by specifications.

549

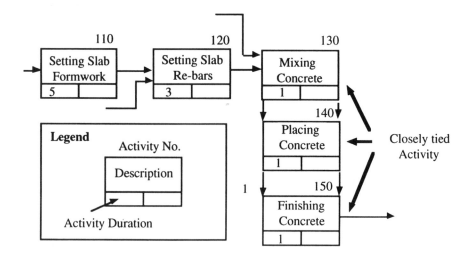

Figure 1 Closely Tied Activity in Precedence Diagram Network

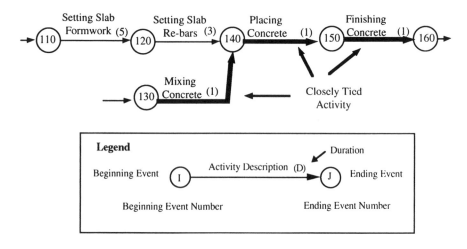

Figure 2 Closely Tied Activity in Arrow Diagram Network

In the event that an activity does not have to start immediately after the completion of its predecessor, it can be called a loosely tied activity. This occurs when two activities are different tasks that do not require a continuous process or operation. A sample of loosely tied activities is shown in Figures 3 and 4. In this sample, re-bar setting must start after formworks are set, but it can wait for a certain period of time. This allowable waiting time is a major characteristic of a loosely tied activity.

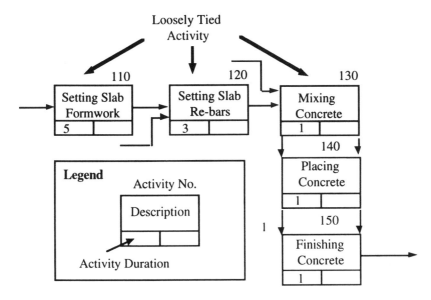

Figure 3 Loosely Tied Activity in Precedence Diagram Network

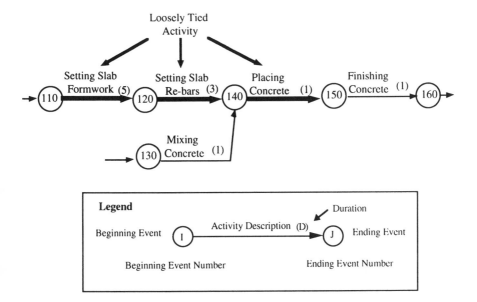

Figure 4 Loosely Tied Activity in Arrow Diagram Network

TIME-COST TRADE-OFF A scheduling technique using the critical path method by which the project duration is shortened with a minimum of added costs.

Time-cost trade-off analysis is accomplished based on a number of special terms *[see Direct cost, Burden (overhead), Crash activity time cost point, Normal activity time cost point, and Cost slope]*. In general, relationships of time and costs are as follows:

1. *Time and direct costs.* When the overall project is shortened, the shortened activities are on the critical paths, and their direct costs increase between normal activity time cost point and crash activity time cost point.
2. *Time and indirect costs.* When the overall project is shortened, its overhead costs decrease.

Therefore, the bottom line of time-cost trade-off is to find the optimum point between normal activity time cost point and crash activity time cost point. Simply, the overall project can be shortened until overhead cost savings are greater than the increased direct costs. Thus the optimum point, called *the least cost point,* is found. This procedure is accomplished by the following steps:

Step 1: Develop the target schedule based on the normal duration and normal cost.
Step 2: Estimate the crash duration and crash cost of each activity.
Step 3: Identify the critical activities.
Step 4: Among the critical activities, identify an activity or activities that can be shortened at the least cost.
Step 5: Shorten the activity or activities by 1 time unit.
Step 6: Calculate the new project duration and cost.
Step 7: Repeat steps 3 to 6 until the overhead cost savings are greater than the increased direct costs.

For the procedures listed above, the following assumptions are made:

1. The planned activity duration can be any point between normal activity duration and crash activity duration.
2. The direct activity costs are linear between normal activity cost and crash activity cost.
3. The overhead costs are constant during the project duration.
4. Unlimited resource are available.

The procedure described above is explained using an example step by step as follows:

1. *Step 1:* • Develop a target schedule based on the normal duration and the normal cost.
   • Activities 1, 3, 7, 9, and 10 are on the critical path.
   • The project duration is 30 days.

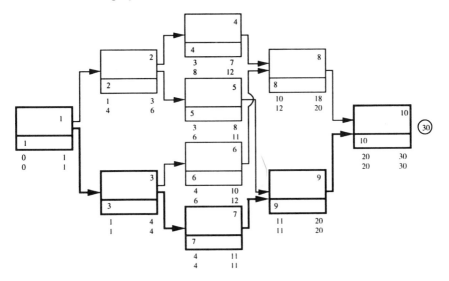

2. *Step 2:* • Estimate the crash duration and the crash cost of each activity.
   • The total cost is $ 7900.

Act.	Duration		Cost $		Δ Cost	Δ Days	Cost Slope	Days Shortened						
	Norm	Crash	Norm	Crash										
1	1	1	100	100	100	-	-							
2	2	2	200	200	80	-	-							
3	3	1	300	400	100	2	50							
4	4	2	400	450	50	2	25							
5	5	3	500	600	100	2	50							
6	6	4	600	650	50	2	25							
7	7	5	700	760	60	2	30							
8	8	6	800	900	100	2	50							
9	9	7	900	980	80	2	40							
10	10	8	1,000	1,100	100	2	50							

Note:

Overhead Cost
= $ 80 / Day

Cost slope
= Cost Increased
/ Days Reduced

Days Cut	▨	
Project Duration	30	
Increased Cost/Day	▨	
Direct Cost	5,500	
Overhead Cost	2,400	
Total Cost	7,900	

3. Steps 3 to 7:

Cycle 1: • Shorten activity 7 by 1 time unit (the least cost activity on the critical path).

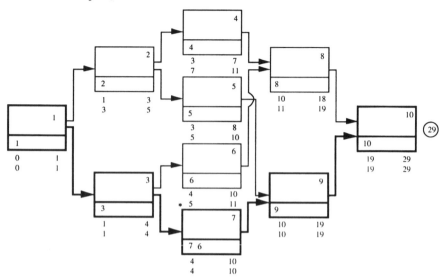

Act.	Duration		Cost $		Δ Cost	Δ Days	Cost Slope	Days Shortened						
	Norm	Crash	Norm	Crash										
1	1	1	100	100	100	-	-							
2	2	2	200	200	80	-	-							
3	3	1	300	400	100	2	50							
4	4	2	400	450	50	2	25							
5	5	3	500	600	100	2	50							
6	6	4	600	650	50	2	25							
7	7	5	700	760	60	2 1	30	1						
8	8	6	800	900	100	2	50							
9	9	7	900	980	80	2	40							
10	10	8	1,000	1,100	100	2	50							

	Days Cut	▨	1	
Overhead Cost = $ 80 / Day	Project Duration	30	29	
	Increased Cost/Day	▨	30	
	Direct Cost	5,500	5,530	
	Overhead Cost	2,400	2,320	
	Total Cost	7,900	7,850	

554

Cycle 2: • The critical path remains the same (activity 1,3,7,9,10).
• Shorten activity 7 by 1 time unit again.

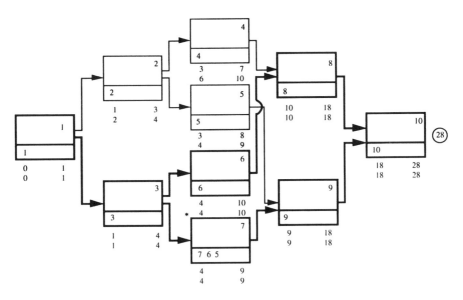

Act.	Duration		Cost $		Δ Cost	Δ Days	Cost Slope	Days Shortened						
	Norm	Crash	Norm	Crash										
1	1	1	100	100	100	-	-							
2	2	2	200	200	80	-	-							
3	3	1	300	400	100	2	50							
4	4	2	400	450	50	2	25							
5	5	3	500	600	100	2	50							
6	6	4	600	650	50	2	25							
7	7	5	700	760	60	2 1 0	30	1	1					
8	8	6	800	900	100	2	50							
9	9	7	900	980	80	2	40							
10	10	8	1,000	1,100	100	2	50							

Overhead Cost = $ 80 / Day

Days Cut	▨	1	1			
Project Duration	30	29	28			
Increased Cost/Day	▨	30	30			
Direct Cost	5,500	5,530	5,560			
Overhead Cost	2,400	2,320	2,240			
Total Cost	7,900	7,850	7,800			

Cycle 3:
- Two critical paths are identified (activity 1,3,7,9,10 and activity 1,3,6,8,10).
- Shorten activities 6 and 9 by 1 time unit (the least cost activities on the critical paths).

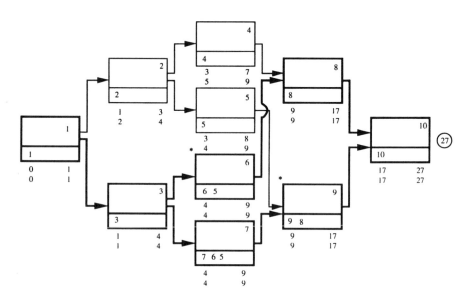

Act.	Duration		Cost $		Δ Cost	Δ Days	Cost Slope	Days Shortened							
	Norm	Crash	Norm	Crash											
1	1	1	100	100	100	-	-								
2	2	2	200	200	80	-	-								
3	3	1	300	400	100	2	50								
4	4	2	400	450	50	2	25								
5	5	3	500	600	100	2	50								
6	6	4	600	650	50	2 1	25		1						
7	7	5	700	760	60	2 1 0	30	1	1						
8	8	6	800	900	100	2	50								
9	9	7	900	980	80	2 1	40		1						
10	10	8	1,000	1,100	100	2	50								

Overhead Cost = $ 80 / Day

Days Cut	▨	1	1	1	
Project Duration	30	29	28	27	
Increased Cost/Day	▨	30	30	65	
Direct Cost	5,500	5,530	5,560	5,625	
Overhead Cost	2,400	2,320	2,240	2,160	
Total Cost	7,900	7,850	7,800	7,785	

556

Cycle 4:
- The critical paths remain the same (activity 1,3,7,9,10 and activity 1,3,6,8,10).
- Shorten activities 6 and 9 again.

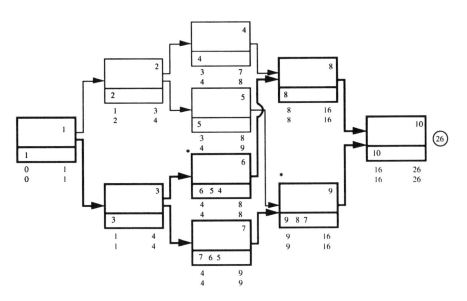

Act.	Duration		Cost $		Δ Cost	Δ Days	Cost slope	Days Shortened							
	Norm	Crash	Norm	Crash											
1	1	1	100	100	100	-	-								
2	2	2	200	200	80	-	-								
3	3	1	300	400	100	2	50								
4	4	2	400	450	50	2	25								
5	5	3	500	600	100	2	50								
6	6	4	600	650	50	2 1 0	25			1	1				
7	7	5	700	760	60	2 1 0	30	1	1						
8	8	6	800	900	100	2	50								
9	9	7	900	980	80	2 1 0	40			1	1				
10	10	8	1,000	1,100	100	2	50								

Overhead Cost = $ 80 / Day

Days Cut	▨	1	1	1	1		
Project Duration	30	29	28	27	26		
Increased Cost/Day	▨	30	30	65	65		
Direct Cost	5,500	5,530	5,560	5,625	5,690		
Overhead Cost	2,400	2,320	2,240	2,160	2,080		
Total Cost	7,900	7,850	7,800	7,785	7,770		

557

Cycle 5: • The critical paths remain the same (activity 1,3,7,9,10 and activity 1,3,6,8,10).
• Shorten activity 10 by 2 time units (the least cost activity on the critical paths).

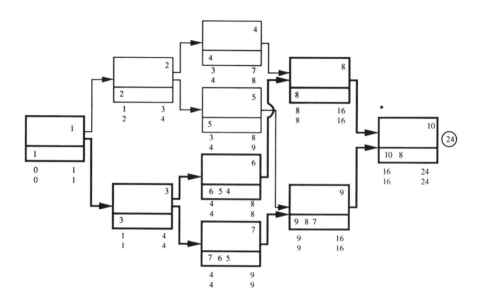

Act.	Duration		Cost $		Δ Cost	Δ Days	Cost Slope	Days Shortened						
	Norm	Crash	Norm	Crash										
1	1	1	100	100	100	-	-							
2	2	2	200	200	80	-	-							
3	3	1	300	400	100	2	50							
4	4	2	400	450	50	2	25							
5	5	3	500	600	100	2	50							
6	6	4	600	650	50	2 1 0	25			1	1			
7	7	5	700	760	60	2 1 0	30	1	1					
8	8	6	800	900	100	2	50							
9	9	7	900	980	80	2 1 0	40			1	1			
10	10	8	1,000	1,100	100	2 0	50					2		

Overhead Cost = $ 80 / Day

Days Cut	▓	1	1	1	1	2	
Project Duration	30	29	28	27	26	24	
Increased Cost/Day	▓	30	30	65	65	50	
Direct Cost	5.500	5.530	5.560	5.625	5.690	5.790	
Overhead Cost	2.400	2.320	2.240	2.160	2.080	1.920	
Total Cost	7.900	7.850	7.800	7.785	7.770	7.710	

558

Cycle 6:
- The critical paths remain the same (activity 1,3,7,9,10 and activity 1,3,6,8,10).
- Shorten activity 3 by 1 time unit (the least cost activity on the critical paths).

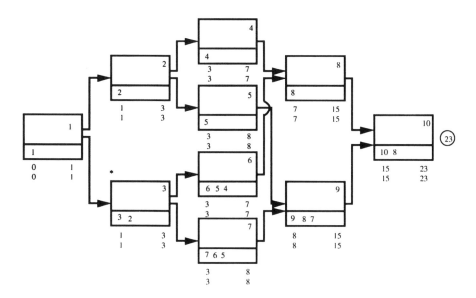

Act.	Duration Norm	Duration Crash	Cost $ Norm	Cost $ Crash	Δ Cost	Δ Days	Cost Slope	Days Shortened						
1	1	1	100	100	100	-	-							
2	2	2	200	200	80	-	-							
3	3	1	300	400	100	2 1	50							1
4	4	2	400	450	50	2	25							
5	5	3	500	600	100	2	50							
6	6	4	600	650	50	2 1 0	25			1	1			
7	7	5	700	760	60	2 1 0	30	1	1					
8	8	6	800	900	100	2	50							
9	9	7	900	980	80	2 1 0	40			1	1			
10	10	8	1,000	1,100	100	2 1 0	50					2		

Days Cut		1	1	1	1	2	1
Project Duration	30	29	28	27	26	24	23
Increased Cost/Day		30	30	65	65	50	50
Direct Cost	5,500	5,530	5,560	5,625	5,690	5,790	5,840
Overhead Cost	2,400	2,320	2,240	2,160	2,080	1,920	1,840
Total Cost	7,900	7,850	7,800	7,785	7,770	7,710	7,680

Overhead Cost = $ 80 / Day

Cycle 7:
- After cycle 6, all activities are on the critical path.
- Shorten activity 3,4,5 by 1 time unit (the least cost activities on the critical paths).
- The calculated total cost is greater than the previous one.
- Stop the cycle.
- The optimum project duration is 23 days, and the optimum total cost is $7680.

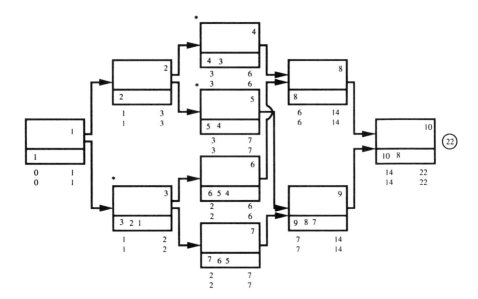

Act.	Duration		Cost $		Δ Cost	Δ Days	Cost Slope	Days Shortened						
	Norm	Crash	Norm	Crash										
1	1	1	100	100	100	-	-							
2	2	2	200	200	80	-	-							
3	3	1	300	400	100	2 1 0	50					1	1	
4	4	2	400	450	50	2 1	25						1	
5	5	3	500	600	100	2 1	50						1	
6	6	4	600	650	50	2 1 0	25		1	1				
7	7	5	700	760	60	2 1 0	30	1	1					
8	8	6	800	900	100	2	50							
9	9	7	900	980	80	2 1 0	40		1	1				
10	10	8	1,000	1,100	100	2 0	50					2		

Overhead Cost = $ 80 / Day

Days Cut	▨	1	1	1	1	2	1	1
Project Duration	30	29	28	27	26	24	23	22
Increased Cost/Day	▨	30	30	65	65	50	50	125
Direct Cost	5,500	5,530	5,560	5,625	5,690	5,790	5,840	5,965
Overhead Cost	2,400	2,320	2,240	2,160	2,080	1,920	1,840	1,760
Total Cost	7,900	7,850	7,800	7,785	7,770	7,710	7,680	7,725

TIME NOW *(see Data date)*

TIME-NOW LINE *(see Status line)*

TIME-PHASED DIAGRAM *(see Time-scaled network diagram)*

TIME-SCALED NETWORK DIAGRAM A CPM network where the length of an activity indicates the duration of that activity as drawn to a calendar scale. The float is usually shown with a dashed line.

A time-scaled CPM diagram represents the logical sequences of project activities with the duration and scheduled start and completion dates on a calendar scale. The development of a time-scaled CPM diagram is the combination of a network diagram and bar chart. The time-scaled CPM plan shows how each activity is carried out in a logical and time order. After the logic diagram is developed, the network analysis must be performed *(see Network analysis)* to identify the early and late start and completion dates before each activity can be drawn in a time-scaled format on a calendar scale.

Each activity in a time-scaled CPM network is plotted as a bar in an early start date and its duration is represented by the bar length. The calculated total float *(see Activity total float)* is represented by the dashed line extended beyond each bar. The logical relationship between two activities is represented by an arrow.

A time-scaled CPM diagram has an advantage over the regular CPM network diagram in that it shows the duration of each activity in a calendar scale which makes the schedule easy to read and understand. It also has the advantage over a regular bar chart that the logical relationship between activities is shown in the diagram so that the impact or flow of one activity to another can be indicated. Figure 1 shows a sample of a time-scaled CPM diagram.

561

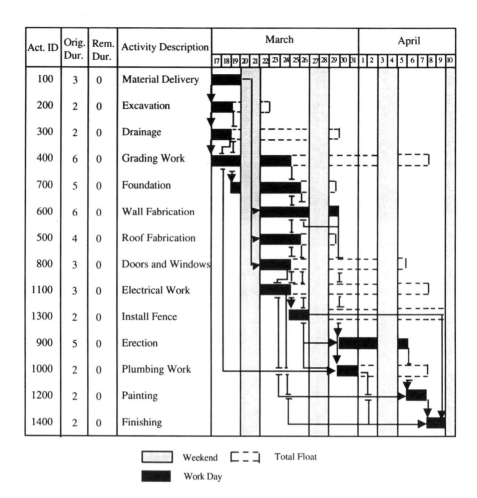

Act. ID	Orig. Dur.	Rem. Dur.	Activity Description	March															April									
				17	18	19	20	21	22	23	24	25	26	27	28	29	30	31	1	2	3	4	5	6	7	8	9	10
100	3	0	Material Delivery																									
200	2	0	Excavation																									
300	2	0	Drainage																									
400	6	0	Grading Work																									
700	5	0	Foundation																									
600	6	0	Wall Fabrication																									
500	4	0	Roof Fabrication																									
800	3	0	Doors and Windows																									
1100	3	0	Electrical Work																									
1300	2	0	Install Fence																									
900	5	0	Erection																									
1000	2	0	Plumbing Work																									
1200	2	0	Painting																									
1400	2	0	Finishing																									

Weekend ☐ Total Float 🔲
Work Day ■

Figure 1 Time-scaled CPM Diagram

TOTAL FLOAT *(see Activity total float)*

TOTAL FLOAT PATH *(see Slack paths)*

TREND ANALYSIS (TRENDING) *(see Progress trend)*

562

U

UNIT OF TIME *(see Calendar unit)*

UNIT OF WORK (WORK UNITS) A deterministic quantification of work adopted as a standard of measurement in a given project (e.g., cubic yards of concrete, 1000 bricks, linear feet of cables).

Work units are used to represent construction output. Project control requires management to identify critical required resources to be measured and monitored during the project execution. During project planning, the resource quantity is estimated and allocated in each work activity. Once the project is begun, actual work units are monitored and compared against the estimated ones in order to identify if the project performance is over or under plan.

Common work units in a construction project are cubic yards of concrete, square feet of paint, and tons of steel. Project planning and scheduling software packages normally allow different types of resources as work units to be defined in the project resource dictionary, as shown in Figure 1, and to be allocated to each activity, as shown in Figure 2, for planning and monitoring purposes.

RESOURCE DICTIONARY COMP

Resource: CONCRETE Units: CY

Description: CONCRETE Driving Resource (Y/N)? N

RESOURCE LIMITS AND PRICES:

Normal limit	Max limit	Through	Price/unit	Through
20	50		0.00	
0	0		0.00	
0	0		0.00	
0	0		0.00	
0	0		0.00	
0	0		0.00	

Commands: Add Curve Del Edit Help Next closeOut Print Return Transfer Usage

Figure 1 Definition of Resources in Work Units (Primavera Project Planner 5.0)

```
                            ACTIVITY FORM                          COMP
    Activity ID:        700                                TF:    0
          Title: FOUNDATION                                PCT:  0.0
    ES: 24MAR93   EF: 30MAR93   Orig. duration:    5     Actual Start:


    LS: 24MAR93   LF: 30MAR93   Rem.  duration:    5     Actual Finish:


Activity Codes:
════════════════════════════════════════════════════════════════════════
RESOURCE SUMMARY:        Resource  1      Resource  2      Resource  3
                         ~~~~~~~~~~~~~~~~~ ~~~~~~~~~~~~~~~~~ ~~~~~~~~~~~~~~~~~
Resource                 CONCRETE
Cost Acct/type
Units per Day                50.00             0.00             0.00
Budget quantity             250.00             0.00             0.00
Resource Lag/Duration    0                 0                0
Percent Complete
Actual qty this Period        0.00             0.00             0.00
Actual qty to date            0.00             0.00             0.00
Quantity to complete        250.00             0.00             0.00
Quantity at completion      250.00             0.00             0.00
Variance (units)              0.00             0.00             0.00
════════════════════════════════════════════════════════════════════════

Commands:Add Delete Edit Help More Next Return autoSort Transfer View Window
Windows :Act.codes Budget Constraints Dates Financial Log Pred Res Succ cUstom
```

Figure 2 Allocation of Resources in Work Units in Activity Data
(Primavera Project Planner 5.0)

UPDATED SCHEDULE A revised schedule reflecting project information at a given data date regarding completed activities, in-progress activities, and changes in logic, cost, and resources required and allocated at an activity level.

During the project planning stage, an as-planned project schedule *(see As-planned project schedule)* is developed to serve as a guideline for project execution. The as-planned project schedule, however, is seldom followed exactly from the start to the completion of a project. After the start of a project, the as-planned schedule therefore needs to be regularly updated or revised to reflect unanticipated or unforeseen problems, change orders, adverse weather conditions, new knowledge, and so on. The updated schedule is required for project control to keep up with project progress and to predict project completion based on current project performance.

A project schedule must be updated periodically or as a specific lapsed time stated in the schedule specification *(see Scheduling specifications)*. The updated schedule should be distributed to all project participants to inform them about the current project status. This schedule should identify the project status as behind or ahead of schedule, the problem areas impeding the project progress, activities causing delays, and in-progress activities.

Updating the project schedule requires activity and network information to be amended or revised at a given data date *(see Data date)*. The following information should be considered for schedule updating:

1. An activity that has been added
2. An activity that has been deleted
3. Changes in logical constraints
4. Changes in activity duration
5. Changes in activity resources such as labor, equipment, and materials
6. Estimated remaining duration and resources for in-progress activities
7. Actual start date of in-progress activities
8. Actual start and completion dates of completed activities

The above-mentioned information is recommended to be recorded on the network diagram *(see Network)* before the new data are input into a computer. The schedule calculation *(see Network analysis)* for the updated schedule is performed based on the remaining incomplete activities at a given data date. All completed activities have no impact on the schedule computation. An updated schedule can be shown in forms of a network diagram, bar chart, and tabular report, as shown in Figures 1 to 4.

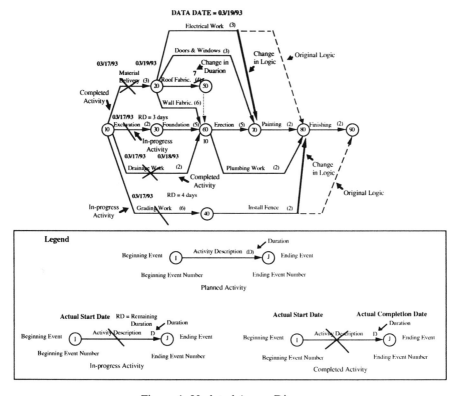

DATA DATE = 03/19/93

Figure 1 Updated Arrow Diagram

567

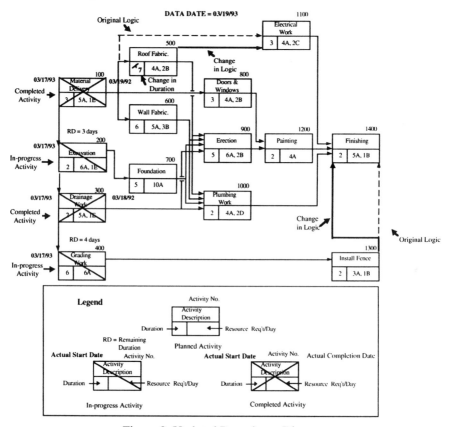

Figure 2 Updated Precedence Diagram

```
CIVIL ENGINEERING DEPARTMENT              PRIMAVERA PROJECT PLANNER

REPORT DATE 24NOV93  RUN NO.  12          PROJECT SCHEDULE REPORT          START DATE 13MAR93  FIN DATE 12APR93
                0:03
CLASSIC SCHEDULE REPORT - SORT BY ES, TF                                   DATA DATE 19MAR93  PAGE NO.    1
```

ACTIVITY ID	ORIG DUR	REM DUR	%	CODE	ACTIVITY DESCRIPTION	EARLY START	EARLY FINISH	LATE START	LATE FINISH	TOTAL FLOAT
100	3	0	100		MATERIAL DELIVERY	17MAR93A	19MAR93A			
300	2	0	100		DRAINAGE WORK	17MAR93A	18MAR93A			
200	2	3	0		EXCAVATION	17MAR93A	23MAR93		23MAR93	0
400	6	4	33		GRADING WORK	17MAR93A	24MAR93		8APR93	11
600	6	6	0		WALL FABRICATION	19MAR93	26MAR93	23MAR93	30MAR93	2
500	4	4	0		ROOF FABRICATION	19MAR93	24MAR93	25MAR93	30MAR93	4
800	3	3	0		DOORS & WINDOWS	19MAR93	23MAR93	2APR93	6APR93	10
1100	3	3	0		ELECTRICAL WORK	19MAR93	23MAR93	6APR93	8APR93	12
700	5	5	0		FOUNDATION	24MAR93	30MAR93	24MAR93	30MAR93	0
1300	2	2	0		INSTALL FENCE	25MAR93	26MAR93	9APR93	12APR93	11
900	5	5	0		ERECTION	31MAR93	6APR93	31MAR93	6APR93	0
1000	2	2	0		PLUMBING WORK	31MAR93	1APR93	7APR93	8APR93	5
1200	2	2	0		PAINTING	7APR93	8APR93	7APR93	8APR93	0
1400	2	2	0		FINISHING	9APR93	12APR93	9APR93	12APR93	0

Figure 3 Updated Tabular Schedule Report (Precedence Diagram)

```
---------------------------------------------------------------------------------------------------------------
CIVIL ENGINEERING DEPARTMENT              PRIMAVERA PROJECT PLANNER

REPORT DATE 23NOV93  RUN NO.  10          PROJECT SCHEDULE REPORT
           22:51                                                          START DATE 13MAR93  FIN DATE 12APR93
BAR CHART BY ES, EF, TF
                                                                          DATA DATE 19MAR93   PAGE NO.   1

                                                                                     DAILY-TIME PER.   1
...........ACTIVITY DESCRIPTION.............    15    22    29    05    12    19    26    03    10    17
ACTIVITY ID  OD   RD  PCT  CODES   FLOAT  SCHEDULE   MAR   MAR   MAR   APR   APR   APR   APR   MAY   MAY   MAY
-----------  ----  ---- ---  ------------  -----  --------  93    93    93    93    93    93    93    93    93    93
-----------------------------------------------------------------------------------------------------------------
DRAINAGE WORK                             CURRENT   . AA* .   .     .     .     .     .     .     .     .
      300    2    0 100                    PLANNED   . EE* .   .     .     .     .     .     .     .     .

MATERIAL DELIVERY                         CURRENT   . AA* .   .     .     .     .     .     .     .     .
      100    3    0 100                    PLANNED   . EEE .   .     .     .     .     .     .     .     .

EXCAVATION                                CURRENT   . AAE..EE   .     .     .     .     .     .     .     .
      200    2    3   0           0        PLANNED   . EE* .    .     .     .     .     .     .     .     .

GRADING WORK                              CURRENT   . AAE..EEE   .     .     .     .     .     .     .     .
      400    6    4  33          11        PLANNED   . EEE..EEE   .     .     .     .     .     .     .     .

DOORS & WINDOWS                           CURRENT   . E..EE   .     .     .     .     .     .     .     .
      800    3    3   0          10        PLANNED   . * EEE   .     .     .     .     .     .     .     .

ELECTRICAL WORK                           CURRENT   . E..EE   .     .     .     .     .     .     .     .
     1100    3    3   0          12        PLANNED   . * EEE   .     .     .     .     .     .     .     .

ROOF FABRICATION                          CURRENT   . E..EEE   .     .     .     .     .     .     .     .
      500    4    4   0           4        PLANNED   . * EEEE   .     .     .     .     .     .     .     .

WALL FABRICATION                          CURRENT   . E..EEEEE .   .     .     .     .     .     .     .
      600    6    6   0           2        PLANNED   . * EEEEE..E   .     .     .     .     .     .     .

FOUNDATION                                CURRENT   . * . EEE..EE   .     .     .     .     .     .     .
      700    5    5   0           0        PLANNED   . E..EEEE .    .     .     .     .     .     .     .

INSTALL FENCE                             CURRENT   . * . EE .   .     .     .     .     .     .     .
     1300    2    2   0          11        PLANNED   . * . EE .   .     .     .     .     .     .     .

PLUMBING WORK                             CURRENT   . * .   . EE .   .     .     .     .     .     .
     1000    2    2   0           5        PLANNED   . * .   .EE   .     .     .     .     .     .

ERECTION                                  CURRENT   . * .   . EEE..EE   .     .     .     .     .     .
      900    5    5   0           0        PLANNED   . * .   .EEEE..E   .     .     .     .     .     .

PAINTING                                  CURRENT   . * .     . EE .   .     .     .     .     .
     1200    2    2   0           0        PLANNED   . * .     .EE   .     .     .     .     .     .

FINISHING                                 CURRENT   . * .     . E..E   .     .     .     .     .
     1400    2    2   0           0        PLANNED   . * .     . EE .   .     .     .     .     .
```

Legend: A = Activity Day as Progressed or Completed
        E = Activity Day as Early Start Schedule

Figure 4 Updated Bar Chart Report (Precedence Diagram)

UPDATING *(see Monitoring)*

UTILIZATION FACTOR (RESOURCE) *(see Resource utilization factor)*

V

VALUE (WORK PERFORMED TO DATE) *(see Budgeted cost of work performed)*

VARIANCE Any actual or potential deviation from an intended or budgeted figure or plan, cost, time, and resources.

The types of variance are shown in Table 1.

TYPE		Unit of Meas.	PURPOSE	REMARKS
Time Variance	Duration Variance	Time	Deviation of actual duration from planned duration	
	Variance of Finish/Start Date	Time	Deviation of actual Finish/Start date from planned Finish /Start date.	
Resource Variance	Variance of Labor Rate	Cost	Deviation of actual labor rate from planned labor rate.	
	Productivity variance	MH /Unit	Deviation of actual man hours/unit of work accomplished from planned man hours/unit of work accomplished.	
	Variance of Material Rate	Cost	Deviation of actual material unit price from planned material unit price.	
	Variance of Material Usage	Q'ty /Unit	Deviation of actual material consumption/unit of work from planned material consumption/unit of work.	
Cost Variance	Cost Variance (CV)	Cost	Performance measure of cost overrun or underrun	*See Cost variance*
	Accounting Variance (AV)	Cost	Performance measure of over budget or under budget	*See Accounting variance*
	Schedule Variance (SV)	Cost	Performance measure of behind schedule or ahead of schedule	*See Schedule variance*
	At Completion Variance (ACV)	Cost	Performance measure of forecast over budget or forecast under budget	*See At-completion variance*

570

W

WBS WORK BREAKDOWN STRUCTURE *(see Contract WBS)*

WORK ITEM A unit of the project breakdown structure (PBS).

Work item is a component of PBS *(see Project breakdown structure)*. A work item is a manageable element at each level. For example, the PBS for a commercial building (Figure 1) shows work items at each level of detail. As shown in the figure, work items are manageable elements at any level of PBS except level 0. Also, the work items at the lower levels have work packages.

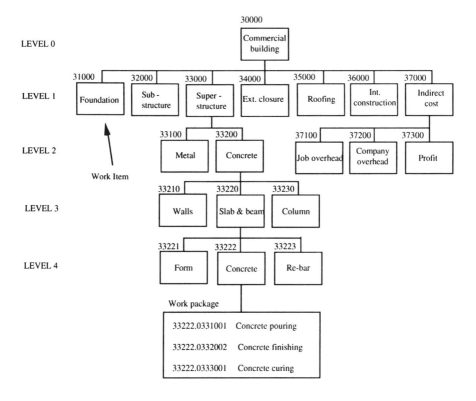

Figure 1 PBS and Its Work Items for Commercial Building

WORK PACKAGE (WP) The unit of a project breakdown structure (PBS) at the lowest developed level. The work package is to be performed by a single organization unit (crew, shop, subcontractors, etc.) and is the base for project estimate, short-interval planning, and collection of expenditures.

The work package is a component of the PBS *(see Project breakdown structure)*. The reason why the work package is needed is that redundancy and inconsistency of data, duplication of data capture, and information lags among different phases are eliminated with effective use of WP throughout the life cycle of a project. For example, the PBS developed for a commercial building is shown in Figure 1. As shown in the figure, Concrete (33222) has a work package. This work package consists of three tasks: Concrete pouring (33222.0331001), Concrete finishing (33222.0332002), and Concrete curing (33222.0332001). Time and cost budgets are assigned to the tasks, and actual time and cost of the tasks are collected.

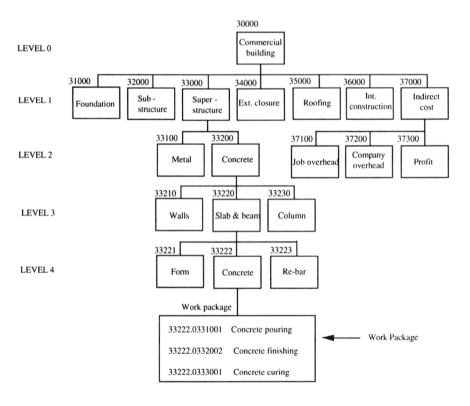

Figure 1 PBS and WP for Commercial Building

WORK UNIT *(see Unit of work)*

572

WORKWEEK The hours or days of work in a calendar week (40 hours/week, 8 hours/day, 5 days/week, etc.).

In most project planning and scheduling software packages, calendar information can customarily be modified to identify specific workdays in a week. Those workdays will serve as a standard work period for the entire project. When a project comprises many activities to be performed in different parts of the country or world, various calendars can be created with specific workdays in a week and be assigned to each activity. For example, a project may require structural component fabrications from another country in which people work only 4 days in a week, Monday to Thursday. For some type of work such as maintenance operations that need not disturb other operations, it is possible to have a workweek consisting of two working days, and they can be Saturday and Sunday. The most common workweek is from Monday to Friday, as shown in Figure 1.

```
→
                          PROJECT CALENDAR                         COMP
═══════════════════════════════════════════════════════════════════════
    Calendar ID: 1                         Title: DAILY  CALENDAR


     Workdays:  X MON   X TUE   X WED   X THU   X FRI     SAT     SUN


───────────────────────────────────────┬───────────────────────────────
   Nonwork periods        Page:    1 |  Exceptions            Page:    1
                                     |
      Start            End           |     Start            End
  ~~~~~~~~~~~~~~~  ~~~~~~~~~~~~~~~    |  ~~~~~~~~~~~~~~~  ~~~~~~~~~~~~~~~
     04JUL93                         |
     25NOV93                         |
     25DEC93                         |
                                     |
                                     |
                                     |
                                     |
                                     |
                                     |
      Scroll using Home, End         |     Scroll using PgUp, PgDn
═══════════════════════════════════════════════════════════════════════
Commands: Add Delete Edit Help More Next Print Return Transfer View Window
Windows : Calendar Global List
```

Figure 1 Project Calendar Information (Primavera Project Planner 5.0)

BIBLIOGRAPHY